Biocatalysis

Qayyum Husain • Mohammad Fahad Ullah
Editors

Biocatalysis

Enzymatic Basics and Applications

Springer

Editors
Qayyum Husain
Department of Biochemistry
Aligarh Muslim University
Aligarh, India

Mohammad Fahad Ullah
Department of Medical Laboratory Technology
FAMS, University of Tabuk
Tabuk, Saudi Arabia

ISBN 978-3-030-25022-5 ISBN 978-3-030-25023-2 (eBook)
https://doi.org/10.1007/978-3-030-25023-2

This Springer imprint is published by the registered company Springer Nature Switzerland AG.
The registered company address is: Gewerbestrasse 11, 6330 Cham, Switzerland

Foreword

Enzymes are known to do wonders for cellular life. If you could peek into an actively grown *E. coli* cell, you would find about 1000 different types of enzymes actively performing their work to keep the little bacterial cell live and vibrant. For an even more complex biological system like our human body, the number of enzymes that are required to carry out coordinated functions essential to life is simply mesmerizing; for example, a human cell utilizes approximately 75,000 enzymes to do its chores.

Enzymes are endowed with necessary powers to build things or take things apart as needed when a cell grows and reproduces. A cell can be simplistically viewed as a microscale biological world filled with enzymes to process all kinds of molecules using fast chemical reactions. Because of the importance of enzymes to cellular life, functional and biochemical investigation of enzymes has been an important discipline of modern science.

However, humans are best known for learning and applying. As a result of our excellent understanding of properties of a diverse array of cellular enzymes, the scientific community has invented many ingenious ways of employing enzymes outside of cells. Enzymes have now been widely used in a variety of industries, such as agriculture, pharmaceuticals, diagnostics, chemical production, biofuels, and consumer products. For example, enzyme-assisted industrial processes have been growing rapidly for the past several decades due to their mild operational conditions, high efficiency, great chemical reaction specificity, and significantly reduced pollution, things that are difficult to achieve with traditional chemical processes.

This book serves to provide a snapshot of current research activities and achievements with some natural and man-made enzymes. It covers several important topics concerning both a basic understanding of enzyme properties and practical applications of enzymes. The book begins with a chapter on synthetic enzymes made of DNA (DNAzymes) by Rothenbroker et al. DNAzymes represent the latest member of macromolecular enzyme family, and the first chapter specifically discusses how DNAzymes can be derived and how they can be employed for the design of biosensors for practical applications.

This is followed by the second chapter of the book, written by Das and Kayastha. It concerns β-amylase, a starch hydrolyzing enzyme widely found in plants, microorganisms, and fungi. The authors provide useful discussion on structure, function, reaction mechanism, and industrial utility of β-amylase. The next chapter, written by Younus, deals with oxidoreductases, a large group of enzymes catalyzing various redox reactions. This chapter places a particular focus on applications of these enzymes in the field of textiles, medicine, food, and chemical production. The chapter by Özgen and Schmidt features Rieske non-heme iron dioxygenases, amazing enzymes that are capable of activating molecular oxygen and producing reactive oxygen species for hydroxylating alkyl substrates. This chapter is rich in information on the basic properties of such enzymes and their potential applications in the industrial synthesis of various natural products, polyfunctionalized metabolites, and pharmaceutical intermediates.

The enzyme stability is highly important to pursuing enzymes for practical applications. The next three chapters of the book cover this topic well. The chapter by Noori et al. presents a detailed discussion on the technique of cross-linked enzyme aggregate (CLEA), which represents an effective way to enhance enzyme performance for industrial applications. This is followed by a chapter from Prakash and Khare who review ongoing research activities aimed to produce highly stable α-amylase for industrial application, with a particular focus on discussing new approaches that take advantage of nanomaterials for enzyme immobilization. The chapter contributed by Husain provides informative discussion on research aimed at utilizing immobilized peroxidases, which are more stable than solution-phase enzymes, for removing industrially produced dyes in polluted water.

The next four chapters deal with biomedical or biopharmaceutical topics. Liu and Liu then review research activities related to screening, optimization, and assembling of key pathway enzymes. These synthetic biology efforts can produce highly effective biological systems for the synthesis of valuable biochemicals. In their chapter, Jha et al. discuss the design of artificial metalloenzymes, an active research area directed at understanding mechanistic insights of natural metalloenzymes and discovering synthetic equivalents for practical applications. This is followed by a chapter written by Chandrasekaran et al. who review the research activities related to secreted phospholipase A2, an enzyme of significant clinical and pharmacological importance. Ullah et al. in their chapter review the clinical significance of enzymes in the diagnosis of cancers, myocardial disorders, and gastrointestinal tract-related impairments in humans.

The final four chapters concern aspects of use of enzymes in the agricultural sector. The chapter by Fernandes provides a brief historic overview concerning uses of enzymes in food and feed, their current applications, and more recent advances in this area. Mehta and Sehgal contribute a chapter on microbial enzymes in food processing with a key focus on modern molecular biology techniques for developing highly efficient expression systems for engineering robust enzymes for food processing. Piotrowska-Długosz then provides a chapter reviewing some historical background and recent activities on two fronts: importance of soil enzymes to agriculture and applications of several types of enzymes in animal farms as feed additives. The final chapter of the book is written by Jamal et al. dealing with the discovery of inhibitors that target proteases from *Helicoverpa armigera* as a new way to control this insect pest.

The editors of the book, Drs. Husain and Ullah, have done a fine job by recruiting acknowledged experts on the chosen topics, and therefore, the information presented in this book should be instantaneously useful to both junior and advanced researchers working in related research areas. Moreover, the topics covered by this are fairly diverse as the book presents in-depth reviews on several important aspects of enzymes, which include fundamental properties of many important enzymes, their industrial or biomedical applications, and general and specific principles guiding enzyme engineering. These topics should be broadly interesting to diverse scientists with either a specific interest in the featured enzymatic systems or a general interest in enzymes.

Department of Biochemistry and Biomedical Sciences, McMaster University, Hamilton, ON, Canada

Yingfu Li

Preface

The Nobel Prize for Chemistry 2018 was awarded to **Prof. Frances H. Arnold** (California Institute of Technology, Pasadena, USA) for *"the directed evolution of enzymes."* His significant work has resulted in the evolution of enzymes to catalyze commercially useful reactions which were previously unknown including catalysis in organic solvents. The idea to archive the complex diversity of catalytic functions in this title *Biocatalysis: Enzymatic Basics and Applications* as a fortunate experience coincided with the enzymes appearing again in the spotlight with the Nobel Prize 2018. This book contains a collection of articles highlighting the significance of enzymes and the related applications. The diversity of the chapters spans a wide range of highly complementary topics with minimal overlap. We commend Springer (Switzerland) for providing the platform for this endeavor and entrusting us with the task of managing and editing the current volume of the compilation that we present before the audience.

Precisely, the volume contains 15 chapters, each focusing on the basic and applied aspects of catalytic functions and their significance. The chapter entitled "DNAzymes: Synthetic enzymes made of DNA" serves to introduce DNAzymes, the synthetic single-stranded DNA molecules with catalytic abilities, with a particular focus on sequence-specific cleavage of a phosphodiester bond located within an all-RNA or a chimeric RNA/DNA molecule. The chapter entitled "β-Amylase: General properties, mechanism and panorama of applications by immobilization on nano-structures" discusses basics, applications, and potential prospects of industrially important β-amylase enzyme, immobilized on various matrices for biotechnological applications. The chapter entitled "Oxidoreductases: Overview and practical applications" describes the oxidoreductases class of enzymes as versatile biological catalysts in a vast range of chemical reactions acting with high specificity, efficiency, and selectivity. The chapter entitled "Rieske non-heme iron dioxygenases: Applications and future perspectives" focuses on current understanding of the structural determinants and the catalytic behavior of dioxygenase-catalyzed reactions and presents several cases where this knowledge has been harnessed to design tailored catalysts for the synthesis of various natural products, polyfunctionalized metabolites, and pharmaceutical intermediates. The chapter entitled "Cross-linked enzyme aggregates: Current developments and applications" presents a detailed discourse on cross-linked enzyme aggregate (CLEA), a new method of immobilization, delivering highly stable, reusable, and an efficient catalyst in terms of performance. The chapter entitled "Immobilization of α-amylases and their analytical applications" summarizes the excellent properties of nanomaterials as effective matrices for enzyme immobilization with enhanced catalytic activity, stability, and reusability, thereby leading to an economically viable and environmentally feasible process. The chapter entitled "Immobilized peroxidase catalyzed decolorization and degradation of industrially important dyes from polluted water" demonstrates enzymatic potential in decolorizing industrially important dyes as effective tools to remove contamination from polluted water. The chapter entitled "Screening, optimization and assembling of key pathway enzymes in metabolic engineering" reports optimization of key pathway enzymes by feedback inhibition removal, catalytic efficiency improvement, and substrate specificity alteration using metabolic engineering approaches to facilitate biochemical production efficiency. The chapter entitled "Designing of artificial metalloenzymes: From concept to applications" discusses the mechanistic and applied potential and the design and synthesis of novel metalloenzymes in an elaborate and well-organized literature. The chapter entitled "Anti-inflammatory and antidote drug discovery with secreted phospholipase A2" describes the current knowledge on phospholipase A2, an enzyme that hydrolyzes the phospholipids at a specific site and initiates the inflammatory/arachidonic acid (AA) pathway, with regard to its biological properties and future perspectives. The chapter entitled "Clinical significance of enzymes in disease and diagnosis" serves to indicate the clinical utilization of enzymes as biomarkers with potential impact in the diagnosis of cancer, myocardial disorders, and gastrointestinal tract-related impairments in humans. The chapter entitled "Enzymes in food and feed industries: where tradition meets innovations" provides a comprehensive description of the role played by enzymes in key processes and goods production in food and feed formulations. The chapter entitled "Microbial enzymes in food processing" describes microbes which offer an important and vast reservoir of biocatalysts for carrying out specific biochemical reactions. The chapter entitled "Significance of enzymes and their application in agriculture" discusses the application of soil enzymes in agronomy as reliable indicators of soil health, fertility, and productivity as affected by differentiated natural and anthropogenic factors. The chapter entitled "Proteinaceous trypsin inhibitors from plants in disarming the insect pest" describes the significance of protease inhibitors as an attractive class of biopesticides against plant pest and pathogen.

We express our gratitude to all the authors for valuable contribution from around the globe. It is indeed their willingness to share their onerous experiences which has facilitated this piece of scientific literature. We appreciate the support of Dr. Beatrice Menz (Senior Editor, Springer–Basel) for working out the procedural framework of our book proposal. We also acknowledge the assistance of Mr. Srinivasan Manavalan (Project Coordinator) and Mr. M. Saravanan (Project Manager) who were instrumental in ensuring the required basics of attractive and meaningful academic production. We are indeed honored to have Prof. Yingfu Li (McMaster University, Canada) introducing the substance of the book in the foreword.

Lastly, we wish that the audience will like the content of this book, and it will serve as a promising literature for the researchers who wish to update their current knowledge on the catalytic properties of enzymes.

Aligarh, India Qayyum Husain
Tabuk, Saudi Arabia Mohammad Fahad Ullah

About the Book

The book entitled *Biocatalysis: Enzymatic Basics and Applications* provides an updated compilation of new insights into enzyme catalysis; and is an attempt to scan the untrodden paths, yet integrated into past knowledge, with critical assessment and futuristic ideas in developing fascinating concepts of catalytic functions. The volume includes a comprehensive review of the literature, which essentially engages a vast extent of basic and applied knowledge with subjective diversity that serves as an element of strength. In the last few decades, molecular engineering and variable dynamics in enzymatic catalysis has contributed enormously to some well-characterized experimental approaches leading to novel biochemical reactions. In concurrence, these simple molecules which were naturally oriented for cellular catalysis have now been modified as a highly efficient molecular tool for manipulating basic cellular reactions and extending to the realm of a broad range of desirable transformations with vast application in organic synthesis, pharmaceuticals, diagnostic assays, food processing and preservation, biotechnology, and industrial processes. Enzymatic traits such as sensitivity of enzyme catalysis to temperature and pH (stability) and narrow substrate range (specificity) have for long limited the use of enzymes in cell-free systems. However, the molecular, structural, and conditional optimization of existing enzymes and the emergence of novel biocatalytic reactions that were previously unknown in nature have resulted in processes that are efficient, are sustainable by design, and have a broad range of applications. The volume shares the experiences of experts working in the area for decades, as these catalytic molecules evolved from basic knowledge to climacteric approaches, some of which finds mention here.

Contents

About the Editors

Qayyum Husain is currently the Dean of the Faculty of Life Sciences, Aligarh Muslim University, Aligarh, India. Dr. Husain has completed his entire education from Aligarh Muslim University and subsequently joined the Department of Biochemistry, Faculty of Life Sciences, Aligarh Muslim University, as a lecturer in January 1987. He was promoted as Reader in 1997 and as a Professor in the same department in 2005. Prof. Husain also served as the Coordinator for M.Sc. Programme in "Enzymes and Fermentation Technology," July 2008 to March 31, 2011. Prof. Husain has supervised 17 Ph.D., one M.Phil., 45 M.Sc., and 19 B.Sc. students for their thesis and dissertations. Presently 7 students are pursuing research under his guidance for the award of Ph.D. degree. Prof. Husain was awarded DAAD (German Academic Exchange Service) Fellowship to work at the University of Konstanz, Germany, June 1992 to September 1994. He was re-invited under the DAAD revisiting Programme to work at the Technical University of Kaiserslautern, December 2008 to January 2009. He has served as a Professor and Head of Biochemistry in the Faculty of Applied Medical Sciences, Jazan University, Jazan, Kingdom of Saudi Arabia, April 1, 2011, to August 13, 2014. He has been selected as an "External Accessor" to evaluate the academic facilities, syllabi, and standard of undergraduate and postgraduate courses run by the Faculty of Biotechnology and Biomolecular Sciences, University of Putra Malaysia, for the duration June 2012 to June 2015.

During his stay in Jazan University, Kingdom of Saudi Arabia, he worked as an active member of Board of Research Project Review, and he was nominated as Chief Editor, to *JazMed Plus* quarterly magazine published by the Faculty of Medicine, Jazan University, Jazan, KSA. Dr. Husain has published more than 130 research and review articles in peer-reviewed and journals of international repute. He has also published 9 book chapters in the books published by International Press such as Academic, Springer, Nova Science, and Wiley Science Publications. His work has been published in some of highly reputed journals and well cited in plenty of books, magazines, and journals of international repute. His present H-index is 39 and i-10 index is 84. His work has total citations 5055 as reported by Googlescholar.com on 12 August, 2018. He has worked as a reviewer for more than 50 journals of international repute. He has been nominated as an examiner to evaluate Ph.D. thesis of several foreign and Indian Universities. He has obtained membership of more than 40 different scientific and universities bodies, i.e., American Chemical Society and Society of Chemical Industry, UK. His work has been presented in more than 100 different seminars, symposium, and conferences at national and international level. He has attended many short-term courses and workshops organized by various reputed institutes and universities. He has delivered numerous invited lectures and speeches at national and international level. Dr. Husain has been actively engaged in organizing several national and international conferences. His main areas of research are environmental biotechnology, enzyme technology, and nanobiotechnology.

Mohammad Fahad Ullah, Ph.D, MRSC, is an Associate Professor of biochemistry in the Department of Medical Laboratory Technology (FAMS) and a research scientist at Prince Fahd Research Chair, University of Tabuk, Saudi Arabia. He received his academic degrees along with a gold medal in M.Sc. (biochemistry) from Aligarh Muslim University, India. Further he worked as a research associate at Experimental Oncology Laboratory, Department of Biomedical and Diagnostic Sciences, University of Tennessee, USA. His research interests include mechanistic insight into the cell signaling and therapeutic pathways against chronic diseases including cancer and diabetes. He is an active member of AACR (USA)/Royal Society of Chemistry (UK) and member of the editorial/reviewer board of a number of scientific journals. Dr. Ullah has more than 9 years of experience in teaching biochemistry to the students of biological and health sciences. His academic works include more than 50 publications in reputed journals and three books entitled *Critical dietary factors in cancer chemoprevention* (Springer-Switzerland), *Illustrated notes on biomolecules* (Partridge-Singapore), and *Nutraceuticals and Natural Product Derivatives: Disease Prevention & Drug Discovery* (John Wiley-UK).

DNAzymes: Synthetic Enzymes Made of DNA

Meghan Rothenbroker, Sandy Zakaria, Devon Morrison, and Yingfu Li

1 In Vitro Selection Technique for Deriving DNAzymes

The traditional role of nucleic acids as a medium for storage and transfer of genetic information for living organisms on Earth has long been understood. Similarly, the role of proteins to act as biological catalysts is also well understood. The ability of nucleic acids to function as catalysts, however, was not realized until the early 1980s (Kruger et al. 1982; Guerrier-Takada et al. 1983). In 1989, Thomas Cech and Sidney Altman were awarded the Nobel Prize in Chemistry "for their discovery of catalytic properties of RNA". Their finding of RNA-based enzymes, or ribozymes, represents a ground-breaking discovery because their work revealed, for the first time, that RNA has what it takes to catalyze complex biochemical reactions. The discovery of natural ribozymes has also served as the motivation to search for man-made ribozymes as well as DNA-based enzymes (DNAzymes), the focus of this chapter. Unlike ribozymes, there is no evidence that DNAzymes exist in nature. DNA's natural state is primarily in double-stranded form, and the presence of the complementary strand makes the double-stranded DNA adopt a simple duplex structure, which takes away DNA's ability to create intricate tertiary structures. In contrast to the structural simplicity of a duplex, single-stranded DNA is highly flexible and can form complex three-dimensional structures that are typically characteristic of enzyme active sites

M. Rothenbroker · D. Morrison
Department of Biochemistry and Biomedical Sciences, McMaster University, Hamilton, ON, Canada

S. Zakaria
Department of Biomedical Engineering, McMaster University, Hamilton, ON, Canada

Y. Li (✉)
Department of Biochemistry and Biomedical Sciences, McMaster University, Hamilton, ON, Canada

Department of Biomedical Engineering, McMaster University, Hamilton, ON, Canada
e-mail: liying@mcmaster.ca

Q. Husain, M. F. Ullah (eds.), *Biocatalysis*,
https://doi.org/10.1007/978-3-030-25023-2_1

(Breaker 1997; Ponce-Salvatierra et al. 2016; Liu et al. 2017). The ability to create intricate tertiary structures is a key characteristic of enzymatic catalysis.

The search for man-made DNAzymes became feasible with the development of the in vitro selection technique by the groups of Larry Gold, Jack Szostak and Gerald Joyce. In vitro selection is a powerful experimental approach that permits the isolation of functional DNA or RNA sequences from pools of DNA or RNA variants. Tuerk and Gold (1990) showed that high-affinity ligands can be isolated from a population of RNA molecules to bind a protein target. They gave their method the acronym "SELEX", which stands for **S**ystematic **E**volution of **L**igands by **EX**ponential enrichment. Ellington and Szostak (1990) showed that RNA molecules that bind organic dyes can be isolated from a random-sequence RNA pool. They called their method "in vitro selection" and the RNA-based recognition elements "aptamers". Robertson and Joyce (1990) showed that a ribozyme that cleaves an RNA molecule can be evolved to cleave a DNA sequence through directed molecular evolution. These studies share two common features: building a pool of RNA molecules, and devising an effective method to enrich molecules that perform the task of interest. These methods are analogous to natural selection as both processes are governed by variation, selection, and amplification. The key difference, however, is that in vitro selection can be conducted in a test tube in a matter of weeks whereas natural selection occurs on a much longer timescale (thousands of years or longer).

Although in vitro selection was initially conceived for deriving functional RNA molecules, it was quickly expanded to isolate DNA counterparts. The first DNA aptamers were reported in 1992: Ellington and Szostak (1992) described the isolation of DNA aptamers for a group of organic dyes, while Bock and coworkers reported DNA aptamers that bind human thrombin (Bock et al. 1992). These studies were followed by the report of the first-ever DNAzyme in 1994 by Breaker and Joyce: from a random-sequence DNA pool, they isolated a DNAzyme capable of carrying out Pb^{2+}-dependent cleavage of an RNA linkage embedded within a DNA sequence (Breaker and Joyce 1994).

In vitro selection for DNAzymes begins with a single-stranded DNA pool that may contain as many as 10^{16} unique sequences. This DNA pool (or DNA library) can be prepared by automated DNA synthesis. Once synthesized and purified, the library is asked to come up with a solution that can speed up a chemical reaction. Let us assume that we are looking for DNAzymes that cut the phosphodiester bond within an RNA substrate using the well-known phosphoester transfer mechanism illustrated in Fig. 1a. In this case, the DNA library will be modified in such a way that each variant in the library will contain an RNA substrate. DNA molecules capable of cleaving the attached RNA substrate by this mechanism will become shorter and can be separated from catalytically inactive variants simply by size. The isolated DNA molecules can be amplified by polymerase chain reaction (PCR) to produce a new library that now will contain increased copies of catalytically active molecules. Prior to starting the next round of selection, it is necessary to convert the double-stranded DNA produced by PCR back into single-stranded DNA to avoid antisense inhibition of catalytically active DNA sequences by their complementary strands. The process

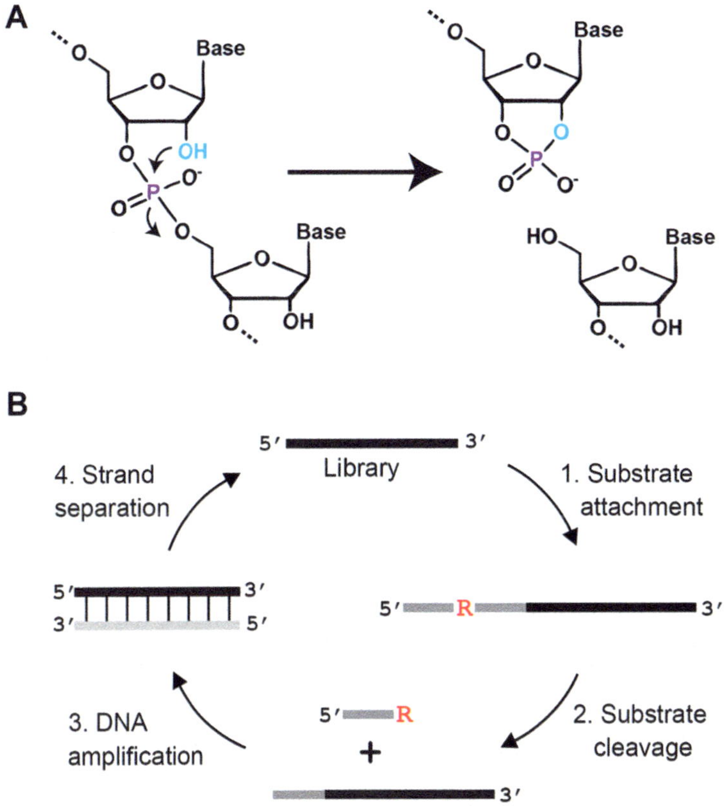

Fig. 1 RNA-cleaving DNAzyme. (**a**) Transesterification reaction catalyzed by RNA-cleaving DNAzyme. (**b**) Four key steps employed in each round to select for RNA-cleaving DNAzymes from a DNA library

of isolation and amplification qualifies as one round of selection. Often, several rounds are required to produce a DNA pool that contains a high percentage of RNA-cleaving DNAzymes (Fig. 1b).

It is common for in vitro selection to generate molecules that do not have the desired catalytic activity. These molecules somehow manifest during the selection process due to a variety of factors. The generation of these molecules, sometimes referred to as "molecular parasites", can be avoided by employing stringent conditions that only permit the selection of catalytically active molecules (Breaker and Joyce 1994). The implementation of these conditions usually occurs after a few initial rounds to give catalytically active DNA sequences a better chance to survive the selection process. Mutations may also be introduced alongside stringent conditions to isolate more proficient DNAzymes. The number of rounds to be conducted is dependent on the activity of the population. Once the population exhibits a satisfactory level of catalytic activity, as defined by the researcher, individual DNAzymes can then be identified by sequencing. The most notable feature of in vitro selection is that it does not systematically screen every individual sequence for activity; rather, only a small fraction of the initial library is tested for activity after a long journey of selective amplification. This feature makes in vitro selection conceptually distinct from screening, which is a one-at-a-time process.

Imperative to the success of any in vitro selection is the ability to choose an effective method, or a combination of methods, that link the survival of molecules to their ability to catalyze the specific reaction. The proper choice of the methods, alongside proper execution, will enable catalytically active DNA sequences to flourish among a background of functionally inert, but potentially "selfish" DNA molecules (Li and Breaker 2001). There are many factors and experimental parameters that can influence the outcome of a DNAzyme selection experiment, which include length of random sequence domain, the sequence diversity of a DNA library, choice of metal ion cofactors, and permissive reaction time (Achenbach et al. 2004). A key advantage of in vitro selection is the flexibility to accommodate different experimental parameters to achieve specific purposes. For example, the permissive reaction time represents an important selection pressure to influence the outcome of in vitro selection, as an extended reaction time would allow diverse DNAzyme sequences to survive whereas a short reaction time would only allow the DNAzymes with high catalytic rates to persist. Therefore, judiciously selecting a specific reaction time can be crucially important to the success of a DNAzyme selection experiment.

In some cases, positive-selection and counter-selection may be applied as an excellent selective pressure to derive DNAzymes with specific properties. For example, if the goal of the experiment is to isolate a DNAzyme that shows high activity towards bacterium A but low activity towards bacteria B, C, and D; bacterium A can be used as the positive-selection target while bacteria B, C, and D can be used as the counter-selection targets (Fig. 2). This strategy will favour the selective amplification of DNAzyme sequences that exhibit the desired selectivity towards bacterium A. It is common to begin applying a counter-selection step after a few rounds of positive-selection have occurred. This is done to give, presumably, a few potential DNAzymes present in the initial library a chance to survive prior to the

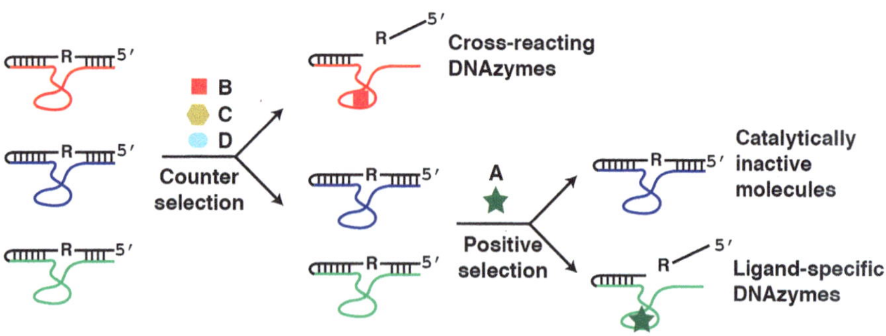

Fig. 2 Selecting RNA-cleaving DNAzymes with high recognition specificity for bacterium A over bacteria B, C and D by employing a counter-selection step with bacteria B, C and D. This is followed by a positive-selection step with bacterium A

introduction of the stringent counter-selection step. If applied too early, the counter-selection may accidently eliminate any of the rare, but desired, DNAzymes.

Recurring small sequence motifs, such as the RNA-cleaving DNAzyme named 8-17, which will be described later, could represent a major challenge for efforts aimed to isolate diverse DNAzyme sequences by in vitro selection. 8-17 has been selected multiple times by different labs from different libraries, due to its small size, ability to permutate, and high frequency in the initial library. The other significant challenge with in vitro selection is the limited amount of sequence space that can be sampled in a single experiment. The sequence space of a random-sequence DNA library is 4^n in size, where n is the number of nucleotides in the random-sequence region. The N_{50} random pool is comprised of 4^{50} or 10^{30} unique sequences; however, it is only possible to produce a synthetic DNA pool containing 10^{13}–10^{16} different sequences. To create a complete sequence library that covers all 4^{50} sequence options would require 3×10^7 kg of DNA (Schlosser et al. 2009). As a result, the sequence space cannot be explored exhaustively.

In vitro selection has given researchers the ability to expand beyond the central dogma of molecular biology through the isolation of single-stranded DNA sequences either recognizing a specific target of choice or catalyzing a specific chemical transformation or performing both functions. Many in vitro selection experiments have generated interesting DNAzymes, which have been further used in wide-ranging applications. In the remaining sections, we will discuss some examples.

2 Representative DNAzymes

2.1 8-17 and 10-23 DNAzymes

The two most studied DNAzymes to date are 8-17 (Fig. 3a) and 10-23 (Fig. 3b), RNA-cleaving DNAzymes isolated by Santoro and Joyce (1997). Both DNAzymes are able to bind and cleave all-RNA substrates as well as chimeric DNA/RNA substrates containing a single RNA unit embedded in a DNA chain. Although both DNAzymes have similar secondary structures, their target substrates are different. It has been shown that 10-23 can be designed to cleave an RNA sequence at any purine-pyrimidine junction; in contrast, the original 8-17 can only cleave an RNA substrate with a G·T wobble base pair adjacent to the cleavage site (Santoro and Joyce 1997). A few years later, our group discovered that there are many 8-17 sequence variants that can collectively cleave any dinucleotide junction of RNA in a chimeric DNA/RNA sequence (Cruz et al. 2004; Schlosser and Li 2010). Furthermore, the investigation of the 8-17 and 10-23 catalytic cores showed that both DNAzymes can retain catalytic activity as long as the key catalytic residues are kept unchanged and the structural integrity is maintained (Wang et al. 2010). Interestingly, investigations of the catalytic core of 8-17 and 10-23 have generated the evidence to suggest that 10-23 is a special variant of 8-17 (Wang et al. 2010). 8-17 has been isolated independently many times by different research groups using different divalent metal ion cofactors that

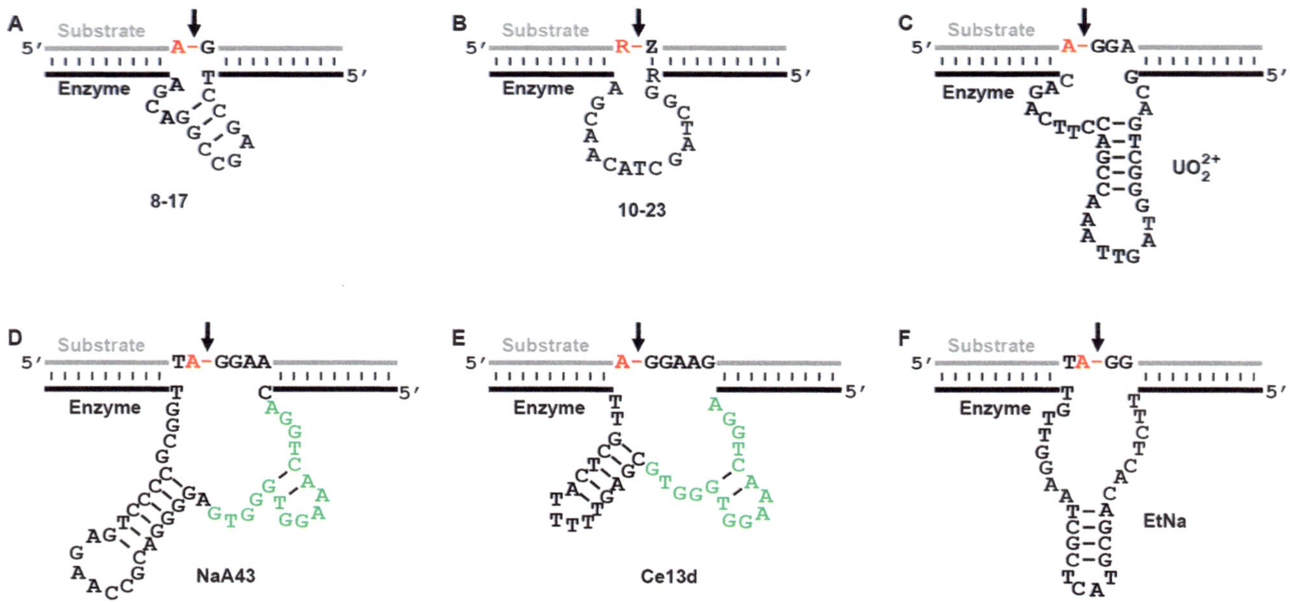

Fig. 3 Proposed secondary structure of six RNA-cleaving DNAzymes. R: A or G; and Z: U or C. The grey bars are the substrate strands that contain a ribonucleotide (shown in red). The arrow indicates cleavage site. Green nucleotides indicate the conserved nucleotides of NaA43 and Ce13d

include Zn^{2+} (Li et al. 2000), Mg^{2+} (Cruz et al. 2004; Faulhammer and Famulok 1996), Ca^{2+} (Faulhammer and Famulok 1996), Mn^{2+} (Cruz et al. 2004),and Cd^{2+} (Kasprowicz et al. 2015).

The functional differences between 8-17 and 10-23 allow each DNAzyme to be used for unique applications. The applications of 8-17 have primarily focused on metal ion sensing and molecular computing, whereas 10-23 has been widely examined for potential therapeutic applications. The first DNAzyme selected was later realized to contain the 8-17 motif, enabling it to detect Pb^{2+} at low concentrations (Breaker and Joyce 1994; Li and Lu 2000). Li and Lu understood how important the development of Pb^{2+} sensors are to human health as Pb^{2+} levels greater than 480 nM in the blood are toxic (Li and Lu 2000). The discovery of the Pb^{2+} DNAzyme has led to the development of sensitive Pb^{2+} sensors using fluorescent, colorimetric, electrochemical, electrochemiluminescent, and electrical techniques (Schlosser and Li 2010; Xiang and Lu 2013). These techniques have also been applied to another metal ion sensor that will be described in Sect. 2.2.

In addition, 8-17 was first demonstrated to be used as a molecular computational device by Stojanovic and colleagues (Stojanovic et al. 2002). For example, logic gates were turned on by effector oligonucleotides that activated the DNAzyme, and this in turn generated a reporting channel which was converted into a Boolean function. More intriguing computing systems were further demonstrated by incorporating 8-17 into logic circuits such as half and full adders, game-playing automata (MAYA) and signal cascades (Lakin et al. 2017). Such computational devices have demonstrated success in viral and bacterial gene detection (Lakin et al. 2017).

Applications for the 10-23 DNAzyme have generally been focused on its therapeutic potential, since 10-23 can be designed to bind and cleave any RNA sequence, as first demonstrated by Santoro and Joyce, who used several 10-23 variants to inactivate HIV-1 *gagypol, env, vpr, tat,* and *ner* mRNAs (Santoro and Joyce 1997). From there, many researchers have shown that 10-23 can be used to down-regulate expression of mRNAspecies associated with viral and bacterial infections, cancer, and cardiovascular, inflammatory and neurological diseases while eliciting no immune response or toxic effects to the host (Fokina et al. 2015). Clinical trials with several versions of 10-23 have been implemented to target carcinomas and inflammation (Cho et al. 2013; Grassi and Grassi 2013; Homburg et al. 2013). The DNAzymes operate by reducing *c-jun* and *GATA-3* expression. A phase-I clinical trial with Dz13, a DNAzyme operating by reducing *c-jun* mRNA expression, has been completed, and the data shows that administration of Dz13 reduced tumor size and the expression of c-Jun in tumor cells as compared to controls (Fokina et al. 2015). A phase-I clinical trial with SB010, a *GATA-3* mRNA targeting DNAzyme, for treating allergic bronchial asthma, has also been completed. SB010 treatment improved lung function compared to controls. Both Dz13 and SB10 were shown to be safe and well tolerated at all administered doses (Fokina et al. 2015).

2.2 UO_2^{2+} DNAzyme

DNAzymes are well suited for metal ion sensing as DNAzymes can be created with a dependency on a specific metal ion. Metal ions can aid in DNAzyme folding and can even play an important role in catalysis. Many RNA-cleaving DNAzymes that show high specificity for metal ions that are detrimental to human health, such as Pb^{2+} (Breaker and Joyce 1994), UO_2^{2+} (Liu et al. 2007), and Hg^{2+}(Hollenstein et al. 2008), have been reported. One notable example is the RNA-cleaving DNAzyme for UO_2^{2+} (Fig. 3c). It is highly sensitive and specific for UO_2^{2+}: a fluorescent sensor constructed with this DNAzyme was able to achieve 1 million-fold selectivity for UO_2^{2+} over other metal ions, and a limit of detection of 45 pM of UO_2^{2+} (Liu et al. 2007).

The detection of uranium has important implications for human health; high exposure to uranium can interfere with normal kidney, brain and liver functions (Nguyen et al. 2017). According to the U.S. Environmental Protection Agency (EPA), it is necessary for the bioavailability of uranyl ions, UO_2^{2+}, in drinking water to be monitored. The EPA determined that the maximum contamination level of UO_2^{2+} in drinking water is 130 nM(Lee et al. 2008). Fluorescence, colorimetric, electrochemical, and electrical signaling methods coupled to the UO_2^{2+} DNAzyme are sensitive enough to meet the EPA's mandated limit of detection. UO_2^{2+} DNAzyme biosensors have been applied to fulfill the need for the development of a simple, easy-to-use, real-time detection and quantification method of UO_2^{2+}. Fluorescent labelling methods, like the covalent coupling of a fluorophore or quencher to the end or internal site of DNA, have demonstrated the lowest detection sensitivity of UO_2^{2+} (45 pM) (Liu et al. 2007). Non-labelling fluorescent methods, like a fluorophore extrinsically bound within the duplex region of the UO_2^{2+} DNAzyme, provides a simpler and more cost-effective system, however, the detection sensitivity is reduced to 3 nM(Xiang et al. 2010). Electrochemical methods have also been shown to detect approximately 1 nM UO_2^{2+} by immobilizing the respective DNAzyme on the surface of gold electrodes (Tang et al. 2013). In the presence of UO_2^{2+}, the immobilized DNAzyme cleaves and releases ferrocene, an electroactive molecule, from the gold electrode. This results in a decrease in current. Another method, using the well-known glucose meter to detect UO_2^{2+}, has been shown to provide a detection limit of 9.1 nM UO_2^{2+} (Xiang and Lu 2011). The UO_2^{2+} DNAzyme and invertase (an enzyme that catalyzes hydrolysis of sucrose into fructose and glucose) complexes are immobilized onto magnetic beads. When UO_2^{2+} is present, the cleaved DNA-invertase complex is released into the solution and the released invertase hydrolyzes sucrose into glucose. The glucose is quantified by a personal glucose meter (PGM) and the readout of the PMG can be used to determine the UO_2^{2+} concentration; the DNA-invertase complexes in solution are proportional to UO_2^{2+} in a test sample. The portability of the PGM allows for onsite detection of UO_2^{2+}. Alternatively, a gold nanoparticle-based (AuNP) system allows for the simple detection of UO_2^{2+} using a color change that can be detected by the naked eye. The color change is influenced by the aggregation (blue/purple) and dispersal (red) state of AuNPs. Both labelling and non-labelling methods of AuNPs provide low detection sensitivity, 50 nM and 1 nM respectively (Lee et al. 2008). Labelled AuNP methods provide a "turn-on sensor" whereby in the presence of UO_2^{2+}, the UO_2^{2+} DNAzyme-labeled AuNP aggregates disassemble and disperse, and generate a color shift from purple to red. Non-labelled AuNP methods provide a "turn-off sensor", where in the presence of UO_2^{2+}, RNAcleavage of the UO_2^{2+}DNAzyme will release a single stranded DNA to be adsorbed onto the surface of the AuNPs. With the addition of NaCl, the AuNPs bound with single-stranded DNA will not aggregate, resulting in a red color. Finally, intracellular detection of UO_2^{2+} in living cells using the UO_2^{2+} DNAzyme-labelled AuNP system provides methods to help understand metal-ion localization and distribution within cells. The AuNP-DNA complexes are easily taken up by HeLa cells, and in the presence of UO_2^{2+}, RNA cleavage will separate the quencher from fluorophore to generate an increase in fluorescence that can be detected (Wu et al. 2013).

2.3 Na^+ DNAzyme

Over the last two decades, extensive research has been made in the development of divalent metal ion-dependent DNAzymes. However, more recently, an interest in isolating monovalent DNAzymes has taken precedence, gearing biosensors towards biological systems and processes in which monovalent ions, such as Na^+, play a critical role (Torabi et al. 2015). Previous attempts to develop monovalent metal ion-activated DNAzymes have only produced DNAzymes with slow catalytic rates, and poor sensitivity and selectivity (Faulhammer and Famulok 1997; Geyer and Sen 1997; Carrigan et al. 2004). Nonetheless, current research has addressed this issue by isolating highly sensitive and selective Na^+-activated RNA-cleaving DNAzymes: NaA43 (Fig. 3d), Ce13d (Fig. 3e), and EtNa (Fig. 3f). NaA43 was isolated by Torabi and colleagues; the DNAzyme was transformed into a fluorescent sensor capable of achieving 10,000-fold selectivity of Na^+ compared to other metal ions, a detection limit of 135 μM, and an observed rate of constant of 0.1 min^{-1} in the presence of 400 mM of Na^+ (Torabi et al. 2015). Interestingly, Ce13D, a different Na^+ DNAzyme with catalytic activity towards the trivalent lanthanide Ce^{3+}, contained the same Na^+ binding motif as NaA43 even though two different and separate in vitro selection experiments were conducted, highlighted by Figs. 3d, e (Torabi and Lu 2015; Zhou et al. 2015b). The K_d of Ce13D for Na^+ was determined to be 20–40 mM (Zhou et al. 2016a, b) and Ce13D was found to be active onlyin the simultaneous presence of Ce^{3+} and Na^+. The isolation of EtNa, by Zhou and colleagues, demonstrated multifaceted functionality of a DNAzyme. EtNa can detect Na^+ in the presence of high amounts of ethanol and DMSO (72%) (Zhou et al. 2015a). Also, in the presence of Na^+, the activity of EtNA is 1000 times greater in ethanol than in water.

The applications of the Na^+ DNAzymes are tailored to intracellular detection and signaling of Na^+, and detection of Na^+ in organic solvents. For intracellular detection, fluorescent NaA43 was modified with alpha-helical cationic polypeptides, to allow for entry into cells, and a photoliable o-nitrobenzyl group, to protect the RNA moiety of NaA43 (Torabi et al. 2015). This modified NaA43 complex was taken up by HeLa cells and with an induced influx of Na^+ from extracellular environment, an increase in fluorescence was observed within 30 min (Torabi et al. 2015). Endogenously, Na^+ within the cell is 9–19 mM, which NaA43 is not able to generate a level of signal. To overcome this issue, a strategy was developed that combined NaA43 with catalytic hairpin assembly (CHA) in order to amplify signal (Wu et al. 2017). In the presence of endogenous Na^+ within HeLa cells, Na^+ induces Na^+ DNAzyme-mediated cleavage and releases a primer-like fragment. The cleaved fragment initiates the CHA amplification and produces a turn-on fluorescence signal. Finally, EtNa was shown to accurately measure, within 5% difference, the alcohol content of commercial alcoholic beverages, thus demonstrating that EtNa is functional in a complex matrix (Zhou et al. 2015a).

3 Engineering DNAzyme-Based Biosensors for Bacterial Pathogen Detection

Our group has been interested in searching for diverse RNA-cleaving DNAzymes and employing them for the development of simple and easy-to-use biosensors (Schlosser and Li 2009; Liu et al. 2017; Morrison et al. 2018). In this section, we will highlight our recent efforts in developing bacteria-responsive RNA-cleaving DNAzymes and using them to devise biosensing systems for the detection of bacterial pathogens. We are particularly interested in developing RNA-cleaving fluorogenic DNAzymes (RFDs) that link catalysis to fluorescence generation. These DNAzymes are designed to cleave an RNA linkage located between a pair of fluorophore/quencher modified nucleotides in a single-stranded DNA sequence; because of this feature, they can be conveniently used as real-time fluorescent sensors (Tram et al. 2012; Ali et al. 2012).

3.1 *In Vitro Selection of Bacteria-Responsive RFDs*

For the duration of human life on Earth, microbes have been ubiquitously present and have played an integral role in our survival. However, in addition to the benefits they provide, bacteria have participated as antagonists as well. Pathogens are often transmitted from contaminated food and water sources, surfaces, the environment, and animal and human vectors (Schürch and Siezen 2010). After an outbreak has occurred, it is critical to identify and isolate the bacterial source, to the level of strain or subtype, in order to prevent further infection and transmission. A DNAzyme probe that is highly specific for a given pathogen can be used as a valuable reporting system. In addition, bacteria have been recently implicated to play huge roles in human health and diseases like cancer, diabetes, obesity, infections and inflammation (Mager 2006; Upadhyaya and Banerjee 2015; John and Mullin 2016). DNAzyme probes can therefore be used to develop cost-effective diagnostics for these diseases.

We have developed an in vitro selection method that can be used to isolate a highly specific bacterium-responsive RFD from random-sequence pools to recognize the bacterium of interest without the need to first identify a specific biomarker for the bacterium. This is done through the implementation of a double-selection strategy comprised of two key steps: a counter-selection in which the DNA pool is incubated with the crude cellular mixture of unintended bacteria to eliminate DNA molecules with cross-reacting activities, followed by a positive-selection where the unreactive fraction of the DNA pool from the counter-selection step is incubated with the crude cellular mixture of the intended bacterium (Fig. 2). We have demonstrated the effectiveness of this approach in bacteria sensors engineered in two different studies that used two different bacteria: *Escherichia coli* (*E. coli*) and *Clostridium difficile* (*C. difficile*). In the study of *E. coli* (Ali et al. 2011), we isolated an RFD, named RFD-EC1, after 20 rounds of selection with the use of *Bacillus subtilis* (*B. subtilis*) as the counter-selection bacterium and *E. coli* as the positive-selection target. We found that RFD-EC1 exhibited strong cleavage activity towards *E. coli* but weak activity towards *B. subtilis* as well as a host of other bacteria (Ali et al. 2011). RFD-EC1 can produce a detectable signal in the presence of as little as 1000 *E. coli* cells without a culturing step, and is able to detect a single colony forming unit with 12 hours of culturing (Ali et al. 2011).

For the work with *C. difficile* (Shen et al. 2016), we have shown that the same selection approach can be used to isolate RFDs that can distinguish BI/027-H, a hypervirulent, antibiotic-resistant strain of *C. difficile*, from other strains of *C. difficile* as well as other species of bacteria. In this study, BI/027-H was used as the positive-selection target while *E. coli*, *B. subtilis* and CD630 (a non-BI/027 strain of *C. difficile*), were used as the counter-selection targets. After 25 rounds of selection, we isolated an RFD, named RFD-CD1, that is not only species-specific but also strain-selective.

The same approach has also been applied to isolate RFDs that can be specifically activated by mammalian cells. More specifically, we have derived a highly selective RFD, named AAI2-5, to recognize MDA-MB-231, a breast cancer cell (He et al. 2014). AAI2-5 was obtained after 25 rounds of selection using the cell lysate of MCF-10A (a normal breast cell line) as the negative-selection target and that of MDA-MB-231 as the positive-selection target. AAI2-5 is able to distinguish MDA-MB-231 from many other human cell lines, including normal breast cells, and other subtypes of breast cancer cells.

The findings from the three studies we have conducted so far indicate that the double-selection strategy can derive highly selective RFDs when suitable cell lines are chosen as negative targets to eliminate the undesired cross-reactivity. The key advantage of this approach is that it does not need to use a specific biomarker for the selection of the DNAzyme; instead, a potential DNAzyme finds its own target during the selection, as illustrated by Fig. 4. In addition, it turns out that RFD-CD1

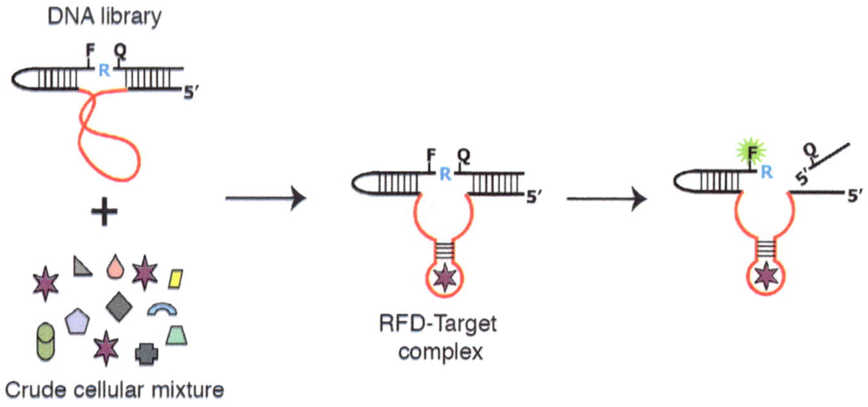

Fig. 4 The approach of the mixed cellular targets for deriving RFDs without using a specific biomarker. In this approach, the DNA library provides many different sequence options whereas many potential targets (represented by different shapes in the graph) may be recruited by RFDs

recognizes a truncated version of TcdC, a transcription factor with a 6-amino acid internal truncation, that is associated with the BI-027-H strain but not with other strains of *C. difficile* (Shen et al. 2016). This is a great example of the remarkable level of specificity that RFDs can achieve.

3.2 DNAzyme-Mediated Colorimetric Detection of Bacterial Pathogens

RFDs can also be employed to set up colorimetric assays, which have the advantage of being inexpensive and easily interpreted by the naked eye. For example, we have developed a colorimetric assay for *E. coli* detection that makes use of the classic litmus test and RFD-EC1 (Tram et al. 2014). The assay uses magnetic beads immobilized with RFD-EC1 that is tagged with the protein enzyme urease (Fig. 5a). When the DNAzyme is activated by *E. coli*, it cleaves itself and falls off the magnetic beads into the solution, carrying along the tagged urease. This solution is collected and used to hydrolyze urea in a new solution containing the pH-sensitive dye phenol red (Fig. 5b). The hydrolysis of urea generates ammonia, causing the pH to rise, accompanied by a color change from yellow to pink (Tram et al. 2014). This simple method can detect as little as 500 *E. coli* cells.

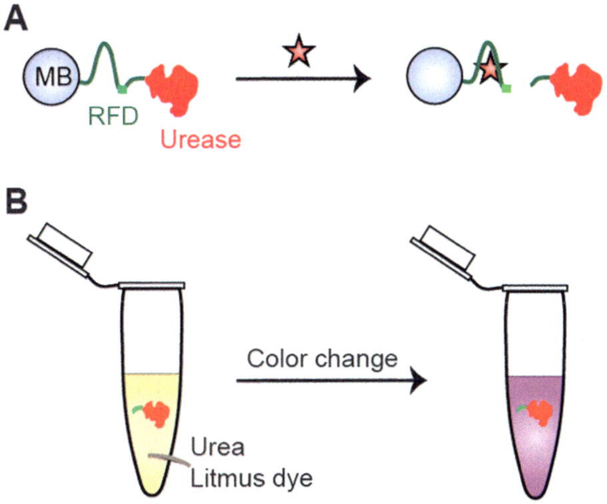

Fig. 5 Bacterial litmus test. (**a**) Cleavage reaction. Star: the bacterium of interest for which an RFD is created. MB: magnetic beads. The cleavage of RFD immobilized on the magnetic bead results in the release of urease to solution. (**b**) Colorimetric reporting. Upon magnetic separation, the released urease is taken to hydrolyze urea in the presence of a litmus dye. The hydrolysis of urea elevates the pH of the test solution, leading to the sharp color change

3.3 Paper Sensors Printed with Bacteria-Activated DNAzymes

Bacteria-responsive DNAzymes can also be used to make paper-based sensors. For example, we have recently shown that a paper sensor for *E. coli* can be produced via inkjet printing using a specially formulated ink containing RFD-EC1 and pullulan (Ali et al. 2017). Pullulan is a polysaccharide with a highly attractive property: upon drying it solidifies into films that stabilize the entrapped biomolecules for a long period of time but readily dissolve in an aqueous solution (Jahanshahi-Anbuhi et al. 2014, 2016; Hsieh et al. 2017). The entrapped molecules have a very long shelf life. For example, when printed onto paper surface with pullulan, RFD-EC1 remains stable for at least 6 months (Hsieh et al. 2017). However, upon the addition of an *E. coli*-containing solution, RFD-EC1 cleaves and generates high levels of a fluorescent signal that can be easily detected. The testing procedure is simple, straightforward, and the results are easy to interpret (Ali et al. 2017). Product stability during ambient storage is a huge advantage for this detection device; it is ideal for transportation and storage in resource-limited regions.

4 Conclusion

Our ability to isolate single-stranded DNA sequences with catalytic activities from random-sequence DNA pools provides endless possibilities of engineering DNA-based enzymes for practical applications. In this chapter, we have only discussed a few examples of RNA-cleaving DNAzymes that have been examined as therapeutics or as sensors for the detection of toxic metal ions and bacterial pathogens. It is certain that more and better DNAzymes will continue to emerge from in vitro selection experiments, which will allow the development of better therapeutics, diagnostics, as well as other useful chemical or biological tools.

References

Achenbach JC, Chiuman W, Cruz RP, Li Y (2004) DNAzymes: from creation in vitro to application in vivo. Curr Pharm Biotechnol 5(4):321–336

Ali MM, Aguirre SD, Lazim H, Li Y (2011) Fluorogenic DNAzyme probes as bacterial indicators. Angew Chem Int Ed 50(16):3751–3754

Ali MM, Aguirre SD, Mok WW, Li Y (2012) Developing fluorogenic RNA-cleaving DNAzymes for biosensing applications. Methods Mol Biol 848:395–418

Ali MM, Brown CL, Jahanshahi-Anbuhi S, Kannan B, Li Y, Filipe CDM, Brennan JD (2017) A printed multicomponent paper sensor for bacterial detection. Sci Rep 7(1):12335

Bock LC, Griffin LC, Latham JA, Vermaas EH, Toole JJ (1992) Selection of single-stranded DNA molecules that bind and inhibit human thrombin. Nature 355(6360):564–566

Breaker RR (1997) DNA enzymes. Nat Biotechnol 15(5):427–431

Breaker RR, Joyce GF (1994) A DNA enzyme that cleaves RNA. Chem Biol 1(4):223–229

Carrigan MA, Ricardo A, Ang DN, Benner SA (2004) Quantitative analysis of a RNA-cleaving DNA catalyst obtained via in vitro selection. Biochemistry 43(36):11446–11459

Cho EA, Moloney FJ, Cai H, Au-Yeung A, China C, Scolyer RA, Yosufi B, Raftery MJ, Deng JZ, Morton SW, Hammond PT, Arkenau HT, Damian DL, Francis DJ, Chesterman CN, Barnetson RSC, Halliday GM, Khachigian LM (2013) Safety and tolerability of an intratumorally injected DNAzyme, Dz13, in patients with nodular basal-cell carcinoma: a phase 1 first-in-human trial (DISCOVER). Lancet 381(9880):1835–1843

Cruz RP, Withers JB, Li Y (2004) Dinucleotide junction cleavage versatility of 8-17 deoxyribozyme. Chem Biol 11(1):57–67

Ellington AD, Szostak JW (1990) In vitro selection of RNA molecules that bind specific ligands. Nature 346(6287):818–822

Ellington AD, Szostak JW (1992) Selection in vitro of single-stranded DNA molecules that fold into specific ligand-binding structures. Nature 355(6363):850–852

Faulhammer D, Famulok M (1996) The Ca^{2+} ion as a cofactor for a novel RNA-cleaving deoxyribozyme. Angew Chem Int Ed 35(23–24):2837–2841

Faulhammer D, Famulok M (1997) Characterization and divalent metal-ion dependence of in vitro selected deoxyribozymes which cleave DNA/RNA chimeric oligonucleotides. J Mol Biol 269 (2):188–202

Fokina AA, Stetsenko DA, François J (2015) DNA enzymes as potential therapeutics: towards clinical application of 10-23 DNAzymes. Expert Opin Biol Ther 15(5):689–711

Geyer CR, Sen D (1997) Evidence for the metal-cofactor independence of an RNA phosphodiester-cleaving DNA enzyme. Chem Biol 4(8):579–593

Grassi G, Grassi M (2013) First-in-human trial of Dz13 for nodular basal-cell carcinoma. Lancet 381(9880):1797–1798

Guerrier-Takada C, Gardiner K, Marsh T, Pace N, Altman S (1983) The RNA moiety of ribonuclease P is the catalytic subunit of the enzyme. Cell 35(3):849–857

He S, Qu L, Shen Z, Tan Y, Zeng M, Liu F, Jiang Y, Li Y (2014) Highly specific recognition of breast tumors by an RNA-cleaving fluorogenic DNAzyme probe. Anal Chem 87(1):569–577

Hollenstein M, Hipolito C, Lam C, Dietrich D, Perrin D (2008) A highly selective DNAzyme sensor for mercuric ions. Angew Chem Int Ed 47(23):4346–4350

Homburg U, Turowska A, Kuhlmann J, Muller A, Renz J, Bille J, Renz H, Garn H (2013) Safety profile and pharmacokinetics of SB010, an inhaled GATA-3-specific DNAzyme, in phase I clinical trials in healthy and asthmatic subjects. Eur Respir J 42:4858

Hsieh PY, Monsur Ali M, Tram K, Jahanshahi-Anbuhi S, Brown CL, Brennan JD, Filipe CD, Li Y (2017) RNA protection is effectively achieved by Pullulan film formation. Chembiochem 18 (6):502–505

Jahanshahi-Anbuhi S, Pennings K, Leung V, Liu M, Carrasquilla C, Kannan B, Li Y, Pelton R, Brennan JD, Filipe CD (2014) Pullulan encapsulation of labile biomolecules to give stable bioassay tablets. Angew Chem Int Ed 53(24):6155–6158

Jahanshahi-Anbuhi S, Kannan B, Leung V, Pennings K, Liu M, Carrasquilla C, White D, Li Y, Pelton RH, Brennan JD, Filipe CDM (2016) Simple and ultrastable all-inclusive pullulan tablets for challenging bioassays. Chem Sci 7(3):2342–2346

John GK, Mullin GE (2016) The gut microbiome and obesity. Curr Oncol Rep 18(7):45

Kasprowicz A, Stokowa-Sołtys K, Wrzesiński J, Jeżowska-Bojczuk M, Ciesiołka J (2015) In vitro selection of deoxyribozymes active with Cd^{2+} ions resulting in variants of DNAzyme 8–17. Dalton Trans 44:8138–8149

Kruger K, Grabowski PJ, Zaug AJ, Sands J, Gottschling DE, Cech TR (1982) Self-splicing RNA: autoexcision and autocyclization of the ribosomal RNA intervening sequence of Tetrahymena. Cell 31(1):147–157

Lakin MR, Stojanovic MN, Stefanovic D (2017) Implementing molecular logic gates, circuits, and cascades using DNAzymes. In: Adamatzky A (ed) Advances in unconventional computing. Emergence, complexity and computation, vol 23. Springer, Charm

Lee JH, Wang Z, Liu J, Lu Y (2008) Highly sensitive and selective colorimetric sensors for uranyl (UO_2^{2+}): development and comparison of labeled and label-free DNAzyme-gold nanoparticle systems. J Am Chem Soc 130(43):14217–14226

Li Y, Breaker RR (2001) In vitro selection of kinase and ligase Deoxyribozymes. Methods 23 (2):179–190

Li J, Lu Y (2000) A highly sensitive and selective catalytic DNA biosensor for lead ions. J Am Chem Soc 122(42):10466–10467

Li J, Zheng W, Kwon AH, Lu Y (2000) In vitro selection and characterization of a highly efficient Zn(II)-dependent RNA-cleaving deoxyribozyme. Nucleic Acids Res 28(2):481–488

Liu J, Brown AK, Meng X, Cropek DM, Istok JD, Watson DB, Lu Y (2007) A catalytic beacon sensor for uranium with parts-per-trillion sensitivity and millionfold selectivity. Proc Natl Acad Sci U S A 104(7):2056–2061

Liu M, Chang D, Li Y (2017) Discovery and biosensing applications of diverse RNA-cleaving DNAzymes. Acc Chem Res 50(9):2273–2283

Mager DL (2006) Bacteria and cancer: cause, coincidence or cure? A review. J Transl Med 4:14

Morrison D, Rothenbroker M, Li Y (2018) DNAzymes: selected for applications. Small Methods 2 (3):1700319

Nguyen PL, Sekhon SS, Ahn JY, Ko JH, Lee L, Cho SJ, Min J, Kim YH (2017) Aptasensor for environmental monitoring. Toxicol Environ Health Sci 9(2):89–101

Ponce-Salvatierra A, Wawrzyniak-Turek K, Steuerwald U, Höbartner C, Pena V (2016) Crystal structure of a DNA catalyst. Nature 529(7585):231–234

Robertson DL, Joyce GF (1990) Selection in vitro of an RNA enzyme that specifically cleaves single-stranded DNA. Nature 344(6265):467–468

Santoro SW, Joyce GF (1997) A general purpose RNA-cleaving DNA enzyme. Proc Natl Acad Sci U S A 94(9):4262–4266

Schlosser K, Li Y (2009) Biologically inspired synthetic enzymes made from DNA. Chem Biol 16 (3):311–322

Schlosser K, Li Y (2010) A versatile endoribonuclease mimic made of DNA: characteristics and applications of the 8-17 RNA-cleaving DNAzyme. Chembiochem 11(7):866–879

Schlosser K, Lam JC, Li Y (2009) A genotype-to-phenotype map of in vitro selected RNA-cleaving DNAzymes: implications for accessing the target phenotype. Nucleic Acids Res 37 (11):3545–3557
Schürch AC, Siezen RJ (2010) Genomic tracing of epidemics and disease outbreaks. Microb Biotechnol 3(6):628–633
Shen Z, Wu Z, Chang D, Zhang W, Tram K, Lee C, Kim P, Salena BJ, Li Y (2016) A catalytic DNA activated by a specific strain of bacterial pathogen. Angew Chem Int Ed 55(7):2431–2434
Stojanovic MN, Mitchell TE, Stefanovic D (2002) Deoxyribozyme-based logic gates. J Am Chem Soc 124(14):3555–3561
Tang Q, Yuan Y, Xiao X, Guo P, Hu J, Ma D, Gao Y (2013) DNAzyme based electrochemical sensors for trace uranium. Microchim Acta 180(11–12):1059–1064
Torabi S, Lu Y (2015) Identification of the same Na^+-specific DNAzyme motif from two in vitro selections under different conditions. J Mol Evol 81(5–6):225–234
Torabi SF, Wu P, Mcghee CE, Chen L, Hwang K, Zheng N, Cheng J, Lu Y (2015) In vitro selection of a sodium-specific DNAzyme and its application in intracellular sensing. Proc Natl Acad Sci U S A 112(19):5903–5908
Tram K, Kanda P, Li Y (2012) Lighting up RNA-cleaving DNAzymes for biosensing. J Nucleic Acids 2012:958683
Tram K, Kanda P, Salena BJ, Huan S, Li Y (2014) Translating bacterial detection by DNAzymes into a litmus test. Angew Chem Int Ed 53(47):12799–12802
Tuerk C, Gold L (1990) Systematic evolution of ligands by exponential enrichment: RNA ligands to bacteriophage T4 DNA polymerase. Science 249(4968):505–510
Upadhyaya S, Banerjee G (2015) Type 2 diabetes and gut microbiome: at the intersection of known and unknown. Gut Microbes 6(2):85–92
Wang B, Cao L, Chiuman W, Li Y, Xi Z (2010) Probing the function of nucleotides in the catalytic cores of the 8−17 and 10−23 DNAzymes by abasic nucleotide and C3 spacer substitutions. Biochemistry 49(35):7553–7562
Wu P, Hwang K, Lan T, Lu Y (2013) A DNAzyme-gold nanoparticle probe for uranyl ion in living cells. J Am Chem Soc 135(14):5254–5257
Wu Z, Fan H, Satyavolu NS, Wang W, Lake R, Jiang J, Lu Y (2017) Imaging endogenous metal ions in living cells using a DNAzyme-catalytic hairpin assembly probe. Angew Chem Int Ed 129(30):8847–8851
Xiang Y, Lu Y (2011) Using personal glucose meters and functional DNA sensors to quantify a variety of analytical targets. Nat Chem 3(9):697–703
Xiang Y, Lu Y (2013) An invasive DNA approach toward a general method for portable quantification of metal ions using a personal glucose meter. Chem Commun 49(6):585–587
Xiang Y, Wang Z, Xing H, Wong NY, Lu Y (2010) Label-free fluorescent functional DNA sensors using unmodified DNA: a vacant site approach. Anal Chem 82(10):4122–4129
Zhou W, Saran R, Chen Q, Ding J, Liu J (2015a) A new Na^+-dependent RNA-cleaving DNAzyme with over 1000-fold rate acceleration by ethanol. Chembiochem 17(2):159–163
Zhou W, Zhang Y, Huang PJ, Ding J, Liu J (2015b) A DNAzyme requiring two different metal ions at two distinct sites. Nucleic Acids Res 44(1):354–363
Zhou W, Ding J, Liu J (2016a) A highly specific sodium aptamer probed by 2-aminopurine for robust Na^+ sensing. Nucleic Acids Res 44(21):10377–10385
Zhou W, Ding J, Liu J (2016b) A selective Na^+ aptamer dissected by sensitized Tb^{3+} luminescence. Chembiochem 17(16):1563–1570

β-Amylase: General Properties, Mechanism and Panorama of Applications by Immobilization on Nano-Structures

Ranjana Das and Arvind M. Kayastha

1 Introduction

The polyglucan homopolysaccharide i.e. starch, is the major constituent of human diet. It is a storage polysaccharide and is composed of single type of monosaccharide called glucose at different anomeric conformation. This starch serves as energy source in all the genera by its breakdown into simple sugar, by a group of enzymes (Husain 2017). β-Amylases are the enzymes hydrolyzing α,1-4 glycosidic linkages in starch and related polysaccharide and are widely distributed in plants as well as microbes. β-Amylase is a member of family 14 of glycosyl hydrolases and catalyzes the hydrolysis of alternate α-1,4-glycosidic linkages in starch and related polysaccharide starting from the non-reducing end. The enzyme releases maltose, continuing until the entire chain is cleaved or the enzyme encounters blockage by a physical or chemical irregularity in the chain. Unlike many other enzymes, the term 'beta' does not imply to the linkage present in the substrate, but the inversion of configuration which results during β-amylase action on starch (Henrissat 1991). β-Amylase is also known to promote various less known reactions which follow other stereochemical paths. Thus, though the enzyme does not hydrolyze maltose (Genghof et al. 1978), it does convert the β-anomer of maltose to maltotetraose to a small extent by a condensation reaction, that is the reversal of maltotetraose hydrolysis (Hehre et al. 1969). β-Amylase further catalyzes the slow irreversible hydration of maltal, an enolic glycosyl donor lacking α- or β-anomeric configuration, to form β-2-deoxymaltose (Kitahata et al. 1991).

The systemic name recommended by the International Union of Biochemistry and Molecular Biology (IUBMB) for β-amylase is "4-α-D-glucan maltohydrolase". It is based on the type of reaction, which the enzyme catalyzes, and on their substrate specificity. Enzyme Commission Number (EC number) assigned to β-amylase is 3.2.1.2 [3: hydrolase (class); 2: glycosylases (substrate catalysis); 1: glycosidase

R. Das · A. M. Kayastha (✉)
School of Biotechnology, Institute of Science, Banaras Hindu University, Varanasi, India

Q. Husain, M. F. Ullah (eds.), *Biocatalysis*,
https://doi.org/10.1007/978-3-030-25023-2_2

(hydrolyze O- and S-glycosyl compounds); 2: serial number]. Chemical Abstracts Service has assigned CAS registry number: 9000-91-3. Since the structure of starch is irregular and intricate, the action of highly peculiar enzyme such a β-amylase had been of great interest in determining a more precise picture of starch structure.

2 History

Kuhn (1924) classified the saccharogenic amylase of malt as a β-amylase on the basis of its ability to convert starch into the β-anomeric form of maltose. Ohlsson (1930), discovered another amylase which yielded β-mannose and named it as β-amylase. Hopkins et al. (1948) were first to study the kinetics of β-amylase on potato amylose. In the 1960s and 1970s, various techniques in the purification of sweet potato β-amylase were improved (Nakayama and Amagase 1963; Takeda and Hizukuri 1969). Thereafter, the enzyme was also successfully immobilized on Agarose and Sepharose gels in the late 1970s (Caldwell et al. 1976a, b). Cloning of the enzyme was successfully carried out in *E. coli* by Yoshida and Nakamura (1991). Amino acid sequence and primary crystal structure of the tetrameric enzyme were determined in 1993 (Cudney and McPherson 1993). The crystal structure was refined in 1995 (Cheong et al. 1995). The evolution of β-amylases was studied by Pujadas et al. (1996), who discovered the enzyme to be an example of stingy divergence illustrated by signature structural motifs. Finally, in 2001, a simple purification method using affinity precipitation was developed (Teotia et al. 2001). Tanaka et al. (2002) studied the effects of guanidine hydrochloride and increased pressure on the activity of the enzyme.

The amylase family is of immense biotechnological significance owing to their large area of applications but less attention has been paid towards β-amylases, which had industrial utility in food and pharmaceutics. So it becomes necessary to create a platform to discuss about the enzyme β-amylase. In this chapter, attention will be paid on molecular aspects, reaction mechanism and recent advancements in applications. Though most of the peculiar features are discussed here, but it is impossible to cover the enormous information available on the subject.

3 Classification: The β-Amylase Family in Plants

Plant genome is encoded by several β-amylase–like proteins (Fulton et al. 2008). In *Arabidopsis*, there are nine genes, which was designated BAM1 to BAM9 by Smith et al. (2004) to provide a unifying nomenclature. Of these nine, only BAM3 is involved in starch degradation till date (Kaplan and Guy 2005; Lao et al. 1999). Sparla et al. (2006) showed that BAM1 (also called TR-BMY and BMY7), like BAM3, encodes an active, chloroplast-targeted β-amylase. BAM1 particularly is regulated *via* thioredoxin mediated reduction (hence the abbreviation TR-BMY;

Sparla et al. 2006). Nonetheless, primary analysis suggested that mutation of BAM1 does not result in excess leaf starch, and its function is not yet defined (Kaplan and Guy 2005). BAM2 (also called BMY9) and BAM4 (also called BMY6) are also predicted as N-terminal chloroplast transit peptides (Lloyd et al. 2005; Smith et al. 2005). However, these predictions are weak, and neither protein has been localized experimentally. Besides, Kaplan and Guy (2005) reported that mutation of BAM2 gene did not result in elevated leaf starch.

To our interest, most of the β-amylase proteins are expected to be localized outside the chloroplast, and their functions are unknown. Amongst the genes, only BAM5 (also called BMY1 and RAM1) has been studied in depth. According to the report of (Laby et al. 2001), 90% of the β-amylase activity in *Arabidopsis* leaves is encoded by the BAM5 locus. Formerly Wang et al. (1995) had reported that the BAM5 protein is localized to the phloem sieve elements. However, mutation of the BAM5 gene did not affect phloem function or leaf starch levels (Laby et al. 2001). To gain more insight into the roles of the β-amylases, the phylogeny of the β-amylase gene family was investigated using reverse genetics to systematically study the functions of those proteins predicted to be targeted to the chloroplast. The results strongly supported the hypothesis that β-amylase is a major enzyme of starch breakdown but also indicated the specialization of BAM family members, including a novel role for a catalytically inactive, β-amylase-like protein in metabolic regulation. The analysis indicated that plant β-amylases can be divided into four major subfamilies (Fig. 1). Each subfamily contained sequences from rice, poplar and *Arabidopsis*, which showed that these subfamilies are conserved in higher plants.

Subfamily I contain β-amylases localized to different cellular compartments. It includes *Arabidopsis* BAM5 and BAM6, together with active β-amylases from soybean and sweet potato (Cheong et al. 1995; Mikami et al. 1993, 1994). BAM5 is an extra-plastidial protein. BAM6 does not carry a predicted transit peptide, and localized to the plastids (Fulton et al. 2008). Therefore, it can be concluded that other proteins present in subfamily I may be targeted to the chloroplasts despite the absence of an apparent transit peptide.

Subfamily II contains BAM1 and BAM3, the two proteins shown to encode active, chloroplastic enzymes involved in transitory starch degradation, particularly at night (Lao et al. 1999; Sparla et al. 2006). Both *Arabidopsis* proteins have putative orthologs in other species, including rice (*Oryza sativa*) and poplar (*Populus sp*).

Subfamily III is comprised of eight sequences which include *Arabidopsis* proteins BAM4 and BAM9 including monocot and dicot sequences. Sequences in this family are more divergent from rest of the three families. Three proteins of this family are predicted to carry transit peptides and there is experimental evidence for the chloroplastic localization of *Arabidopsis* BAM4. However, *Arabidopsis* BAM9 also localizes to the chloroplasts despite the lack of a predicted transit peptide (Fulton et al. 2008), suggesting that other members of subfamily III may be chloroplastic. The proteins found in subfamily III cannot be expected to possess catalytic activity (except for two sequences: putative rice protein and banana β-amylase), because they do not contain the catalytic residue Glu^{380} (Kang et al. 2004; Totsuka and Fukazawa 1996).

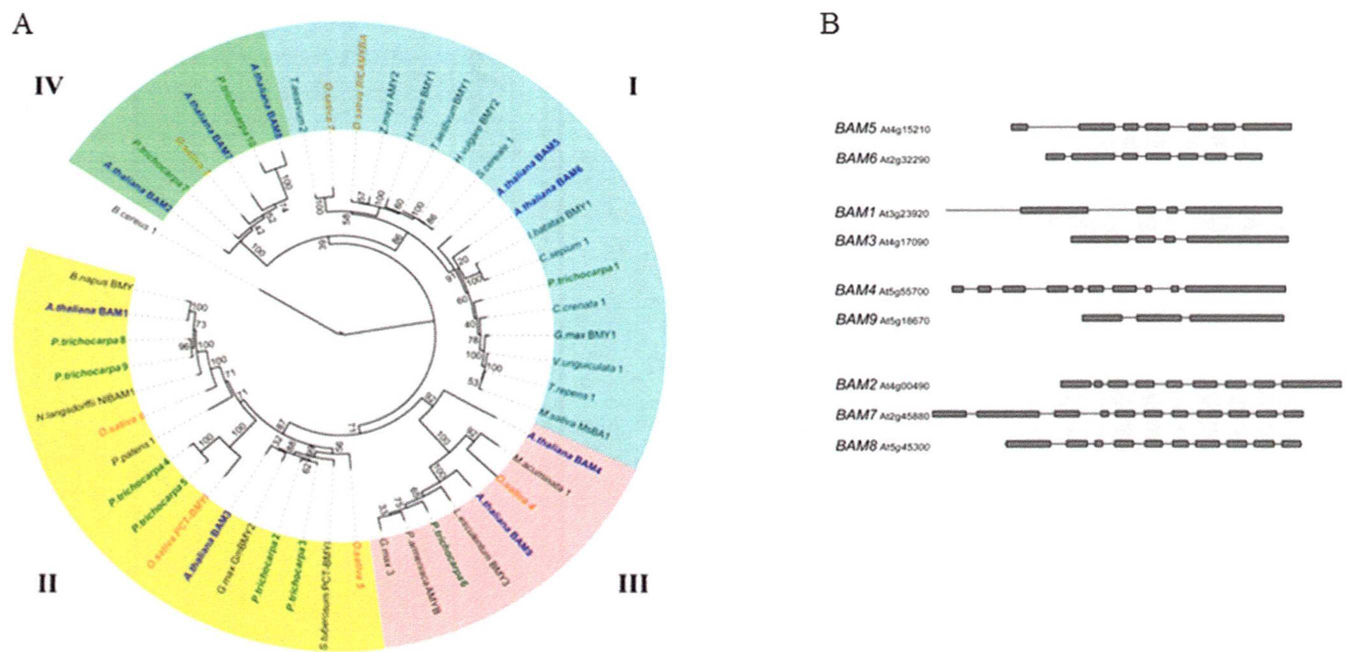

Fig. 1 (**a**) Phylogram showing four families of plant β-amylase. The core β-amylase domains of 48 plant β-amylases, corresponding to amino acids 17 to 439 of the soybean protein, were aligned and used to generate maximum-likelihood tree presented using Tree Of Life software. β-Amylase from *Bacillus cereus*, served as an outlier to root the tree. The robustness of the tree was derived from 100 bootstrap replicates. *Arabidopsis* proteins are shown in blue, rice proteins in orange, and poplar proteins in green. (**b**) Intron-exon structures of the genes encoding the *Arabidopsis* BAM family. Gene names and AGI identifiers are shown on the left. Exons are shown as grey boxes; introns are shown as black lines. Sequence conservation between genes is indicated by grey shading (Adapted from Fulton et al. 2008)

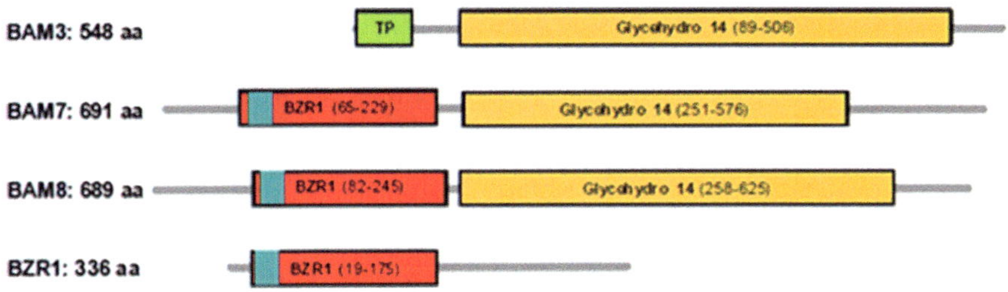

Fig. 2 Domain structures of the two *Arabidopsis* BZR1-BAMs, BAM7 and BAM8 (Adapted from Reinhold et al. 2011)

Subfamily IV contains chloroplastic (*Arabidopsis* BAM2) and non-chloroplastic β-amylase sequences (*Arabidopsis* BAM7 and BAM8), which localize to the nucleus. BAM2 has very low specific activity inside the cell and thus, its physiological function could not be assigned until today. BAM7 and 8 differ from other *Arabidopsis* β-amylase family in carrying an additional N-terminal domain of the BZR1-types. Other BZR1-BAM sequences are encoded by rice, poplar and many other plant genomes which are present in subfamily IV. Localization of both the proteins in the nucleus was based on their amino acid sequence, determined by WoLF PSORT and BaCelLo protein localization predictor (Horton et al. 2007; Pierleoni et al. 2006). BAM2 and BAM7 are putative paralogs, residing within a segment of chromosomal duplication (Fulton et al. 2008).

The phylogenetic relations between the *Arabidopsis* β-amylase sequences can also be speculated by the gene structures of BAM1-9 (Figs. 1b and 2). The intron-exon structure between the glucosyl hydrolase encoding sequences of BAM2, BAM7, and BAM8 are highly conserved except for the fact that BAM7 and BAM8 are extended by a 5′ sequence encoding the BZR1-type domain (shown in red color) whereas BAM2 encodes an N-terminal transit peptide (TP shown in green color in figure). Interestingly, BAM7 is predicted to encode a large 5′ UTR (spanning the first and second exon), which is not the case for BAM8.

4 Structural Characteristics of β-Amylase

The 3D structures derived from X-ray crystallography of β-amylases are known for sweet potato (Cheong et al. 1995), soybean (Mikami et al. 1993, 1994), barley (Mikami et al. 1999a), *Bacillus cereus* (Mikami et al. 1999b) etc. The structure of soybean β-amylase is composed of $(\beta/\alpha)_8$ core domain and a C-terminal loop region. The catalytic site of the soybean β-amylase (SBA) lies in a sac like structure of 18 Å deep cleft, and allows the endwise hydrolysis of unbranched amylose and maltosaccharide chains. Two molecules of glucose are known to bind in tandem in the active site pocket of SBA, occupying the subsites 1–2 and 3–4, where two catalytic residues, Glu^{186} and Glu^{380}, are situated between subsites 3 and 4. The open-close movement of the flexible loop (residues 96–103 in SBA) is known to play an important role in the catalytic procedure. The aforementioned amino acid residues are important for catalysis and binding of the substrate and well-conserved between plant and bacterial enzymes, however only 29% identical residues are found between *B. cereus* β-amylase and SBA in the core domain. The flexible loop in the closed position helps to protect the reaction center from solvent, and assists an ordered water molecule adjacent to Glu^{380} to provide the steric outcome of the hydrolysis/hydration reaction. This structural arrangement explains the role of α-cyclodextrin and maltose as competitive inhibitors of β-amylase (Mikami et al. 1993, 1994).

Chemical evaluation revealed that one disulfide bridge and one sulfhydryl group exist in *B. cereus* β-amylase and *B. polymyxa* β-amylase, whereas plant enzyme contains several sulfhydryl groups (Uozumi et al. 1991). Cys^{91} in *B. cereus* β-amylase forms the disulfide bridge near the flexible loop. This loop contains the aspartic acid and valine residues which interact with glucose residues at subsites 1 and 4, respectively. The disulfide bridge of *B. cereus* β-amylase (Cys^{90}–Cys^{99}) is thought to stabilize the flexible loop (residues 91–97), without affecting the open-close movements. The structure of the core domain of *B. cereus* β-amylase is similar to that of soybean β-amylase, except for the two novel maltose binding sites in the L4 region and in the C-terminal domain. The property of *B. cereus* β-amylase to digest raw starch can be attributed to these two maltose-binding sites (Oyama et al. 1999).

5 Distribution and Functional Properties of β-Amylase

β-Amylases are not ubiquitous in nature and have been purified from a number of plant, bacterial, fungal sources, but absent in animal kingdom. β-Amylase is synthesized by various plants like soybean (Mikami et al. 1994), sweet potato (Cheong et al. 1995), barley (Mikami et al. 1999b), rye seeds (Sadowski et al. 1993), alfalfa (*Medicago sativa*) tap roots (Doehlert et al. 1982), *Arabidopsis* sp. (Laby et al. 2001), rhizome of *Calystegia sepium* (Van Damme et al. 2001). Various bacterial strains (aerobic and anaerobic) like *B. cereus*, *B. megaterium*, *B. polymyxa*, *Clostridium thermosulfurogenes*, *Thermoactinomyces sp.* (Ray 2004; Ray and Nanda 1996) etc. were found to synthesize β-amylases. Cloning and amino acid sequencing was done in many strains of *Bacillus sp.* among which the cloned β-amylase gene from *Bacillus cereus var. mycoides* was found to have 1638 bp for coding (Yamaguchi et al. 1996). Miyake et al. (2002) explained the mechanism of the chemical rescue of catalytic activity of β-amylase from *B. cereus var. mycoides* by azide. A β-amylolytic strain of *Clostridium thermosulfurogenes SV2* was found to produce pullulanase and another neopullulanase showing β-amylolytic activity. It was also found to be present in *B. stearothermophilus* (Kamasaka et al. 2002).

Early work on highly active crude enzyme preparations led to the opinion that these enzymes are simple proteins, with no evidence for requirement of cofactors, metal ions or non-protein active groups (Bernfeld 1955). β-Amylase is a unique enzyme in having wide range of operational parameters. The ranges of parameters

Table 1 Operational parameters and their range for efficient hydrolysis by β-amylase

Operational parameters	Range
pH optimum	4.5–6.5
Temperature optimum	30–60/75 °C
K_m (starch)	1.25–13.6 mg/mL
Specific activity	5–4664 μmol/min/mg
pI	4.45–6.4
Mol. Wt.	19.6–315 kDa
Turnover number	299–6540 s^{-1}

have been given in the Table 1. As we can see the optimum pH range for β-amylase is 4.5–6.2 for most β-amylases. However, pH optima as low as 3.5 for *Hordeum vulgare* (Yoshigi et al. 1994) and as high as 8–10 for *Halobacillus species* (Li and Yu 2011) have also been reported. The stability range of pH varies in between 2 and 9 with approximately 35 and 75% of maximum activity, respectively, for the two pH values. Plant β-amylases have an optimum temperature upto a maximum of 65 °C but microbial enzyme can go beyond 75 °C. Mostly the range is 50–65 °C for both plant and bacterial enzyme. Approximate molecular mass of the enzyme ranges in between 50 and 65 kDa. The protein as small as 19.6 kDa has been reported from *Triticum aestivum* (Rexova et al. 1967) and as large as 315 kDa from *Ipomoea batatas* with enzyme modified with periodate-oxidized maltohexaose has been reported (Ann et al. 1990). K_m for most of the enzymes lies between 1 and 3 mg/mL. Turnover number which is a measure of the catalytic activity of the enzyme ranges between 200 and 400 s^{-1}. Highest turnover number for *Curculigo pilosa*, 6540 s^{-1}, has been reported (Dicko et al. 1999).

β-Amylases from the starchless seeds of fenugreek, led to a new finding in association of the enzyme with protophloem, establishing its role in germination. The enzyme had a high specific activity of 732 units/mg (Srivastava and Kayastha 2014). Specific activity of β-amylase have been reported to be as low as 5 units/mg (Viksø-Nielsen et al. 1997), to as high as 4664 units/mg (Chang et al. 1996). Recently, a novel antioxidative, native starch digesting β-amylase was purified from peanuts which could have potential employment in food and pharmaceutical industries (Das and Kayastha 2018). Apart from seeds, the enzyme is also purified from other parts of the plant like leaves of ramie (He et al. 2017), stems of *Abrus precatorius* (Sagu et al. 2015), roots of *Paederia foetida* (Sottirattanapan et al. 2017). Production of the enzyme by a new strain *Paenibacillus chitinolyticus* CKS1 was also performed using molasses and sugar beet pulp (Mihajlovski et al. 2016).

Various metallic salts and chemical reagents have differential effect on the activity of β-amylase. Divalent cations like Ca^{2+}, Mg^{2+}, Zn^{2+}, Fe^{3+} and EDTA are known to increase the activity of purified β-amylase and provide stability. Iodoacetic acid, N-ethylmaleimide, urea and Fe^{2+} reduce the activity moderately. The enzyme is completely inhibited by the presence of heavy metals such as Cu^{2+}, Pb^{2+}, Sn^{2+}, Hg^{2+} and by SH-inhibitor such as PCMB, the activity being restored by thiol compounds, such as cysteine. This indicates that a -SH group exists in the molecular structure of the enzyme, as in other plant β-amylases. Plant β-amylases have been reported to require free sulfhydryl groups for activity and simultaneously inhibited by heavy metals as well as other binding reagents (Thoma et al. 1971; Tkachuk and Tipples 1966).

6 Action Pattern

The endwise action of β-amylase is well-established but substantial discussion has arisen over whether the enzyme degrades linear starch molecules (amylose) by a single chain or multi-chain mechanism (Fig. 3). The enzyme degrades one substrate molecule completely before it attacks a second molecule, in single chain mechanism. In the multi chain mechanism, enzyme having removed one maltose unit from the substrate, attacks a second chain molecule, degrading all the substrate molecules more or less homogeneously. Validation, for or against these theories has in general brought about with some measurement of the change in the average DP (degree of polymerization) of the substrate. Narrow DP distribution with no change in overall DP of the substrate, suggests the mechanism to be as single chain. The multi chain mechanism proceeds through decrease in average DP of the substrate during the course of reaction. Elucidation regarding the DP of the substrate has generally been

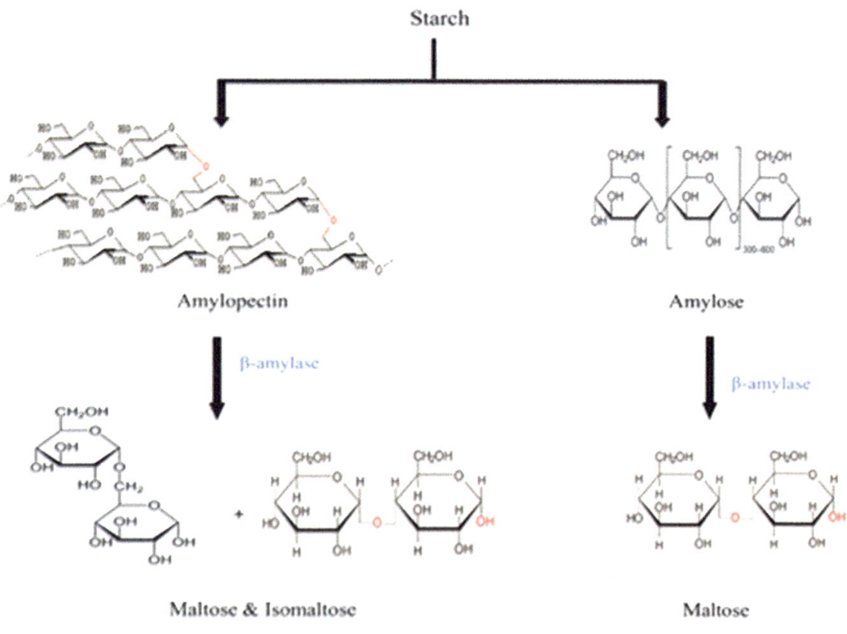

Fig. 3 Demonstrates hydrolysis of starch molecules to maltose by β-amylase

drawn from the iodine spectrum or from physical methods. In some cases, the action pattern has been studied using oligosaccharides as substrate (Youngquist 1962).

Swanson (1948) gave the view that since no shift was found in the iodine spectrum during β-amylase hydrolysis of the substrate, the enzyme must be proceeding in a single chain fashion. In the same year, Cleveland and Kerr (1948) also reached the climax that the mechanism is of the single chain type. Later, French et al. (1950) appended to this theory by using maltoheptaose as a substrate. The hydrolysate was examined by electrophoresis and no appreciable amount of maltopentaose was found. Whilst this was not a very sensitive method for maltopentaose, the results proposed that the enzyme was acting by a single chain mechanism. Hopkins et al. (1948) also implicated a multi chain action from similar information. French (1961) had reported that it is not possible to differentiate between the two theories when the distribution of the substrate chain length is the most probable distribution. Husemann and Pfannemüller (1961) had reported that the action of β-amylase on a synthetic polysaccharide with a moderately sharp chain length distribution seems to differ from the action of the enzyme on natural amylose. This authenticates French's theory.

β-Amylase also proceeds by multiple attack pattern of action, which is intermediary between single and multi chain attack. In this pattern, the enzyme having released the product ones, remains attached to the substrate and may degrade it further, unless the enzyme and partially degraded substrate segregate from each other. The probable number of hydrolytic actions during a successful enzyme-substrate meet is under the control of kinetic constants for the reaction. For multi chain mechanism, this number would be strictly zero, and for single chain, it would reach the number of approximate linkages per amylose molecules. The basic idea of multiple attack, likely to single chain mechanism, is the belief that end-wise attack (whether by synthesis or degradation) on a straight chain polymer of consistent structure drives to a homolog of the original substrate molecule.

7 Reaction Mechanism

Enzymologists have been attracted by two most peculiar reaction mechanisms of β-amylase. One is the specificity of enzyme to produce only maltose as the end product with traces of isomaltose. Secondarily, the fact that maltose produced in β-anomeric form, requires inversion of configuration accompanied by the hydrolysis reaction. These problems can be divided into two groups, namely, the components involved in the binding of enzyme with the substrate, and the chemistry of the catalytic step. Let us shed some more light on the mechanism of hydrolytic reaction on various polysaccharides by the enzyme β-amylase.

In the case of high molecular weight substrates like amylose and amylopectin, β-amylase exhibits hydrolytic phenomenon of multiple attack; i.e. the enzyme releases several products effectively from a single enzyme-substrate complex without dissociation, through multiple or repetitive attack involving many branching reaction paths (Bailey and French 1957; French and Youngquist 1963; Thoma et al. 1970).

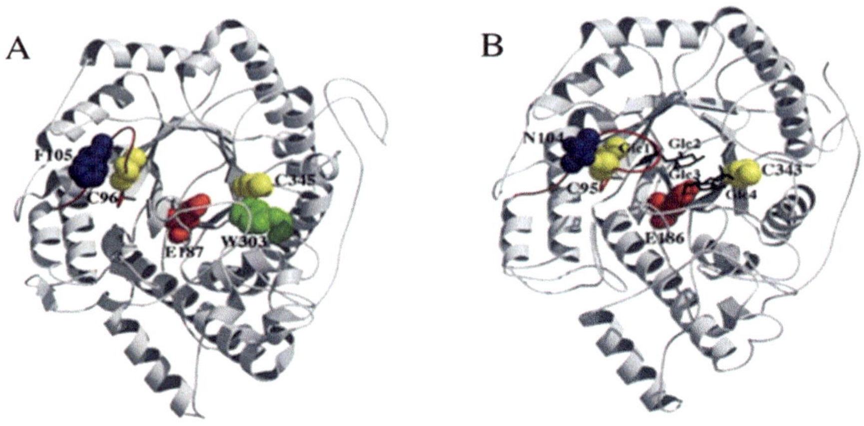

Fig. 4 β-Amylase structure of (**a**) sweet potato β-amylase and (**b**) soybean β-amylase complex with β-maltose (Adapted from Tanaka et al. 2002)

From the results of X-ray structure analysis of the mutant enzyme, it was clarified that one carboxyl residue plays an important role in the multiple attack. The action of multiple attack needs the force of enzyme sliding on the substrate. Also, for the multiple attack, it is important that the enzyme and substrate have the characteristic of a stable productive substrate-enzyme complex through formation of hydrogen bond between the non-reducing end of the substrate and the carboxyl residue of the β-amylase (Ishikawa et al. 2007).

As demonstrated by X-ray crystallographic study, the active site is located in the cleft of the enzyme, and the substrate binding site at the non-reducing end is blocked by a part of the β-barrel structure of the enzyme. The structure of soybean β-amylase that was resolved by crystal structure complexed with β-maltose showed that Glu^{186} and Glu^{380} are catalytic residues, as previously discussed in this chapter. The glucose binding sites on both sides of these residues are postulated as subsites 1 and 2, and subsites 3 and 4. The Cys^{95} residue (conserved as Cys^{96} in sweet potato β-amylase) located on a flexible loop (Fig. 4), forms part of the active site and Cys^{345} (Cys^{343} in sweet potato β-amylase) is in the vicinity of subsites 3 and 4.

In the soybean β-amylase, Glu^{186} plays role as a general acid and Glu^{380} plays role as a general base and the hydrolysis of glycosidic bond proceeds as a general acid-base catalysis mechanism. The carboxyl group of Glu^{186} is positioned on the hydrophilic surface of the glucose, and donates a proton to the glycosidic oxygen. The carboxyl group of Glu^{380} is located on the hydrophobic surface of the glucose residue at the subsite1 and activates an incoming water molecule. After cleavage of the glycosidic bond, the deprotonated Glu^{186} is stabilized by Thr^{342} (Kang et al. 2004; Mikami et al. 1994). The maltose produced has the reducing glucose in the β-form.

After resolving the crystal structure, it was found that SBA shares 67% amino acid identity and has the same β/α barrel as sweet potato β-amylase (Mikami et al. 1993, 1994). The structural difference between α-cyclodextrin (α-CD)/β-amylase and β-maltose/β-amylase complex indicated that saccharide binding induces a significant local conformational change, which induces the release and removal of the reaction product. Furthermore, within the α-CD/β-amylase complex, only one glucose residue in α-CD binds near subsite 4 and the exact position of the glucose residue is shifted by half of one residue towards the reducing end from subsite 4. This positional flexibility of glucose binding on subsite 4 may be responsible for starch slipping in a single-chain attack mechanism. Using high pressure, it was found that a local conformation in the vicinity of subsites 3 and 4 and the large amplitude of this local fluctuation might be responsible for the positional flexibility of glucose binding (Tanaka et al. 2002).

8 β-Amylase Engineering

Traditional mutation or cloning can be used to increase the enzyme yield or to improve the strain quality for enhanced production of β-amylase. Multistep mutagenesis was reported to multiply the yield of β-amylases in *Saccharomyces fibuligera* (Ray and Chakraverty 1998) and MNNG (1-Methyl-3-nitro-1-nitrosoguanidine) was used as a potent mutagen. A raw starch digesting β-amylase was produced from a mutant strain of *A. nidulans MNU82* (Chatterjee et al. 1992), formed by multistep mutagenesis using successive UV and MNNG treatment.

9 Role of β-Amylase in Starch Degradation

Main function of the enzyme is starch degradation which aids in seed germination process and ripening of the fruits, imparting sweet flavor. Starch is the main energy storage component in plants, present in the form of granules. Starch granules are highly organized semi-spherical structures that, depending on the plant source, display different morphology and composition. The variation in the granules occurs from the proportions and packaging of amylose and amylopectin—the two types of glucose polymers bound by glycosidic bonds. These two components of starch also differ in structure and their properties.

β-Amylase proceeds hydrolysis of the gelatinized starch from the non-reducing end, in vitro, producing maltose which finds application in food and pharmaceutics. Under in vivo condition, β-amylase play an important function in the hydrolysis of starch reserves in germinating cereal endosperm and has been extensively studied in this context (Ziegler 1999). Simultaneously high activities of β-amylase are also found in other plant tissues, including leaves (Doehlert et al. 1982). In pea and *Arabidopsis* for example, the activity of β-amylase exceeds the activities of other glucan-metabolizing enzymes by many fold (Zeeman et al. 1998).

It has been demonstrated from the antisense repression of a chloroplast-targeted isoform in potato, that β-amylase function in transitory starch breakdown comes.

The leaves of transformed plants have reduced β-amylase activity and exhibit decreased starch metabolization during night, compared with the wild-type plants. Consequently, the leaf starch contents at the end of the night are significantly increased. Likewise, observations were also made in *Arabidopsis* mutants, where specific β-amylase isoforms were eliminated by insertional mutagenesis. Till date, the importance of multiple isoforms of chloroplastic β-amylase in *Arabidopsis* is not clear. A possibility of functional redundancy of individual isoforms may have arisen through gene duplication. Preferably, each gene may have a distinct, tissue-specific expression pattern, or encode a protein with distinct catalytic and/or regulatory properties (Scheidig et al. 2002; Zeeman et al. 2004).

Majority of the β-amylase activity in *Arabidopsis* leaves is extraplastidial and its exact function is unclear. Their activity is attributable to a single isoform encoded at the RAM1 locus (reduced amylase). This isoform has been well-characterized and is reported to be localized in phloem sieve elements. Mutation of the RAM1 gene eradicates 90–95% of the total β-amylase activity in the leaf, but has no apparent effect on starch metabolism or phloem function, or any other phenotypic consequences (Laby et al. 2001). Soybean mutants have also been reported in which β-amylase activity is curtailed without any noticeable effects on starch metabolism or growth. It can be assumed that these mutants also lack extra plastidial isoforms of β-amylase. Cell fractionation studies on protoplasts from pea or wheat leaves have evidenced the presence of β-amylase in the vacuole. Again, the function of these β-amylase isoforms remains to be understood (Hildebrand and Hymowitz 1981; Ziegler and Beck 1986).

Experiments using β-amylase for starch granules hydrolysis in *Arabidopsis* showed an increase under conditions of simultaneous starch phosphorylation. This effect was reported for the *Arabidopsis* enzymes BAM1 and BAM3 and also for PCT-BMY1 (Scheidig et al. 2002), the potato ortholog of BAM3. In contrast, the activity of a commercially available barley β-amylase (*Hordeum vulgare*; Megazyme) was insignificantly affected by a simultaneous phosphorylation of the granules by glucan water dikinase (GWD). It was hypothesized that protein-protein interaction between BAM and GWD is important. During granule breakdown, binding of GWD to starch is strongly enhanced (Kötting et al. 2005; Ritte et al. 2000). Therefore, it can be concluded that GWD probably accelerates starch breakdown by targeting other proteins to starch.

It has been suggested that maltose, the product of β-amylase is exported to the cytosol during hydrolytic cleavage. Various studies have shown that export of maltose and its further breakdown product, glucose, occurs from isolated chloroplast (Servaites and Geiger 2002). This is further evidenced by a recent discovery of a maltose translocator (Fig. 5) in the chloroplast membrane. Mutation in the maltose translocator resulted in starch excess phenotype and eventually an elevated maltose content (Niittylä et al. 2004). As soon as the maltose is out of the cytosol, it is further metabolized to glucose and/or sucrose and maltodextrins by the activity of cytosolic glucosyltransferases during transitory starch degradation.

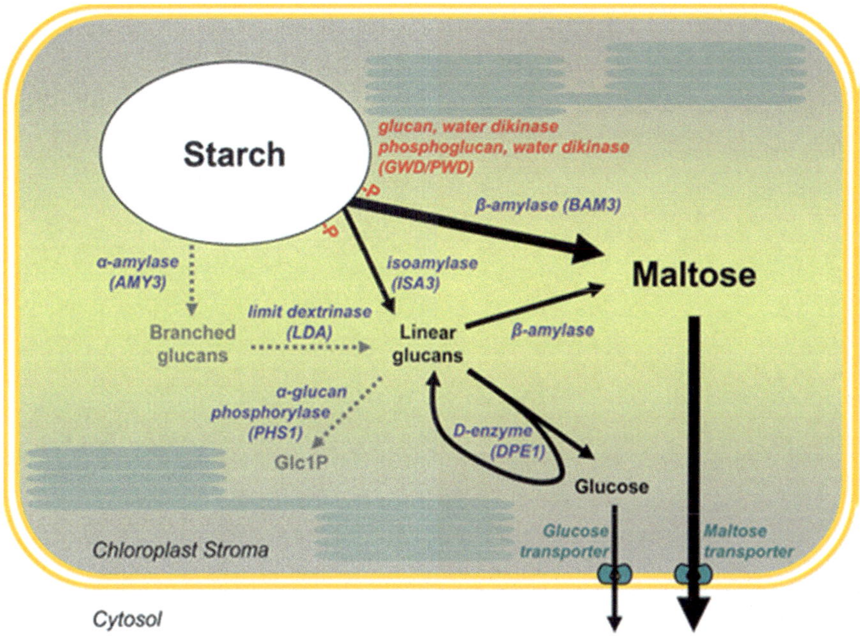

Fig. 5 Model representing starch degradation pathways in *Arabidopsis* leaves. Maltose being the major product of starch degradation, released by the action of β-amylases and glucose occurs as a minor product. Glucose and maltose are exported from the chloroplast *via* specific transporters. (Adapted from Zeeman et al. 2007)

10 Granular Starch Hydrolysis by β-Amylase

β-Amylases are generally incapable of hydrolyzing native starch granules. The effect of β-amylase in hydrolysis of A-type and B-type spherulitic polycrystalline amylose has also been examined and it was demonstrated that the pattern of enzyme attack depends on the crystal type, and that enzyme has a specific mode of attack on B-type amylose even in chemically homogeneous spherulites of a single crystalline type (Williamson et al. 1992). In barley, β-amylase was only significantly correlated with hydrolysis of boiled soluble starch and it did not contribute significantly to hydrolysis of native starch granules (Sun and Henson 1991). β-Amylase from *Curculigo pilosa* showed 4.5 and 6.2% degradation for native corn starch and wheat starch, respectively (Dicko et al. 1999). Recently, a novel β-Amylase from *Arachis hypogaea* has been reported to act efficiently on native potato and corn starch with 40 and 10% hydrolysis, respectively (Das and Kayastha 2019). This novel enzyme was found to corrode the surface of starch granules, leaving broken granules to smaller particles at later stage of digestion (Fig. 6). Such property of some plant β-amylases can be assumed to be based on extra sugar binding domains. Exact mechanism of attachment is still unclear. Bacterial β-amylases are more efficient

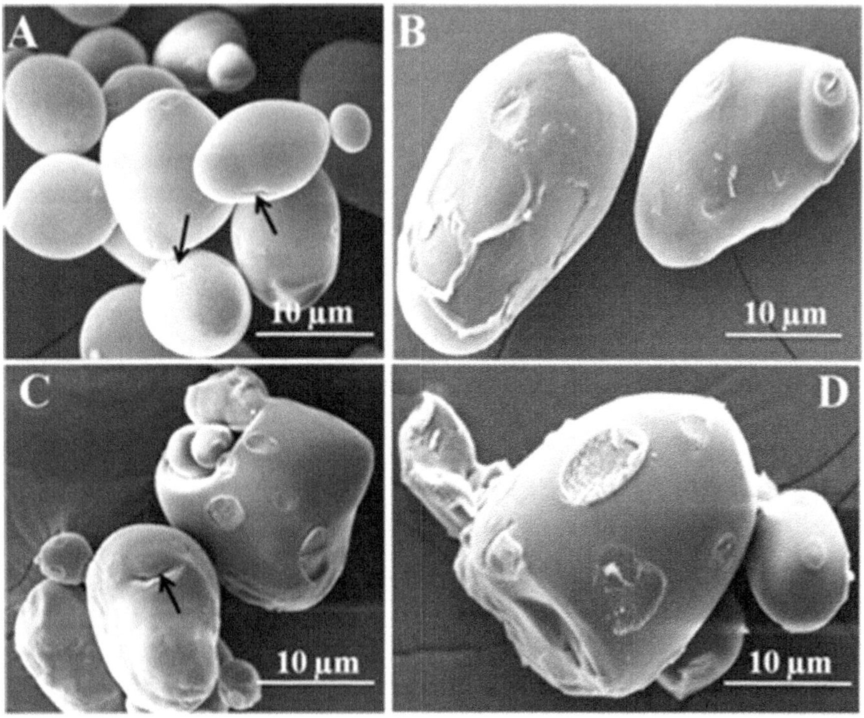

Fig. 6 SEM micrographs (×2000) for (**a**) native, (**b**) 4 h hydrolyzed, (**c**) 16 h hydrolyzed, and (**d**) 24 h hydrolyzed potato starch by β-amylase from peanut. Arrow represents the weak point from where hydrolysis starts (Adapted from Das and Kayastha 2019)

than plant β-amylase in hydrolysis of native starch granule i.e. they have raw starch-binding ability owing to their C-terminal starch-binding-domain (SBD) that resembles SBDs of cyclodextrin glucosyltransferase (CGTase) and glucoamylase. From the crystal structure of the catalytic site mutant E172A ($Glu^{172} \rightarrow Ala$) of β-amylase from *Bacillus cereus var. mycoides* complexed with maltopentose (G5), it was found that there exist three sugar-binding sites apart from the active site (Demirkan et al. 2005; Oyama et al. 2003).

11 Applications and Recent Advancements in Immobilization of β-Amylase for Enhanced Use

β-Amylase is used in mashing and brewing process owing to its hydrolysis of cereal grain starch and production of maltose. It is used in the production of malt as an additive in foodstuff. β-Amylase is utilized in the production of maltose rich syrup. Maltose gains importance as a sweetener in candy, confectionery, "tsukundani" (food boiled in soy sauce), ice cream and other related foods because, compared to glucose, it has a full-boiled taste, lower Maillard reaction rate, and is resistant to crystallization. Since maltose has a refined and mild type of sweetness and a low level of coloring, it is particularly indispensable as a sweetener for Japanese confectionery. High maltose syrup acts as a stabilizer and improves the shelf-life of the product. It helps to reduce the freezing point thereby decreasing the manufacturers freeze time and improving the freezer capacity. High purity maltose obtained by crystallizing maltose syrups is utilized in pharmaceutical industry for the manufacture antibiotics, vaccines, maltitol etc. Anhydrous crystalline maltose is used as a desiccant (Nehete et al. 1992).

β-Amylase is also known to inhibit retrogradation of starch. The enzyme shortens the α-1,4-linkage in the straight chain starch molecule utilizing its exo-type activity and ultimately reducing intermolecular association of the straight chain portion of amylose, which is known as the main cause for starch retrogradation (Das and Kayastha 2018). In addition, the moisture holding capacity of maltose produced by the action of enzyme, contribute to the softness of starch-containing foods. Thus, β-amylase is used in preventing retrogradation of rice cake. β-Amylase present in wheat flour is considered to be effective in the inhibition of retrogradation during the process of baking, although the barley-derived enzyme is used on some occasions, its use is limited because of the issue of thermostability.

Immobilization is mandatory for enzymes in industry, amplifying their usefulness. Moreover, immobilization protects the features that under unfavorable condition incline towards deformation, thereby lowering its activity. Immobilized enzymes are used in various reactor configurations, permitting an easy control of the reaction, avoiding contamination of the product by the enzyme (this is highly relevant in food technology), and permitting their reuse over many reaction cycles (Khan et al. 2011; Barbosa et al. 2013). Thus, a proper immobilization system should provide a strong enough immobilization so as to avoid enzyme release that may contaminate the product and result in loss of enzyme (and catalytic activity). Also, immobilization and stability are closely related terms, as only stable enough biocatalysts could be reused (Schmid et al. 2001; Schoemaker et al. 2003). β-Amylase, being one of the mainstay in starch based industries, display great specificity and are not permanently modified by their participation in reactions which, in turn, becomes cost-effective to use them repeatedly. Here, immobilization becomes an important aspect in enzymes technology. Therefore, the enzyme is attached to the reactor in a way that, they can be used again, after the products have been removed. They have been immobilized on a number of matrices which do not modify its functional property and stability quotient.

For reasonable and efficient usage of β-amylase, the enzyme has been immobilized on various conventional supports like chitosan coated PVC, chitosan/ PVP blend (Srivastava et al. 2015), acrylic carriers (Bryjak 2003), phenyl boronate agarose (Viera et al. 1988), etc. Immobilization is also well documented on a variety of carriers in literatures: adsorbed on raw starch (Rani et al. 1994), entrapment in alginate (Kokufuta et al. 1986), immobilization on ion exchanger (Deleyn and Stouffs 1990), silica beads (Germain and Crichton 1988), porous chitosan beads (Yoshida et al. 1989), agarose (Viera et al. 1988), etc. The enzyme immobilized on chitosan beads has been used in application of semi continuous production of maltose. Recently, a new β-amylase was isolated from *Pergularia tomentosa* and immobilized onto a matrix of titanium dioxide-based hybrid materials incorporated to cellulose acetate butyrate and copolymer of acrylonitrile and acrylamide, which was found to be a highly active biocatalyst at pH 7.0, 70 °C and thermal-stability at 60 °C (Lahmar et al. 2018). These studies demonstrate that it is appropriate to use the aforementioned biocatalyst for further applications in order to minimize the quantity of requested matrices retaining enzyme activity and might provide an interesting tool for an eco-friendly immobilization of plants amylases.

Nowadays, significant efforts are made to obtain immobilized forms of the enzyme on various support systems, which involves a variety of new materials, including nanostructures. Nanostructures have gained attention particularly for features such as high surface to volume ratio, desired aqueous suspending ability, small size, abundant oxygen containing surface functionalities and designable function (Singh and Kayastha 2014; Singh et al. 2014; Torres-Salas et al. 2011). β-amylase from various sources have been reported to be immobilized on graphene oxide (Srivastava et al. 2014), graphene oxide carbon nanotubes, iron oxide nanoparticles (Fe_3O_4), molybdenum sulfide (MoS_2) (Das et al. 2017, 2018a, b), modified magnetic nanoparticles $ZnFe_2O_4@SiO_2\text{-}NH_2$ (Rasouli et al. 2016). Recent studies have shown that magnetic nano particles possess a great potential for the immobilization of enzyme that include enzyme reusability using an external magnetic field. Moreover, contrary to porous compounds, magnetic nanoparticles discard limitation for diffusion of substrates and products. The betterment of β-amylase immobilization in terms of pH, temperature and stability would confer wider range of application for maltose production and accordingly, suitable for food and pharmaceuticals industries.

12 Future Perspective

β-Amylases are one of the major enzymes necessitate in starch based industries, and accordingly efforts should be made to explore plant, cereal, biocompatible new microorganisms, etc. for the efficient production of β-amylases at lower production cost, with high purity. In future three dimensional structures of more thermostable and kinetically favorable β-amylases will be determined in order to get a clear picture of structure-function relationship and molecular mechanism of this multidomain protein. Combined with site-directed mutagenesis, the role of important residues for catalysis and adaptive parameters of raw starch digesting, plant based β-amylases will add on useful information. Moreover, future fast kinetic experiments on different β-amylases shall provide a detailed mechanism of folding and unfolding transitions.

Lastly, nano-structured metal-oxide based amylase sensors should be developed with high sensitivity, fast response time, and stability/shelf-life for various biotechnological applications. In future, efforts should be made to construct different maltose biosensors using these ultrasensitive nanoparticles and explore its implementations in pharmaceutical and industrial processes. The sensors will be schemed by co-immobilization of β-amylase, α-amylase and α-glucosidase on suitable supports. Besides, each step of manufacture process of proposed maltose biosensor shall be monitored employing spectroscopic and microscopic tools like scanning electron microscopy (SEM), transmission electron microscopy (TEM) with energy dispersive X-ray analysis (EDX), X-ray photoelectron spectroscopy (XPS), fourier transform infrared spectrometry (FTIR), Raman spectroscopy, fluorescence imaging, atomic force microscopy (AFM) and surface enhanced Raman scattering (SERS) techniques. Moving forward in the future, food, pharmaceutics and starch based industries will be the major targets for these biosensors and biocatalysts.

References

Ann YG, Iizuka M, Yamamoto T, Minamiura N (1990) Preparation and some properties of active monomer of sweet potato β-amylase. Agric Biol Chem 54(3):769–774

Bailey J, French D (1957) The significance of multiple reactions in enzyme-polymer systems. J Biol Chem 226(1):001–014

Barbosa O, Torres R, Ortiz C, Berenguer-Murcia Á, Rodrigues RC, Fernandez-Lafuente R (2013) Heterofunctional supports in enzyme immobilization: from traditional immobilization protocols to opportunities in tuning enzyme properties. Biomacromolecules 14(8):2433–2462

Bernfeld P (1955) Amylases, alpha and beta. Methods Enzymol 1:149–158

Bryjak J (2003) Glucoamylase, α-amylase and β-amylase immobilisation on acrylic carriers. Biochem Eng J 16(3):347–355

Caldwell KD, Axén R, Bergwall M, Olsson I, Porath J (1976a) Immobilization of enzymes based on hydrophobic interaction. III. Adsorbent substituent density and its impact on the immobilization of β-amylase. Biotechnol Bioeng 18(11):1605–1614

Caldwell KD, Axén R, Wall MB, Porath J (1976b) Immobilization of enzymes based on hydrophobic interaction. I. Preparation and properties of a β-amylase adsorbate. Biotechnol Bioeng 18 (11):1573–1588

Chang C, Liou HY, Tang HL, Sung HY (1996) Activation, purification and properties of beta-amylase from sweet potatoes (*Ipomoea batatas*). Biotechnol Appl Biochem 24:13–18

Chatterjee B, Ghosh A, Das A (1992) Starch digestion and adsorption by β-amylase of *Emericella nidulans* (*Aspergillus nidulans*). J Appl Microbiol 72(3):208–213

Cheong CG, Eom SH, Chang C, Shin DH, Song HK, Min K, Moon JH, Kim KK, Hwang KY, Suh SW (1995) Crystallization, molecular replacement solution, and refinement of tetrameric β-amylase from sweet potato. Proteins: Struct Funct Bioinf 21(2):105–117

Cleveland F, Kerr R (1948) The action of beta-amylase on corn amylose. Cereal Chem 25 (2):133–139

Cudney R, McPherson A (1993) Preliminary crystallographic analysis of sweet potato beta amylase. J Mol Biol 229(1):253–254

Das R, Kayastha AM (2018) An antioxidant rich novel β-amylase from peanuts (*Arachis hypogaea*): its purification, biochemical characterization and potential applications. Int J Biol Macromol 111:148–157

Das R, Kayastha AM (2019) Enzymatic hydrolysis of native granular starches by a new β-amylase from peanut (*Arachis hypogaea*). Food Chem 276:582–590

Das R, Mishra H, Srivastava A, Kayastha AM (2017) Covalent immobilization of β-amylase onto functionalized molybdenum sulfide nanosheets, its kinetics and stability studies: a gateway to boost enzyme application. Chem Eng J 328:215–227

Das R, Talat M, Srivastava ON, Kayastha AM (2018a) Covalent immobilization of peanut β-amylase for producing industrial nano-biocatalysts: a comparative study of kinetics, stability and reusability of the immobilized enzyme. Food Chem 245:488–499

Das R, Ranjan R, Sinha N, Kayastha AM (2018b) Comparative characterization of peanut β-amylase immobilization onto graphene oxide and graphene oxide carbon nanotubes by solid-state NMR. J Phys Chem 122:19259–19265

Deleyn F, Stouffs R (1990) Immobilised β-amylase in the production of maltose syrups. Starch – Stärke 42(4):158–160

Demirkan ES, Mikami B, Adachi M, Higasa T, Utsumi S (2005) α-Amylase from *B. amyloliquefaciens*: purification, characterization, raw starch degradation and expression in *E. coli*. Process Biochem 40(8):2629–2636

Dicko MH, Searle-van Leeuwen M, Beldman G, Ouedraogo O, Hilhorst R, Traore A (1999) Purification and characterization of β-amylase from *Curculigo pilosa*. Appl Microbiol Biotechnol 52(6):802–805

Doehlert DC, Duke SH, Anderson L (1982) Beta-amylases from alfalfa (*Medicago sativa L.*) roots. Plant Physiol 69(5):1096–1102

French D (1961) Action pattern of β-amylase. Nature 190(4774):445–446

French D, Youngquist R (1963) The mode of action of β-amylase on starch oligosaccharides. Starch-Stärk 15(12):425–431

French D, Levine ML, Pazur J, Norberg E (1950) Studies on the schardinger dextrins. IV. The action of soy bean beta amylase on amyloheptaose. J Am Chem Soc 72(4):1746–1748

Fulton DC, Stettler M, Mettler T, Vaughan CK, Li J, Francisco P, Gil M, Reinhold H, Eicke S, Messerli G (2008) β-Amylase, a noncatalytic protein required for starch breakdown, acts upstream of three active β-amylases in *Arabidopsis* chloroplasts. Plant Cell 20(4):1040–1058

Genghof DS, Brewer CF, Hehre EJ (1978) Preparation and use of α-maltosyl fluoride as a substrate by beta amylase. Carbohydr Res 61(1):291–299

Germain P, Crichton RR (1988) Characterization of a chemically modified β-amylase immobilized on porous silica. J Chem Technol Biotechnol 41(4):297–315

He L, Park SH, Dang NDH, Duong HX, Duong TPC, Tran PL, Park JT, Ni L, Park KH (2017) Characterization and thermal inactivation kinetics of highly thermostable ramie leaf β-amylase. Enzym Microb Technol 101:17–23

Hehre EJ, Okada G, Genghof DS (1969) Configurational specificity: unappreciated key to understanding enzymic reversions and de novo glycosidic bond synthesis: I. reversal of hydrolysis by α-, β-and glucoamylases with donors of correct anomeric form. Arch Biochem Biophys 135:75–89

Henrissat B (1991) A classification of glycosyl hydrolases based on amino acid sequence similarities. Biochem J 280(2):309–316

Hildebrand D, Hymowitz T (1981) Role of β-amylase in starch metabolism during soybean seed development and germination. Physiol Plant 53(4):429–434

Hopkins R, Jelinek B, Harrison L (1948) The action of β-amylase on potato amylose. Biochem J 43 (1):32–38

Horton P, Park KJ, Obayashi T, Fujita N, Harada H, Adams-Collier C, Nakai K (2007) WoLF PSORT: protein localization predictor. Nucleic Acids Res 35(2):W585–W587

Husain Q (2017) Nanomaterials as novel support for the immobilization of amylolytic enzymes and their applications: a review. Biocatalysis (DeGruyter) 3(1):37–53

Husemann VE, Pfannemüller B (1961) About the different behavior of synthetic and natural amylose versus phosphorylase and β-amylase. Macromol Chem Phys 43:156–159

Ishikawa K, Nakatani H, Katsuya Y, Fukazawa C (2007) Kinetic and structural analysis of enzyme sliding on a substrate: multiple attack in β-amylase. Biochemistry 46(3):792–798

Kamasaka H, Sugimoto K, Takata H, Nishimura T, Kuriki T (2002) B*acillus stearothermophilus neopullulanase* selective hydrolysis of amylose to maltose in the presence of amylopectin. Appl Environ Microbiol 68(4):1658–1664

Kang YN, Adachi M, Utsumi S, Mikami B (2004) The roles of Glu^{186} and Glu^{380} in the catalytic reaction of soybean *β*-amylase. J Mol Biol 339(5):1129–1140

Kaplan F, Guy CL (2005) RNA interference of *Arabidopsis* beta-amylase prevents maltose accumulation upon cold shock and increases sensitivity of PSII photochemical efficiency to freezing stress. Plant J 44(5):730–743

Khan MJ, Khan FH, Husain Q (2011) Application of immobilized *Ipomoea batata* β amylase in the saccharification of starch. J Appl Biol Sci 5(2):33–39

Kitahata S, Chiba S, Brewer CF, Hehre EJ (1991) Mechanism of maltal hydration catalyzed by beta-amylase: role of protein structure in controlling the steric outcome of reactions catalyzed by a glycosylase. Biochemistry 30(27):6769–6775

Kokufuta E, Sodeyama T, Katano T (1986) Initiation–cessation control of an enzyme reaction using pH-sensitive poly(styrene) microcapsules with a surface-coating of poly(iminoethylene). J Chem Soc Chem Commun 1(9):641–642

Kötting O, Pusch K, Tiessen A, Geigenberger P, Steup M, Ritte G (2005) Identification of a novel enzyme required for starch metabolism in *Arabidopsis* leaves. The phosphoglucan, water dikinase. Plant Physiol 137(1):242–252

Kuhn R (1924) About the constitution of the starch and the different modes of action of the amylases. Berichte der Deutschen Chemischen Gesellschaft (A and B Series) 57(10):1965–1968

Laby RJ, Kim D, Gibson SI (2001) The ram1 mutant of *Arabidopsis* exhibits severely decreased *β*-amylase activity. Plant Physiol 127(4):1798–1807

Lahmar I, Radeva G, Marinkova D, Velitchkova M, Belghith H, Ben Abdallah F, Yotova L, Belghith K (2018) Immobilization and topochemical mechanism of a new *β*-amylase extracted from *Pergularia tomentosa*. Process Biochem 64:143–151

Lao NT, Schoneveld O, Mould RM, Hibberd JM, Gray JC, Kavanagh TA (1999) An *Arabidopsis* gene encoding a chloroplast-targeted *β*-amylase. Plant J 20(5):519–527

Li X, Yu HY (2011) Extracellular production of beta-amylase by a halophilic isolate, *Halobacillus sp.* LY9. J Ind Microbiol Biotechnol 38(11):1837–1843

Lloyd JR, Kossmann J, Ritte G (2005) Leaf starch degradation comes out of the shadows. Trends Plant Sci 10(3):130–137

Mihajlovski KR, Radovanović NR, Veljović ĐN, Šiler-Marinković SS, Dimitrijević-Branković SI (2016) Improved *β*-amylase production on molasses and sugar beet pulp by a novel strain *Paenibacillus chitinolyticus* CKS1. Ind Crop Prod 80:115–122

Mikami B, Hehre EJ, Sato M, Katsube Y, Hirose M, Morita Y, Sacchettini JC (1993) The 2.0-ANG resolution structure of soybean beta-amylase complexed with alpha-cyclodextrin. Biochemistry 32(27):6836–6845

Mikami B, Degano M, Hehre EJ, Sacchettini JC (1994) Crystal structures of soybean beta-amylase reacted with beta-maltose and maltal: active site components and their apparent roles in catalysis. Biochemistry 33(25):7779–7787

Mikami B, Yoon HJ, Yoshigi N (1999a) The crystal structure of the sevenfold mutant of barley *β*-amylase with increased thermostability at 2.5 Å resolution. J Mol Biol 285(3):1235–1243

Mikami B, Adachi M, Kage T, Sarikaya E, Nanmori T, Shinke R, Utsumi S (1999b) Structure of raw starch-digesting *Bacillus cereus* *β*-amylase complexed with maltose. Biochemistry 38 (22):7050–7061

Miyake H, Otsuka C, Nishimura S, Nitta Y (2002) Catalytic mechanism of *β*-amylase from *Bacillus cereus var. mycoides*: chemical rescue of hydrolytic activity for a catalytic site mutant (Glu^{367} → Ala) by azide. J Biochem 131(4):587–591

Nakayama S, Amagase S (1963) An improved method for the purification of sweet potato *β*-amylase. J Biochem 54(4):375–377

Nehete PN, Shah NK, Ramamurthy V, Kothari RM (1992) An optimized protocol for the production of high purity maltose. World J Microbiol Biotechnol 8(4):446–450

Niittylä T, Messerli G, Trevisan M, Chen J, Smith AM, Zeeman SC (2004) A previously unknown maltose transporter essential for starch degradation in leaves. Science 303(5654):87–89

Ohlsson E (1930) Über die beiden komponenten der malzdiastase, besonders mit rücksicht auf die mutarotation der bei der hydrolyse der stärke gebildeten produkte. Hoppe-Seyler′ s Zeitschrift für Physiologische Chemie 189(1-2):17–63

Oyama T, Kusunoki M, Kishimoto Y, Takasaki Y, Nitta Y (1999) Crystal structure of β-amylase from *Bacillus cereus var. mycoides* at 2.2 Å resolution. J Biochem 125(6):1120–1130

Oyama T, Miyake H, Kusunoki M, Nitta Y (2003) Crystal structures of β-amylase from *Bacillus cereus var. mycoides* in complexes with substrate analogs and affinity-labeling reagents. J Biochem 133(4):467–474

Pierleoni A, Martelli PL, Fariselli P, Casadio R (2006) BaCelLo: a balanced subcellular localization predictor. Bioinformatics 22(14):e408–e416

Pujadas G, Ramírez FM, Valero R, Palau J (1996) Evolution of β-amylase: patterns of variation and conservation in subfamily sequences in relation to parsimony mechanisms. Proteins: Struct Funct Bioinf 25(4):456–472

Rani RR, Jana SC, Nanda G (1994) Saccharification of indigenous starches by β-amylase of *Bacillus megaterium*. World J Microbiol Biotechnol 10(6):691–693

Rasouli N, Sohrabi N, Zamani E (2016) Influence of a novel magnetic recoverable support on kinetic, stability and activity of beta-amylase enzyme. Phys Chem Res 4(2):271–283

Ray RR (2004) β-Amylases from various fungal strains. Acta Microbiol Immunol Hung 51 (1-2):85–95

Ray RR, Chakraverty R (1998) Extracellular β-amylase from *Syncephalastrum racemosum*. Mycol Res 102(12):1563–1567

Ray RR, Nanda G (1996) Microbial β-amylases: biosynthesis, characteristics, and industrial applications. Crit Rev Microbiol 22(3):181–199

Reinhold H, Soyk S, Šimkovà K, , Hostettler C, Marafino J, Mainiero S, Vaughan CK, Monroe JD, Zeeman CS (2011). β-Amylase-like proteins function as transcription factors in *Arabidopsis*, controlling shoot growth and development. Plant Cell 23: 1391–1403

Rexova L, Kopec Z, Kiel B (1967) Isolation and certain properties of wheat β-amylase. Collect Czech Chem C 32(2):678–684

Ritte G, Lorberth R, Steup M (2000) Reversible binding of the starch-related R1 protein to the surface of transitory starch granules. Plant J 21(4):387–391

Sadowski J, Rorat T, Cooke R, Delseny M (1993) Nucleotide sequence of a cDNA clone encoding ubiquitous beta-amylase in rye (*Secale cereale L.*). Plant Physiol 102(1):315

Sagu ST, Nso EJ, Homann T, Kapseu C, Rawel HM (2015) Extraction and purification of beta-amylase from stems of *Abrus precatorius* by three phase partitioning. Food Chem 183:144–153

Scheidig A, Fröhlich A, Schulze S, Lloyd JR, Kossmann J (2002) Downregulation of a chloroplast-targeted β-amylase leads to a starch-excess phenotype in leaves. Plant J 30(5):581–591

Schmid A, Dordick JS, Hauer B, Kiener A, Wubbolts M, Witholt B (2001) Industrial biocatalysis today and tomorrow. Nature 409(6817):258–268

Schoemaker HE, Mink D, Wubbolts MG (2003) Dispelling the myths-biocatalysis in industrial synthesis. Science 299(5613):1694–1697

Servaites JC, Geiger DR (2002) Kinetic characteristics of chloroplast glucose transport. J Exp Bot 53(374):1581–1591

Singh K, Kayastha AM (2014) Optimal immobilization of α-amylase from wheat (*Triticum aestivum*) onto DEAE-cellulose using response surface methodology and its characterization. J Mol Catal B Enzym 104:75–81

Singh N, Srivastava G, Talat M, Raghubanshi H, Srivastava ON, Kayastha AM (2014) *Cicer* α-galactosidase immobilization onto functionalized graphene nanosheets using response surface method and its applications. Food Chem 142:430–438

Smith SM, Fulton DC, Chia T, Thorneycroft D, Chapple A, Dunstan H, Hylton C, Zeeman SC, Smith AM (2004) Diurnal changes in the transcriptome encoding enzymes of starch metabolism provide evidence for both transcriptional and post-transcriptional regulation of starch metabolism in *Arabidopsis* leaves. Plant Physiol 136(1):2687–2699

Smith AM, Zeeman SC, Smith SM (2005) Starch degradation. Annu Rev Plant Biol 56:73–98

Sottirattanapan P, Nantachai K, Daduang S, Funahashi T, Yamada M (2017) Purification and characterization of amylase from roots of *Paederia foetida Linn.* Biocatal Agric Biotechnol 10:329–335

Sparla F, Costa A, Schiavo FL, Pupillo P, Trost P (2006) Redox regulation of a novel plastid-targeted β-amylase of *Arabidopsis*. Plant Physiol 141(3):840–850

Srivastava G, Kayastha AM (2014) β-Amylase from starchless seeds of *Trigonella foenum-graecum* and its localization in germinating seeds. PLoS One 9(2):e88697

Srivastava G, Singh K, Talat M, Srivastava ON, Kayastha AM (2014) Functionalized graphene sheets as immobilization matrix for fenugreek β-amylase: enzyme kinetics and stability studies. PLoS One 9(11):e113408

Srivastava G, Roy S, Kayastha AM (2015) Immobilisation of fenugreek β-amylase on chitosan/PVP blend and chitosan coated PVC beads: a comparative study. Food Chem 172:844–851

Sun Z, Henson CA (1991) A quantitative assessment of the importance of barley seed α-amylase, β-amylase, debranching enzyme, and α-glucosidase in starch degradation. Arch Biochem Biophys 284(2):298–305

Swanson MA (1948) Studies on the structure of polysaccharides IV. Relation of the iodine color to the structure. J Biol Chem 172(2):825–837

Takeda Y, Hizukuri S (1969) Improved method for crystallization of sweet potato beta-amylase. Biochim Biophys Acta 185(2):469–471

Tanaka N, Kajimoto S, Mitani D, Kunugi S (2002) Effects of guanidine hydrochloride and high pressure on subsite flexibility of β-amylase. Biochim Biophys Acta Protein Struct Mol Enzymol 1596(2):318–325

Teotia S, Khare S, Gupta M (2001) An efficient purification process for sweet potato beta-amylase by affinity precipitation with alginate. Enzym Microb Technol 28(9-10):792–795

Thoma JA, Brothers C, Spradlin JE (1970) Subsite mapping of enzymes. Studies on *Bacillus subtilis* amylase. Biochemistry 9(8):1768–1775

Thoma JA, Spradlin JE, Dygert S (1971) 6 plant and animal amylases. In: The enzymes, vol 5. Elsevier, New York, pp 115–189

Tkachuk R, Tipples K (1966) Wheat beta-amylases. II. Characterization. Cereal Chem 43:62–79

Torres-Salas P, del Monte-Martinez A, Cutiño-Avila B, Rodriguez-Colinas B, Alcalde M, Ballesteros AO, Plou FJ (2011) Immobilized biocatalysts: novel approaches and tools for binding enzymes to supports. Adv Mater 23(44):5275–5282

Totsuka A, Fukazawa C (1996) Functional analysis of Glu^{380} and Leu^{383} of soybean β-amylase: a proposed action mechanism. Eur J Biochem 240(3):655–659

Uozumi N, Matsuda T, Tsukagoshi N, Udaka S (1991) Structural and functional roles of cysteine residues of *Bacillus polymyxa* beta-amylase. Biochemistry 30(18):4594–4599

Van Damme EJ, Hu J, Barre A, Hause B, Baggerman G, Rouge P, Peumans WJ (2001) Purification, characterization, immunolocalization and structural analysis of the abundant cytoplasmic β-amylase from *Calystegia sepium* (hedge bindweed) rhizomes. FEBS J 268(23):6263–6273

Viera FB, Barragan BB, Busto BL (1988) Reversible immobilization of soybean β-amylase on phenylboronate–agarose. Biotechnol Bioeng 31(7):711–713

Viksø-Nielsen A, Christensen TM, Bojko M, Marcussen J (1997) Purification and characterization of β-amylase from leaves of potato (*Solanum tuberosum*). Physiol Plant 99(1):190–196

Wang Q, Monroe J, Sjolund RD (1995) Identification and characterization of a phloem-specific β-amylase. Plant Physiol 109(3):743–750

Williamson G, Belshaw NJ, Self DJ, Noel TR, Ring SG, Cairns P, Morris VJ, Clark SA, Parker ML (1992) Hydrolysis of A-and B-type crystalline polymorphs of starch by α-amylase, β-amylase and glucoamylase. Carbohydr Polym 18(3):179–187

Yamaguchi T, Matsumoto Y, Shirakawa M, Kibe M, Hibino T, Kozaki S, Takasaki Y, Nitta Y (1996) Cloning, sequencing, and expression of a β-amylase gene from *Bacillus cereus var. mycoides* and characterization of its products. Biosci Biotechnol Biochem 60(8):1255–1259

Yoshida N, Nakamura K (1991) Molecular cloning and expression in *Escherichia coli* of cDNA encoding the subunit of sweet potato β-amylase. J Biochem 110(2):196–201

Yoshida M, Oishi K, Kimura T, Ogata M, Nakakuki T (1989) Continuous production of maltose using a dual immobilized enzyme system. Agric Biol Chem 53(12):3139–3142

Yoshigi N, Okada Y, Sahara H, Koshino S (1994) Expression in *Escherichia coli* of cDNA encoding barley *β*-amylase and properties of recombinant *β*-amylase. Biosci Biotechnol Biochem 58(6):1080–1086

Youngquist RW (1962) Beta-amylase action on high molecular weight maltosaccharides. Retrospective Theses and Dissertations 2033

Zeeman SC, Northrop F, Smith AM, Rees TA (1998) A starch-accumulating mutant of *Arabidopsis thaliana* deficient in a chloroplastic starch-hydrolysing enzyme. Plant J 15(3):357–365

Zeeman SC, Smith SM, Smith AM (2004) The breakdown of starch in leaves. New Phytol 163 (2):247–261

Zeeman SC, Smith SM, Smith AM (2007) The diurnal metabolism of leaf starch. Biochem J 401:13–28

Ziegler P (1999) Cereal beta-amylases. J Cereal Sci 29(3):195–204

Ziegler P, Beck E (1986) Exoamylase activity in vacuoles isolated from pea and wheat leaf protoplasts. Plant Physiol 82(4):1119–1121

Oxidoreductases: Overview and Practical Applications

Hina Younus

1 Overview of Oxidoreductases

Oxidoreductases are a large group of enzymes that are present in different biomes of life (microbes, plants, and animals). In the enzyme commission (EC) number classification of enzymes, they are classified as EC 1. They comprise about one third of the enzymatic activities that are registered in BRaunschweig Enzyme Database (BRENDA) (Sellés Vidal et al. 2018). These enzymes catalyze the exchange of electrons between the donor and acceptor molecules, in reactions involving electron transfer, proton/hydrogen extraction, hydride transfer, oxygen insertion, or other important steps (Toone 2010; Husain 2017). In general, two half reactions i.e. one oxidative and one reductive occur, and at least two substrates i.e. one reducing and one oxidizing are activated or transformed (Laskar et al. 2017).

$$A^- + B \rightarrow B^- + A$$

Shown below is an example of an oxidoreduction reaction catalyzed by the oxidoreductase enzyme, aldehyde dehydrogenase (ALDH) which oxidizes the substrate aldehydes to carboxylic acids, while the other substrate (co-enzyme) NAD^+ or $NADP^+$ is reduced to NADH or NADPH, respectively (Fig. 1).

The different types of oxidoreductases are oxidases, dehydrogenases, hydroxylases, oxygenases, peroxidases and reductases (Nicholas and Lewis 1999) (Fig. 2). When molecular oxygen acts as an acceptor of hydrogen or electrons, the enzymes oxidases are involved. Whereas, dehydrogenases are the enzymes which oxidize a substrate by transferring hydrogen to an acceptor molecule that is either nicotinamide adenine dinucleotide/nicotinamide adenine dinucleotide phosphate ($NAD^+/NADP^+$) or a flavin co-enzyme (Alam et al. 2016). Peroxidases catalyze the reduction of

H. Younus (✉)
Interdisciplinary Biotechnology Unit, Aligarh Muslim University, Aligarh, Uttar Pradesh, India
e-mail: hyounus.cb@amu.ac.in

Q. Husain, M. F. Ullah (eds.), *Biocatalysis*,
https://doi.org/10.1007/978-3-030-25023-2_3

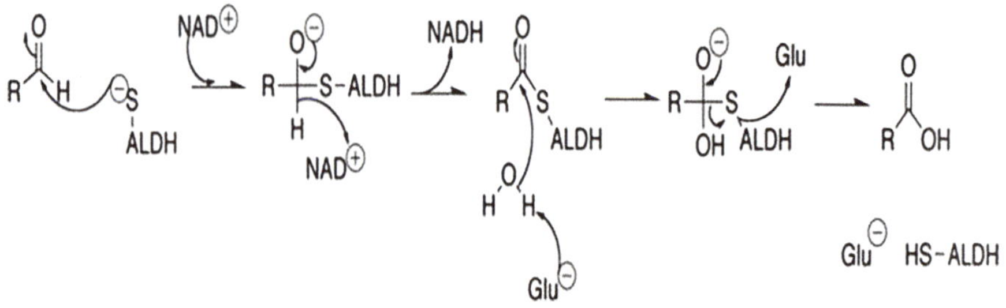

Fig. 1 Catalytic mechanism of aldehyde dehydrogenase (ALDH)

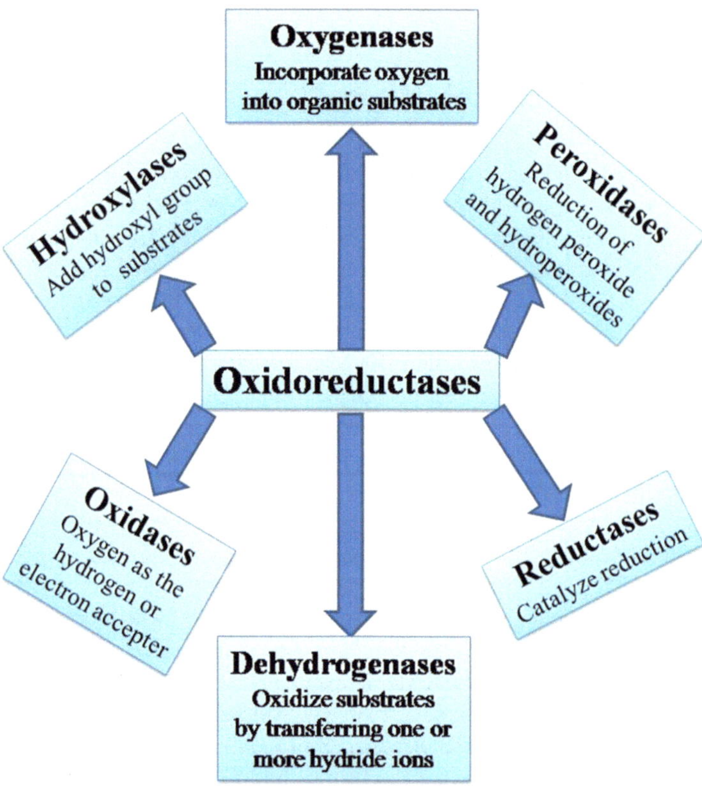

Fig. 2 The classification of oxidoreductases

hydrogen peroxide and are located in the peroxisomes. The enzymes hydroxylases do the addition of hydroxyl groups to their substrates. The oxygenases incorporate oxygen into organic substrates from molecular oxygen. Reductases catalyze reduction reactions, and in most cases they act like oxidases.

Oxidoreductases perform important roles in both aerobic and anaerobic metabolism. They have a wide range of substrates, both organic (e.g. alcohols, amines and ketones) and inorganic (e.g. small anions such as sulphite and metals such as mercury). These enzymes have various redox-active centres for performing their physiological functions (Webb 1992). These centres are protected by the polypeptide backbone of oxidoreductases since they are very labile in nature. The polypeptide backbone of the enzyme also helps in selectivity, reactivity, redox potency, stability, and inhibition-resistance. The common redox centres include amino acid residues (e.g. tyrosine/cysteine), metal ions or complexes (e.g. Cu, Mo, Fe, Fe-S cluster, or heme), and coenzymes [e.g. flavin mononucleotide (FMN), flavin dinucleotide (FAD), pterin, and pyrroloquinoline quinone (PQQ)]. Table 1 shows the various classes of oxidoreductases and their redox-active centres.

2 Practical Applications of Oxidoreductases

Oxidoreductase based catalysis fits well with the development of highly efficient, sustainable, and environment-friendly industries since they are biodegradable, specific in nature, and energy saving. These enzymes can incorporate different cofactors such as heme, flavin and metal ions to catalyze redox reactions, hence they use a variety of electron acceptors and a large number of electron-donating substrates, and this leads to the production of many products of industrial interest. These enzymes are currently utilized in the field of textiles, medicine, food and for chemical synthesis (May and Padgette 1983). Below are some important practical uses of these enzymes.

3 Lignocellulosics Biotransformation

Lignin can be removed from wood by the process of delignification. Chlorine- or oxygen-based chemical oxidants are conventionally used for pulp delignification. Although these are very effective, however they can cause serious problems in by-products disposal or cellulose fiber-strength loss. The enzyme based delignification systems are better alternatives. The oxidoreductases like laccases, peroxidases, and others participate in the natural delignification by the lignolytic white-rot fungi. Various laccases have been reported to be capable of degrading both natural and synthetic lignin (Balakshin et al. 2001; Leonowicz et al. 2001; Shleev et al. 2003). These enzymes oxidize the phenolic components of lignin either directly or indirectly or, in the presence of a redox mediator. As a result, radicals can be generated in lignin, which can result in aliphatic/aromatic C–C bond cleavage and depolymerization. Laccases are also utilized in the biosynthesis of adhesives (Vert et al. 2001). To initiate or enhance the cross-linking efficiency, the enzyme can be used in three ways: to directly oxidize wood particles/pulp to form radicals for cross-linking, to functionalize wood particles/pulp with small compounds (e.g aromatic, isocyanate, carboxyl or acrylamide substances) which act as cross-linking agents,

Table 1 Oxidoreductases and their cofactors

Oxidoreductases	Cofactors	References
1. Oxidases		
1.1. Co-factor free oxidases (Urate oxidase)	Thr–Lys	Imhoff et al. (2003)
1.2. Thiol oxidases	Fe, Cu or Cys + FAD	Fomenko and Gladyshev (2012), Thorpe et al. (2002)
1.3. Cu-containing oxidases	Cu	Claus (2004)
1.4. Flavin containing oxidases Carbohydrate oxidases Amine oxidases	FMN or FAD	Massey (2000), Bannwarth et al. (2004), and Pollegioni et al. (2002)
1.5. Multi-redox-center oxidases	Cu, Tyr, Flavin or Heme	Kagan and Li (2003), Whittaker (2003)
2. Peroxidases		
2.1. Heme peroxidases	Heme	Sharp et al. (2003)
2.2. Catalases	Heme	Veitch (2004)
2.3. Haloperoxidases	Ser-His-Asp-Heme	van de Velde et al. (2001)
2.4. Other peroxidases	Cys or Se-Cys	Rhee et al. (2001), Sun et al. (2004)
3. Oxygenases/hydroxylases		
3.1. Cofactor-free oxygenases	Amino acids	Fetzner (2002)
3.2. Fe-containing oxygenases Nonheme-Fe monooxygenases Cytochrome P450	Fe or Heme	Bugg (2001), Que and Watanabe 2001
3.3. Cu-containing oxygenases Polyphenol oxygenases Dopamine β-monooxygenases	Cu	Durán et al. (2002), Que and Watanabe (2001), and Yamazaki et al. (2004)
3.4. Multicenter oxygenases	Fe, Mo, Pterin, Flavin, Fe-S	Bugg (2001), Poulos (2005)
3.5. Flavin containing oxygenases	FMN or FAD	Massey (2000)
4. Dehydrogenases/reductases		
4.1. Flavin-containing dehydrogenases	FMN or FAD	Yagi et al. (2001)
4.2. Quinone-containing dehydrogenases	PQQ	Anthony (2004)
4.3. Zn-containing dehydrogenases	Amino acid and Zn	Forrest and Gonzalez (2000), Leskovac et al. (2002)

(continued)

Table 1 (continued)

Oxidoreductases	Cofactors	References
4.4. Multi-redox-center dehydrogenases	Flavin, PQQ, Heme, Fe-S, Mo	Cameron and Aust (2001), Oubrie (2003), and Xu et al. (2001)
4.5. Aldo-ketoreductases	Tyr–His–Asp–Lys	Ellis (2002)
4.6. Cu-containing reductases	Cu	Suzuki et al. (2000)
4.7. Flavin-containing reductases	FMN or FAD	Massey (2000)

and to transform isolated lignin (usually a by-product of pulping), phenolic polysaccharide, protein, or starch into radical-rich and nontoxic adhesives (Vert et al. 2001). These applications of laccases do not only replace toxic or expensive chemical adhesives, but also transform wastes such as lignin from paper industries into value-added products. Further optimization is required to commercialize these enzymatic bio-adhesives for particle-/linerboard manufacturing, so that they can be competitive against the inexpensive chemical counterparts.

4 Carbohydrates Derivatization

Carbohydrates can be utilized as a renewable and inexpensive raw materials, precursors, building blocks, or additives for various industrial products. Earlier, useful organic acid such as lactic acid has been produced from sugars by whole cell fermentation method (Ghaffar et al. 2014). Using the oxidoreductase enzymes, sugars used in our daily life such as glucose and sucrose can be modified into other useful products. D-glucose has been converted to D-glucosone by glucose 2-oxidase (Karmali and Coelho 2011). The cheese process industries have produced lactose as a by-product, which was converted to lactobionic acid by a lactose oxidase (Koka et al. 2004). The lactobionic acid is utilized as a valuable food additive, chelator, acidulant, and a polymer precursor (Gutiérrez et al. 2012).

The glucose molecule and other carbohydrates have many chiral centres, which can be functionalized by specific carbohydrate oxidoreductases. The derivatized carbohydrates may act as useful precursors for useful industrial or pharmaceutical products. For instance, galactomannan has been oxidized by galactose oxidase, which converted its C6–OH into a highly reactive aldehyde, making the polymer susceptible to various functionalizations, including the use as a paper additive (Brady et al. 2002).

5 Food Improvement Applications

Oxidoreductases have a high potential for use in the food industry. The substrates of many these enzymes, i.e. carbohydrates, unsaturated fatty acids, phenolics, and thiol-containing proteins, are components of various foods and beverages. The modification of these by oxidoreductases may result in new functions, improvement of quality, or in cost reduction (Kirk et al. 2002). Oxygen (O_2) is sometimes detrimental to the quality or storage of food/beverage because of the unwanted oxidation reactions. Therefore, oxidases can be used as O_2^- scavengers for better food packaging (Andersson et al. 2002).

For bread making applications, glucose oxidase has been commercialized. Addition of the enzyme to dough can lead to various physicochemical changes including cross-linking of wheat albumin, globulin, and to some extent, glutenin (Primo-Martin et al. 2003; Rasiah et al. 2005). Thereafter, the dough shows improved viscoelastic/rheological characteristics, and the baked bread has better crumb, larger volume or other properties. This effect is thought to be caused by the H_2O_2 generated by the enzyme. However, the action of the enzyme is not superior to that of chemical oxidative additives, bromate and azodicarbonamide (Kohajdová et al. 2009). Therefore, there is a need to discover or develop other carbohydrate oxidases for this application. The lipoxygenase enzyme is a promising candidate for the baking applications (Casey et al. 1999). The enzyme can provide dough-strengthening and bread whitening effects by modifying the endogenous lipids/unsaturated fatty acids (and their emulsification property) and forming oxidative peroxide. However, the addition of the enzyme to certain foods may result in loss of flavour or depletion of endogenous antioxidants.

The use of oxidases is also being extended to other wheat products, such as noodles, pasta, and cakes (Hoeegh 2004). Similarly as in the case of bread, these enzymes modify flour components directly or indirectly through the production of active oxygen species. Laccases can be used for the modification of the colour appearance of food or beverages. In one interesting case involving the ripe-olive processing, laccase replaces the conventional dye solution and oxidatively polymerizes the various phenolics (such as oleuropein) in olive, resulting in the darkening of colour and debittering (Lal et al. 1998). Some oxidoreductases, for example glucose oxidase and catalase, have been shown to improve the preservation of shrimp and fish (Dondero et al. 1993). It is thought that the active oxygen species generated by the enzymes act as bacteriocides, and hence help in keeping these food items fresh. Further, the disulfide-reducing enzymes may be useful in the recovery of proteins from the low-value sources (such as fishery by-products) for reconstructed meat products (Chung et al. 2000).

6 Beverage Improvement Applications

Turbidity, browning, and haze formation may occur during processing or storage of fruit juices, beer and wine and this is a major problem for the beverage industry. The phenolic compounds are mainly involved in this process. Commonly, the undesirable phenolics are adsorbed and removed by various fining agents such as gelatine and bentonite, which usually have low specificity, may affect the colour or aroma of the drink, and can pose disposal problems (Schmid and Urlacher 2007). The laccases and other oxidases can be utilized to remove or modify the problematic phenolic saccharides and hence improve the clarity, colour appearance, taste, flavour, aroma, or stability of fruit juices or fermented alcohol beverages (Minussi et al. 2002). After the laccase treatment, the oxidized and polymerized (or precipitated) unwanted phenolic substances can be removed by silicate fining or filtration methods.

7 Dairy Applications

Oxidoreductases may also be applied for dairy applications. For example, a carbohydrate oxidase was utilized to convert lactose during cheese-making (Miwa et al. 2004). Lactose which is found in the whey fraction is generally discarded as a cheese-making by-product. The product of the enzymatic oxidation, lactobionic acid is valuable since it is widely used as a food additive, drug formulant, acidulant, chelator, and polymer precursor (Gutiérrez et al. 2012). By the convertion of lactose in situ to lactobionic acid may have many benefits in terms of added value and improved quality.

8 Environment Protection Applications

The oxidoreductases have an immense potential in environment-protection applications, along with the hydrolases, transferases, and lyases (Ahuja et al. 2004). Many microbes can utilize polycyclic aromatic hydrocarbons, chlorinated compounds, and other pollutants as their carbon, nitrogen, or energy sources. And a significant part of their metabolism is performed by the oxidoreductases. Degradation reactions catalyzed by these enzymes include electron transfer, H extraction, and O insertion, which can result in ring opening, mineralization, depolymerization, and other transformations (Kirk and Farrell 1987).

Oxidoreductases have immense applications in biodegradation. These enzymes are applied to degrade many substances such as undesirable contaminants, byproducts, or discarded materials (Husain 2006; Husain and Husain 2008; Husain and Ulber 2011). Some of these uses have already been described above. Laccase has been shown to oxidize and degrade lipids such as trilinolein and methyl linoleate

(Karlsson et al. 2001; Zhang et al. 2002). These unsaturated fatty compounds are not the typical substrates of the enzyme. The products of this conversion include hydroperoxides and epoxides. It appears that laccase promotes the initial pentadienyl and subsequent peroxy radical formation (Karlsson et al. 2001). This reaction is of interest due to the occurrence of the fatty compounds in wood and food, which may get involved in laccase catalyzed delignification and food modification, respectively. A heme peroxidase has been used for transforming the discarded dyes from the effluent of dye houses prior to its discharge into the environment (Wesenberg et al. 2003). The oxidation of dyes by peroxidases, or laccases, may lead to bleaching, polymerization, or degradation to facilitate downstream treatments (Matto and Husain 2009; Ali et al. 2016, 2017, 2018; Ali and Husain 2018). These enzymes can also be utilized to treat effluents from pulp/cotton mills, breweries, or food processing plants (Shaffiqu et al. 2002; Husain 2010). Laccase can also degrade plastic waste having olefin units (Sivan 2011). The oxidation of the olefin units by the enzyme, preferably in the presence of small redox mediators, can initiate a radical chain reaction, leading to the disintegration of the plastic. Laccases, peroxidases, and oxygenases are being considered as biocatalysts for degrading hazardous coal substances, particularly the sulphur-containing components (Fakoussa and Hofrichter 1999). This study is of interest in terms of lowering the pollution around coal mines and emission of acid rain causing agents from the power plants. The oxidoreductases may also be used to reduce the odour emitted from places such as garbage disposal sites, livestock farms, or pulp mills (Abe and Hiramoto 2004; Xu and Kaplan 2004). Various oxidases can oxidatively degrade ammonia, sulphide, thiol, amine, short-chain fatty acids, or other volatile organic compounds that result in malodour (Burton 2003a).

Oxidoreductases are used for biodetoxification and biodecontamination. Numerous pesticides, xenobiotics, coal substances, and other industrial products derived from polycyclic, aromatic, halogenated hydrocarbons and other organic compounds are toxic environmental pollutants (Husain et al. 2009, 2011; Karim and Husain 2009, 2010, 2011). Oxidoreductases can be used to detoxify and remove them and this is an active area of research. Laccases and peroxidases have been utilized to transform (often in the presence of redox mediators) various xenobiotics, polycyclic aromatic hydrocarbons, and other pollutants found in the industrial wastes and contaminated soil/water (Abe and Hiramoto 2004; Junghanns et al. 2005; Lai and Lin 2005; Minussi et al. 2002). The redox potential of these compounds is too high for laccases to directly oxidize them through electron transfer. Hence, the use of redox mediators allows other reactions, such as H extraction, to occur. The laccase catalyzed reaction can result in either direct degradation or polymerization/immobilization. Examples of direct degradation by laccases include dechlorination of chlorophenols, cleavage of aromatic rings, and mineralization of polycyclic aromatic hydrocarbons (Madhavi and Lele 2009). The process includes polymerization among pollutants themselves or copolymerization with other nontoxic substances (e.g., humic materials). The polymerized pollutants often become insoluble or immobilized, thus facilitating their easy removal by adsorption, sedimentation, or filtration methods. Chlorinated compounds like tetrachloroethene and polychlorinated biphenyl are one of the major biodetoxification targets, which contaminate the underground water. The oxidoreductases, particularly mono and dioxygenases, are strong dehalogenating agents (Furukawa 2000; Schultz et al. 2001). The ethenemonooxygenase and biphenyl dioxygenase enzymes are promising candidates for large-scale industrial applications in this field.

Advancements made in the sequencing of the genomes of environmentally relevant microbes and the direct DNA sequencing of soil microbes are revealing new metabolic enzymes/pathways potentially beneficial for bioremediation. By protein engineering, we can also create new enzymes/functions for biodetoxification (Mason et al. 1997). For example, a dioxygenase that oxidizes arene to vicinal arenediols has been engineered. This recombinant enzyme extends its reactivity from toluene, naphthalene, and biphenyl to polychlorinated biphenyls, trichloroethylene, and dioxin (Gibson and Parales 2000).

9 Organic Synthesis Applications

The biocatalysis by oxidoreductases has high potential for efficient, asymmetric synthetic applications (Alphand et al. 2003; Breuer et al. 2004; Roberts 2004). This is mainly due to their broad versatility, high specificity, high stereo-selectivity, low requirement for stringent/extreme conditions, and environment-friendliness. The oxidoreductases are capable of doing many types of diverse reactions. They can transform a wide range of aliphatic/aromatic molecules and functionalize inert hydrocarbons (by hydroxylation, sulfoxidation, epoxidation, etc). These enzymes can carry out regio-, enantio- (on racemic substrates); enantiotopo- (on prochiral substrates); and chemo-selective reactions. They can build useful synthons from inexpensive and renewable biomaterials and therefore save on energy and other resources, and relieve negative environmental impact (Wong and Whitesides 1994). Although only a few oxidoreductases have been commercialized in this field as yet, however significant growth in using oxidoreductase-based biocatalysis is projected for the coming years.

NAD(P)H-dependent dehydrogenases are very important for asymmetric/chiral synthesis (Stewart 2001). By catalyzing reversible dehydrogenation reactions, these enzymes can transform prochiral aldehydes/ketones to chiral alcohol, hydroxyl acids, amino acids, etc., with high regio- or enantio-selectivity (with enantio excess >90 to 95%). The hydroxy acid dehydrogenases can catalyze the synthesis of chiral hydroxy acids from aliphatic, linear/branched, or aromatic 2-oxo acids, that are valuable synthon (Santaniello et al. 1992). The amino acid dehydrogenases can catalyze the reductive amidation of 2-keto acids to either D- or L-amino acids, useful for making rare/unnatural amino acids with selected aliphatic, aromatic, linear, or branched side chains (Santaniello et al. 1992). This kind of synthesis is almost unattainable by conventional fermentation amino acid-production methods. The chiral synthesis catalyzed by dehydrogenases has higher yield than that done by hydrolases, however it is more difficult and expensive due to its dependency on NAD(P).

Further, the oxygenases and hydroxylases are also good biocatalysts for asymmetric synthesis. They can functionalize inert molecules (such as hydrocarbons) by the introduction of active O atoms (Shul'pin 2016). The flavin-containing monoxygenases/hydroxylases, such as phydroxybenzoate 3-hydroxylase, can catalyze highly regio- and stereo-selective arene hydroxylation reaction (Jadan et al. 2004). Baeyer-Villiger monooxygenase is of particular interest, due to its ability to catalyze the nucleophilic oxygenation of ketone (Baeyer-Villiger reaction) (Alphand et al. 2003). It can further catalyze the nucleophilic oxygenation of boron, as well as the electrophilic oxygenation of sulfur, nitrogen, selenium, or phosphorus, and with high enantio- or enantiotopo-selectivity (Branchaud and Walsh 1985). The enzymatic methods are more advantageous than the traditional chemical ones (using oxidants like m-chloroperbenzoic acid or trifluoroperacetic acid), because the latter are toxic and poorly enantioselective.

The iron-containing dioxygenases can convert arenes into vicinal, cis-arene diols, a task which is not attainable by conventional chemical synthesis. More than 300 vicinal arene diols have been identified from microbial oxidation of aromatic hydrocarbons, that provides a useful pool for selecting desirable synthons for various industrial/medicinal substances, such as postaglandins and other hypertensive agents (Jadan et al. 2004). Lipoxygenases can be used to synthesize many natural products and valuable building blocks since they have high regio- and stereo-selectivity towards their unsaturated fatty acid substrates (Nanda and Yadav 2003). Non-aqueous media, immobilization, and protein engineering have been applied to optimize the biocatalysis by these enzymes for asymmetric synthesis. Chloroperoxidase and cytochrome P450 enzymes can be utilized for asymmetric olefin epoxidation, allylic hydroxylation, and sulfoxidation, due to their high-valentferryl-oxo species (Lütz et al. 2004). A major hurdle in the commercialization of chloroperoxidase is that it is unstable in the presence of medium to high level H_2O_2, the oxidizing substrate. This issue may be resolved by in situ H_2O_2 generation with either a secondary enzyme (e.g., glucose oxidase) or an electrode. As for cytochrome P450, the large-scale production of this enzyme is difficult and remains a challenge (Conesa et al. 2002).

10 Nanomaterial and Polymer Synthesis Applications

The oxidoreductases that are capable of generating radicals are potential biocatalysts for polymer synthesis (Alvarez et al. 2003; Xu and Kaplan 2004). Phenol oxidases (e.g. tyrosinase) are of use in making polyphenolic polymers from monomeric phenols (e.g. catechol). The carbohydrate oxidases (e.g. galactose oxidase) are utilized to modify oligo-/polymeric carbohydrates to facilitate cross-linking. The amino acid-based polymers are preferred because of their biocompatibility and biodegradability. Horseradish peroxidise (HRP) is able of catalyzing tyrosine polymerization. When nano-patterned on a defined surface, the enzymatic polymerization creates well-organized nanofeatures that are potentially useful for specialty membrane/circuitry (Xu and Kaplan 2004). Immobilized HRP enzyme can direct the synthesis of polyaniline film of controlled thickness, a property that is highly useful for coating the surface of biosensors (Alvarez et al. 2003). This property is not attainable by the chemically synthesized polyaniline since it is insoluble and uncompressible. HRP is also utilized to synthesize phenol/aniline polymer via oxidative dehydrogenation (Durán and Esposito 2000; Jin et al. 2001). The immobilized peroxygenases may be applied to catalyze the generation of fatty epoxides, important polymer precursors, plasticizers, or plastic stabilizers (Burton 2003b).

11 Medicinal and Other Synthetic Applications

The oxidoreductases are also important in medicinal synthesis. Laccases can be utilized to synthesize a large number of complex medicinal agents, such as triazolo (benzo)cycloalkylthiadiazines, cephalosporin antibiotics, vinblastine, penicillin X dimer, and dimerized vindoline (Mikolasch et al. 2007; Sagui et al. 2009). The laccases can also be applied to synthesize various functional organic compounds including polymers with specific mechanical/optical/electrical properties, flavoring agents, textile dyes, cosmetic pigments, and pesticides (Ose et al. 2003). The application of oxidoreductases may lead to the development of new industrial synthetic methods. For example, Baeyer-Villiger monooxygenase can catalyze a useful ring-expansion reaction by transforming a cyclic ketone to the corresponding lactone (Alphand et al. 2003). Macrophomate synthase enzyme can catalyze Diels-Alder reaction (Sedmera et al. 2004). Sometimes when the oxidoreductase acts on its substrate, it may induce a secondary reaction on other parts of the substrate, leading to a new type of biocatalysis (Kimoto and Yamamoto 2004). By using oxidoreductases, we can catalyze reactions that are not easily feasible. For example, chloroperoxidase and cytochrome P450 enzymes can functionalize inert hydrocarbons through hydroxylation (Bell et al. 2003). Enone reductase can hydrogenate unsaturated bonds to convert ketones to hydrocarbons (Colonna et al. 2004). The so-called "old yellow enzyme" obtained from yeast which is a FMN-containing enzyme, can catalyze the reduction (by NADPH) of the olefinic $>C=C<$, not carbonyl $>C=O$, the site of cyclohexenone (Massey 2000).

Many new intermediates/synthons can be synthesized using oxidoreductases (Amao and Watanabe 2004; Pilone and Pollegioni 2002; Srivastava et al. 2004). Methane monooxygenase can convert methane to methanol which is a useful synthon. The xylose reductase enzyme can produce xylitol from glucose, which is a natural sweetener (Mayer et al. 2002). The FAD-containing vanillyl-alcohol oxidase can form vanillin from creosol, whose ~2000 tons worldwide annual consumption depends mostly on artificial synthesis (Van den Heuvel et al. 2004).

12 Concluding Remarks

In the present world, where environmental aspect is of uttermost consideration, oxidoreductases may be utilized as advantageous biocatalysts to replace the toxic/ expensive chemicals, save on energy/resources consumption, create novel functionalities, or reduce detrimental impacts on the environment. The oxidoreductases enzymes have the highest potential for the production of building blocks of the polymers, sustainable chemicals/materials from plant biomass in lignocellulose biorefineries. However, for the large scale production of chemicals, the industries are not yet employing the enzymatic oxidation reactions to a large extent. And this is mainly due to the lack of enzymes with the desired selectivity, commercial availability and compatibility with the stringent process conditions e.g. high concentrations of the substrate, use of solvents, and strong oxidative conditions. Currently, these enzymes are mostly employed in specific segments of the chemical industry and frequently in the form of whole-cell catalysts, unlike as isolated enzymes in medium and large scale biotransformations. The main problems for the industrial implementation of oxidative enzymatic biocatalysts have been addressed via protein engineer ing and the optimization of the process using state-of-the-art technologies. Many of these enzymes have been engineered, combining rational and computational design with directed evolution, to achieve the selectivity, catalytic efficiency and stability properties required for their industrial use (Martínez et al. 2017). The application of these enzymes to new manufacturing areas is important for the future growth of industrial oxidoreductase biocatalysts.

References

Abe K, Hiramoto T (2004) Deodorization method. In: Google Patents

Ahuja SK, Ferreira GM, Moreira AR (2004) Utilization of enzymes for environmental applications. Crit Rev Biotechnol 24:125–154

Alam MF, Laskar AA, Choudhary HH, Younus H (2016) Human salivary aldehyde dehydrogenase: purification, kinetic characterization and effect of ethanol, hydrogen peroxide and sodium dodecyl sulphate on the activity of the enzyme. Cell Biochem Biophys 74:307–315

Ali M, Husain Q (2018) Guar gum blended alginate/agarose hydrogel as a promising support for the entrapment of peroxidase: stability and reusability studies for the treatment of textile effluent. Int J Biol Macromol 116:463–471

Ali M, Ahmad M, Husain Q (2016) Redox mediated decolorization and detoxification of direct blue 80 by partially purified ginger (Zingiber officinale) peroxidase. Int J Environ Agri Res 2 (4):19–30

Ali M, Husain Q, Ahmad M (2017) Enhanced catalytic activity and stability of ginger peroxidase immobilized on amino functionalized silica coated titanium dioxide nanocomposite. Water Air Soil Poll 222:28

Ali M, Husain Q, Sultana S, Ahmad M (2018) Immobilization of peroxidase on polypyrrole-cellulose-graphene oxide nanocomposite via non-covalent interactions for the degradation of reactive blue 4 dye. Chemosphere 202:198–207

Alphand V, Carrea G, Wohlgemuth R, Furstoss R, Woodley JM (2003) Towards large-scale synthetic applications of Baeyer-Villiger monooxygenases. Trends Biotechnol 21:318–323

Alvarez S, Manolache S, Denes F (2003) Synthesis of polyaniline using horseradish peroxidase immobilized on plasma-functionalized polyethylene surfaces as initiator. J Appl Pol Sci 88:369–379

Amao Y, Watanabe T (2004) Photochemical and enzymatic synthesis of methanol from HCO_3^- with dehydrogenases and Zinc Porphyrin. Chem Lett 33:1544–1545

Andersson M, Andersson T, Adlercreutz P, Nielsen T, Hornsten EG (2002) Toward an enzyme-based oxygen scavenging laminate. Influence of industrial lamination conditions on the performance of glucose oxidase. Biotechnol Bioeng 79:37–42

Anthony C (2004) The quinoprotein dehydrogenases for methanol and glucose. Arch Biochem Biophys 428:2–9

Balakshin MY, Evtuguin DV, Pascoal Neto C, Cavaco-Paulo A (2001) Polyoxometalates as mediators in the laccase catalyzed delignification. J Mol Catal B Enzymatic 16:131–140

Bannwarth M, Bastian S, Heckmann-Pohl D, Giffhorn F, Schulz GE (2004) Crystal structure of pyranose 2-oxidase from the white-rot fungus Peniophora sp. Biochemistry 43:11683–11690

Bell SG, Orton E, Boyd H, Stevenson JA, Riddle A, Campbell S, Wong LL (2003) Engineering cytochrome P450cam into an alkane hydroxylase. Dalton Transac 11:2133–2140

Brady R, Cheng HN, Haandrikman A, Moore A, Kuo PK, McNabola W, Wheeler C, Xu ZF, Riehle R, Nguyen T (2002). Reduced molecular weight galactomannans oxidized by galactose oxidase. In: Google Patents

Branchaud BP, Walsh CT (1985) Functional group diversity in enzymic oxygenation reactions catalyzed by bacterial flavin-containing cyclohexanone oxygenase. J Am Chem Soc 107:2153–2161

Breuer M, Ditrich K, Habicher T, Hauer B, Kesseler M, Sturmer R, Zelinski T (2004) Industrial methods for the production of optically active intermediates. Angew Chem 43:788–824

Bugg TD (2001) Oxygenases: mechanisms and structural motifs for O 2 activation. Curr Opin Chem Biol 5:550–555

Burton SG (2003a) Oxidizing enzymes as biocatalysts. Trends Biotechnol 21:543–549

Burton SG (2003b) Laccases and phenol oxidases in organic synthesis - a review. Curr Org Chem 7:1317–1331

Cameron MD, Aust SD (2001) Cellobiose dehydrogenase–an extracellular fungal flavocytochrome. Enzym Microb Technol 28:129–138

Casey R, West SI, Hardy D, Robinson DS, Wu Z, Hughes RK (1999) New frontiers in food enzymology: recombinant lipoxygenases. Trends Food Sci Technol 10:297–302

Chung YC, Ho ML, Chyan FL, Jiang ST (2000) Utilization of freeze-dried mackerel (Scomber australasicus) muscle proteins as a binder in restructured meat. Fisheries Sci 66:130–135

Claus H (2004) Laccases: structure, reactions, distribution. Micron 35:93–96

Colonna S, Pironti V, Carrea G, Pasta P, Zambianchi F (2004) Oxidation of secondary amines by molecular oxygen and cyclohexanone monooxygenase. Tetrahedron 60:569–575

Conesa A, Punt PJ, Van den Hondel CA (2002) Fungal peroxidases: molecular aspects and applications. J Biotechnol 93:143–158

Dondero M, EgaÑA W, Tarky W, Cifuentes A, Torres JA (1993) Glucose oxidase/catalase improves preservation of Shrimp (Heterocarpus reedi). J Food Sci 58:774–779

Durán N, Esposito E (2000) Potential applications of oxidative enzymes and phenoloxidase-like compounds in wastewater and soil treatment: a review. Appl Catal B Environ 28:83–99

Durán N, Rosa MA, D'Annibale A, Gianfreda L (2002) Applications of laccases and tyrosinases (phenoloxidases) immobilized on different supports: a review. Enzym Microb Technol 31:907–931

Ellis EM (2002) Microbial aldo-keto reductases. FEMS Microbiol Lett 216:123–131

Fakoussa R, Hofrichter M (1999) Biotechnology and microbiology of coal degradation. Appl Microbiol Biotechnol 52:25–40

Fetzner S (2002) Oxygenases without requirement for cofactors or metal ions. App Microbiol Biotechnol 60:243–257
Fomenko DE, Gladyshev VN (2012) Comparative genomics of thiol oxidoreductases reveals widespread and essential functions of thiol-based redox control of cellular processes. Antioxid Redox Signal 16:193–201
Forrest GL, Gonzalez B (2000) Carbonyl reductase. Chem Int 129:21–40
Furukawa K (2000) Engineering dioxygenases for efficient degradation of environmental pollutants. Curr Opin Biotechnol 11:244–249
Ghaffar T, Irshad M, Anwar Z, Aqil T, Zulifqar Z, Tariq A, Kamran M, Ehsan N, Mehmood S (2014) Recent trends in lactic acid biotechnology: a brief review on production to purification. J Radiat Res Appl Sci 7:222–229
Gibson DT, Parales RE (2000) Aromatic hydrocarbon dioxygenases in environmental biotechnology. Curr Opin Biotechnol 11:236–243
Gutiérrez LF, Hamoudi S, Belkacemi K (2012) Lactobionic acid: a high value-added lactose derivative for food and pharmaceutical applications. Int Dairy J 26:103–111
Hoeegh L (2004) Method of improving the hydration of pasta and preparation of pasta products. In: Google Patents
Husain Q (2006) Potential applications of the oxidoreductive enzymes in the decolorization and detoxification of textile and other synthetic dyes from polluted water: a review. Crit Rev Biotechnol 26:201–221
Husain Q (2010) Peroxidase mediated decolorization and remediation of wastewater containing industrial dyes: a review. Rev Environ Sci Biotechnol 9(2):117–140
Husain Q (2017) High yield immobilization and stabilization of oxidoreductases using magnetic nanosupports and their potential applications: an update. Current Catal 6(3):168–187
Husain M, Husain Q (2008) Applications of redox mediators in the treatment of organic pollutants by using oxidoreductive enzymes: a review. Crit Rev Environ Sci Technol 38(1):1–42
Husain Q, Ulber R (2011) Immobilized peroxidase as a valuable tool in the remediation of aromatic pollutants and xenobiotic compounds: a review. Crit Rev Environ Sci Technol 41(8):770–804
Husain Q, Husain M, Kulshrestha Y (2009) Remediation and treatment of organo-pollutants mediated by peroxidases: a review. Crit Rev Biotechnol 29(2):94–119
Husain Q, Karim Z, Rizvi ZA (2011) Oxidative degradation and polymerization of methyl parathion, a pesticide by using fenugreek seeds (*Trigonella foenum graecum* L) peroxidase. Environ Prog Sustain Energy 30(2):392–398
Imhoff RD, Power NP, Borrok MJ, Tipton PA (2003) General base catalysis in the urate oxidase reaction: evidence for a novel Thr-Lys catalytic diad. Biochemistry 42:4094–4100
Jadan AP, Moonen MJH, Boeren S, Golovleva LA, Rietjens IMCM, Van Berkel WJH (2004) Biocatalytic potential of p-Hydroxybenzoate hydroxylase from Rhodococcus rhodnii 135 and Rhodococcus opacus 557. Adv Synth Catal 346:367–375
Jin Z, Su Y, Duan Y (2001) A novel method for polyaniline synthesis with the immobilized horseradish peroxidase enzyme. Synth Metals 122:237–242
Junghanns C, Moeder M, Krauss G, Martin C, Schlosser D (2005) Degradation of the xenoestrogen nonylphenol by aquatic fungi and their laccases. Microbiology 151:45–57
Kagan HM, Li W (2003) Lysyl oxidase: properties, specificity, and biological roles inside and outside of the cell. J Cell Biochem 88:660–672
Karim Z, Husain Q (2009) Redox-mediated oxidation and removal of aromatic amines from synthetic water by partially purified bitter gourd (*Momordica charantia*) peroxidase. Int Biodeterior Biodegradation 63(5):587–593
Karim Z, Husain Q (2010) Removal of anthracene from polluted water by immobilized peroxidase from *Momordica charantia* in batch process as well as in a continuous spiral-bed reactor. J Mol Catal B Enzym 66(3–4):302–310
Karim Z, Husain Q (2011) Removal of benzidine from polluted water by soluble and immobilized peroxidase in batch processes and continuous horizontal bed-reactor. Environ Technol 32:83–91

Karlsson S, Holmbom B, Spetz P, Mustranta A, Buchert J (2001) Reactivity of Trametes laccases with fatty and resin acids. Appl Microbiol Biotechnol 55:317–320
Karmali A, Coelho J (2011) Bioconversion of D-glucose into D-glucosone by glucose 2-oxidase from Coriolus versicolor at moderate pressures. Appl Biochem Biotechnol 163:906–917
Kimoto N, Yamamoto H (2004). Novel enone reductases isolated from Kluyveromyces lactis, methods for producing same, and methods for selectively reducing a carbon-carbon double bond of an Alpha, Beta-unsaturated ketone using the reductases. In: Google Patents
Kirk TK, Farrell RL (1987) Enzymatic "combustion": the microbial degradation of lignin. Annu Rev Microbiol 41:465–501
Kirk O, Borchert TV, Fuglsang CC (2002) Industrial enzyme applications. Curr Opin Biotechnol 13:345–351
Kohajdová Z, Karovičová J, Schmidt Š (2009) Significance of emulsifiers and hydrocolloids in bakery industry. Acta Chimica Slovaca 2:46–61
Koka R, Mehnert D, Fritsch R, Steffan W, Habermeier P, Bradbury A, Wolfschoon-Pombo A, Rose M (2004) Process for manufacturing cheeses and other dairy products and products thereof. In: Google Patents
Lai YC, Lin SC (2005) Application of immobilized horseradish peroxidase for the removal of p-chlorophenol from aqueous solution. Process Biochem 40:1167–1174
Lal GK, Turksma H, Van PK. PAM, Winkel C (1998). Debittering of olive oil. In: Google Patents
Laskar AA, Alam MF, Younus H (2017) In vitro activity and stability of pure human salivary aldehyde dehydrogenase. Int J Biol Macromol 96:798–806
Leonowicz A, Cho NS, Luterek J, Wilkolazka A, Wojtas-Wasilewska M, Matuszewska A, Hofrichter M, Wesenberg D, Rogalski J (2001) Fungal laccase: properties and activity on lignin. J Basic Microbiol 41:185–227
Leskovac V, Trivic S, Pericin D (2002) The three zinc-containing alcohol dehydrogenases from baker's yeast, Saccharomyces cerevisiae. FEMS Yeast Res 2:481–494
Lütz S, Steckhan E, Liese A (2004) First asymmetric electroenzymatic oxidation catalyzed by a peroxidase. Electrochem Commun 6:583–587
Madhavi V, Lele S (2009) Laccase: properties and applications. BioRes 4:1694–1717
Martínez AT, Ruiz-Dueñas FJ, Camarero S, Serrano A, Linde D et al (2017) Oxidoreductases on their way to industrial biotransformations. Biotechnol Adv 35(6):815–831
Mason J, Briganti F, Wild J (1997) Protein engineering for improved biodegradation of recalcitrant pollutants. Persp Bioremed (Springer) 19:107–118
Massey V (2000) The chemical and biological versatility of riboflavin. Biochem Soc Trans 28:283–296
Matto M, Husain Q (2009) Decolorization of textile effluent by bitter gourd peroxidase immobilized on concanavalin a layered calcium alginate-starch beads. J Hazard Mater 164(2–3):1540–1546
May SW, Padgette SR (1983) Oxidoreductase enzymes in biotechnology: current status and future potential. Nat Biotech 1:677–686
Mayer G, Kulbe KD, Nidetzky B (2002) Utilization of xylitol dehydrogenase in a combined microbial/enzymatic process for production of xylitol from D-glucose. Appl Biochem Biotechnol 98:577–589
Mikolasch A, Niedermeyer THJ, Lalk M, Witt S, Seefeldt S, Hammer E, Schauer F, Gesell Salazar M, Hessel S, Jülich WD (2007) Novel cephalosporins synthesized by amination of 2, 5-dihydroxybenzoic acid derivatives using fungal laccases II. Chem Pharma bulletin 55:412–416
Minussi RC, Pastore GM, Duran N (2002) Potential applications of laccase in the food industry. Trends Food Sci Technol 13:205–216
Miwa N, Kumazawa Y, Nakagoshi H, Sakaguchi S (2004). Method for modifying raw material milk and dairy product prepared by using the modified raw material milk. In: Google Patents
Nanda S, Yadav JS (2003) Lipoxygenase biocatalysis: a survey of asymmetric oxygenation. J Mol Catal B: Enzymatic 26:3–28

Nicholas C, Lewis S (1999) Fundamentals of enzymology: the cell and molecular biology of catalytic proteins. Oxford University Press, Oxford
Ose T, Watanabe K, Mie T, Honma M, Watanabe H, Yao M, Oikawa H, Tanaka I (2003) Insight into a natural Diels-Alder reaction from the structure of macrophomate synthase. Nature 422:185–189
Oubrie A (2003) Structure and mechanism of soluble glucose dehydrogenase and other PQQ-dependent enzymes. Biochim Biophys Act (BBA) Prot Proteom 1647:143–151
Pilone MS, Pollegioni L (2002) D-amino acid oxidase as an industrial biocatalyst. Biocatal Biotransformation 20:145–159
Pollegioni L, Diederichs K, Molla G, Umhau S, Welte W, Ghisla S, Pilone MS (2002) Yeast D-amino acid oxidase: structural basis of its catalytic properties. J Mol Biol 324:535–546
Poulos TL (2005) Structural and functional diversity in heme monooxygenases. Drug Metab Dispos 33:10–18
Primo-Martin C, Valera R, Martinez-Anaya MA (2003) Effect of pentosanase and oxidases on the characteristics of doughs and the glutenin macropolymer (GMP). J Agric Food Chem 51:4673–4679
Que L, Watanabe Y (2001) Oxygenase pathways: Oxo, peroxo, and superoxo. Science 292:651–653
Rasiah IA, Sutton KH, Low FL, Lin HM, Gerrard JA (2005) Crosslinking of wheat dough proteins by glucose oxidase and the resulting effects on bread and croissants. Food Chem 89:325–332
Rhee SG, Kang SW, Chang TS, Jeong W, Kim K (2001) Peroxiredoxin, a novel family of peroxidases. IUBMB Life 52:35–41
Roberts SM (2004) Biocatalysts in synthetic organic chemistry. Tetrahedron 60:499–500
Sagui F, Chirivì C, Fontana G, Nicotra S, Passarella D, Riva S, Danieli B (2009) Laccase-catalyzed coupling of catharanthine and vindoline: an efficient approach to the bisindole alkaloid anhydrovinblastine. Tetrahedron 65:312–317
Santaniello E, Ferraboschi P, Grisenti P, Manzocchi A (1992) The biocatalytic approach to the preparation of enantiomerically pure chiral building blocks. Chem Rev 92:1071–1140
Schmid RD, Urlacher V (2007) Modern biooxidation: enzymes, reactions and applications. Wiley, Weinheim
Schultz A, Jonas U, Hammer E, Schauer F (2001) Dehalogenation of chlorinated hydroxybiphenyls by fungal laccase. Appl Environ Microbiol 67:4377–4381
Sedmera P, Halada P, Peterbauer C, Volc J (2004) A new enzyme catalysis: 3,4-dioxidation of some aryl β-d-glycopyranosides by fungal pyranose dehydrogenase. Tetrahedron Lett 45:8677–8680
Sellés Vidal L, Kelly CL, Mordaka PM, Heap JT (2018) Review of NAD(P)H-dependent oxidoreductases: properties, engineering and application. Biochim Biophys Acta 1866(2):327–347
Shaffiqu T, Roy JJ, Nair RA, Abraham TE (2002) Degradation of textile dyes mediated by plant peroxidases. Appl Biochem Biotechnol 102:315–326
Sharp KH, Moody PCE, Raven EL (2003) A new framework for understanding substrate binding and functional diversity in haem peroxidases. Dalton Trans 22:4208–4215
Shleev SV, Khan IG, Gazaryan IG, Morozova OV, Yaropolov AI (2003) Novel laccase redox mediators: spectral, electrochemical, and kinetic properties. App Biochem Biotechnol 111:167–184
Shul'pin GB (2016) New trends in oxidative functionalization of carbon–hydrogen bonds: a review. Catalysts 6:50
Sivan A (2011) New perspectives in plastic biodegradation. Curr Opin Biotechnol 22:422–426
Srivastava S, Spite M, Trent JO, West MB, Ahmed Y, Bhatnagar A (2004) Aldose reductase-catalyzed reduction of aldehyde phospholipids. J Biol Chem 279:53395–53406
Stewart JD (2001) Dehydrogenases and transaminases in asymmetric synthesis. Curr Opin Chem Biol 5:120–129
Sun Y, Li T, Chen H, Zhang K, Zheng K, Mu Y, Yan G, Li W, Shen J, Luo G (2004) Selenium-containing 15-mer peptides with high glutathione peroxidase-like activity. J Biol Chem 279:37235–37240

Suzuki S, Kataoka K, Yamaguchi K (2000) Metal coordination and mechanism of multicopper nitrite reductase. Acc Chem Res 33:728–735
Thorpe C, Hoober KL, Raje S, Glynn NM, Burnside J, Turi GK, Coppock DL (2002) Sulfhydryl oxidases: emerging catalysts of protein disulfide bond formation in eukaryotes. Arch Biochem Biophys 405:1–12
Toone EJ (2010) Advances in enzymology and related areas of molecular biology, Protein evolution, vol 75. Wiley, Hoboken, NJ
Van de Velde F, Van Rantwijk F, Sheldon RA (2001) Improving the catalytic performance of peroxidases in organic synthesis. Trends Biotechnol 19:73–80
Van den Heuvel RHH, Van den Berg WAM, Rovida S, Van Berkel WJH (2004) Laboratory-evolved vanillyl-alcohol oxidase produces natural vanillin. J Biol Chem 279:33492–33500
Veitch NC (2004) Horseradish peroxidase: a modern view of a classic enzyme. Phytochemistry 65:249–259
Vert M, Chiellini E, Gil H, Braunegg G, Buchert J, Gatenholm P, van den Zee M (2001) Biorelated polymers: sustainable polymer science and technology. Kluwer, New York
Webb EC (1992) Enzyme nomenclature 1992. Recommendations of the Nomenclature Committee of the International Union of Biochemistry and Molecular Biology on the nomenclature and classification of enzymes. Academic, San Diego, CA
Wesenberg D, Kyriakides I, Agathos SN (2003) White-rot fungi and their enzymes for the treatment of industrial dye effluents. Biotechnol Adv 22:161–187
Whittaker JW (2003) Free radical catalysis by galactose oxidase. Chem Rev 103:2347–2364
Wong CH, Whitesides GM (1994) Enzymes in synthetic organic chemistry, vol 12. Academic, Burlington
Xu P, Kaplan DL (2004) Horseradish peroxidase catalyzed polymerization of tyrosine derivatives for Nanoscale surface patterning. J Macromol Sci Part A 41:1437–1445
Xu F, Golightly EJ, Duke KR, Lassen SF, Knusen B, Christensen S, Brown KM, Brown SH, Schülein M (2001) Humicola insolens cellobiose dehydrogenase: cloning, redox chemistry, and "logic gate"-like dual functionality. Enzym Microb Technol 28:744–753
Yagi T, Seo BB, Di Bernardo S, Nakamaru-Ogiso E, Kao MC, Matsuno-Yagi A (2001) NADH dehydrogenases: from basic science to biomedicine. J Bioenerget Biomem 33:233–242
Yamazaki SI, Morioka C, Itoh S (2004) Kinetic evaluation of catalase and peroxygenase activities of tyrosinase. Biochemistry 43:11546–11553
Zhang X, Eigendorf G, Stebbing D, Mansfield S, Saddler J (2002) Degradation of trilinolein by laccase enzymes. Arch Biochem Biophys 405:44–54

Rieske Non-Heme Iron Dioxygenases: Applications and Future Perspectives

Fatma Feyza Özgen and Sandy Schmidt

1 Introduction

Nature has developed manifold creative solutions for C–H-bond functionalization reactions (Dong et al. 2018; Lewis et al. 2011). For many of these reactions no 'classical' chemical counterpart is known. In particular, the catalytic asymmetric dihydroxylation of alkenes has attracted considerable attention due to the facile further transformation of the chiral diol products into valuable derivatives (Bataille and Donohoe 2011; Chang et al. 2004; Gally et al. 2015; Punniyamurthy et al. 2005; Zaitsev and Adolfsson 2006) making them important building blocks for the pharmaceutical and chemical industry. Rieske non-heme iron dioxygenases (ROs) represent very promising biocatalysts to promote catalytic asymmetric dihydroxylation reactions of alkenes since they are the only enzymes known to catalyze the stereoselective formation of vicinal *cis*-diols in one step. These enzymes are soluble multicomponent systems that harness the reductive power of NAD(P)H for oxygen activation. Due to their versatility, ROs are considered as the non-heme analogues of cytochrome P450 monooxygenases and, in addition to their relaxed substrate specificity, these enzymes can catalyze various oxidation reactions (Boyd and Sheldrake 1998; Gally et al. 2015; Halder et al. 2018; Parales and Resnick 2004; Sydor et al. 2011).

ROs are depending on a complex electron transport chain very similar to those having been observed in P450 monooxygenases (Fig. 1a). The electron transfer proceeds from a reductase to a ferredoxin (Fd) and further to the terminal oxygenase where the hydroxylation reaction takes place (Mason and Cammack 1992). Due to that, the majority of synthetic applications of ROs relies on recombinant, whole cell catalysts. Reactions were usually performed with whole cells due to the reported instability of isolated ROs (Catterall and Williams 1971; Sauber et al. 1977) as well

F. F. Özgen · S. Schmidt (✉)
Institute of Molecular Biotechnology, Graz University of Technology, Graz, Austria
e-mail: f.oezgen@tugraz.at; s.schmidt@tugraz.at

Q. Husain, M. F. Ullah (eds.), *Biocatalysis*,
https://doi.org/10.1007/978-3-030-25023-2_4

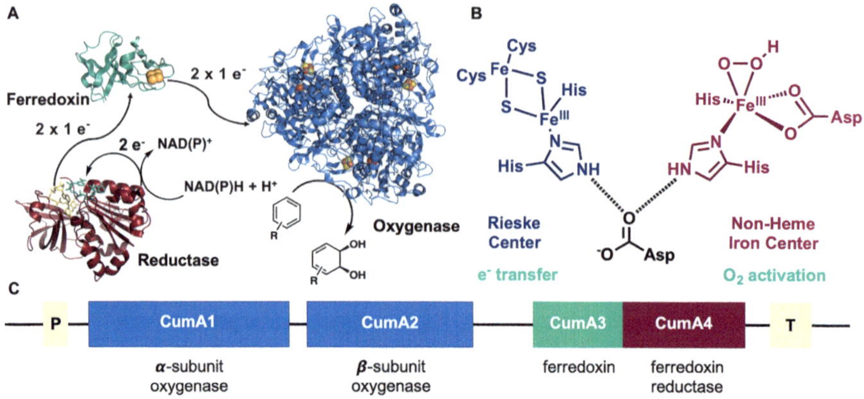

Fig. 1 (**a**) RO electron transfer chain from NAD(P)H via a reductase (FdR) and a ferredoxin (Fd) to the terminal oxygenase (Oxy) component. (**b**) Electron transfer from the Rieske center (blue) to the catalytic mononuclear iron (magenta) of the adjacent subunit proceeds via a conserved aspartate residue (black). (**c**) Arrangement of the RO genes in an operon exemplified for cumene dioxygenase from *Pseudomonas fluorescens* IP01; *P* promoter, *T* terminator

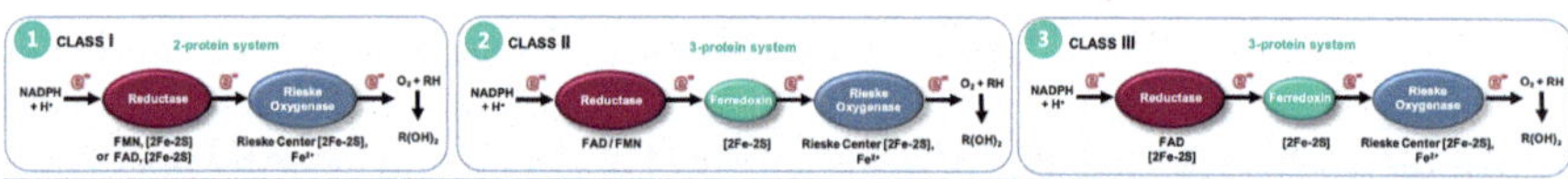

System	Class	Reductase	Ferredoxin	Oxygenase	Model Enzyme
2-component	IA	FMN, [2Fe-2S]	none	[2Fe-2S], Fe	Phthalate dioxygenase
	IB	FAD, [2Fe-2S]	none	[2Fe-2S], Fe	Benzoate dioxygenase
3-component	IIA	FAD	[2Fe-2S]	[2Fe-2S], Fe	Pyrazon dioxygenase
	IIB	FAD	[2Fe-2S]	[2Fe-2S], Fe	Cumene dioxygenase
	III	FAD, [2Fe-2S]	[2Fe-2S]	[2Fe-2S], Fe	Naphtalene dioxygenase

Fig. 2 Classification of ROs by their components (magenta: reductase, green: ferredoxin, blue: oxygenase). Class I: two-component dioxygenases, flavin and iron–sulfur cluster in the reductase, Class IA: FMN, Class IB: FAD. Class II: three-component dioxygenases, electron-transport chain with a flavoprotein and a separate ferredoxin, Class IIA: plant-type ferredoxin, Class IIB: Rieske-type iron–sulfur cluster in the ferredoxin. Class III: three-component dioxygenases, electron-transfer chain with iron–sulfur flavoprotein and ferredoxin

as to provide an efficient in-situ cofactor regeneration system (Schmidt et al. 2015). In contrast to cytochrome P450s, ROs are less well understood. ROs were first identified as enzymes involved in the degradation of aromatic compounds by *Pseudomonas putida* (Axcell and Geary 1975; Gibson et al. 1968) and are characterized by their unique [2Fe–2S] cluster in that one of the two Fe atoms is coordinated by two histidine residues rather than two cysteine residues (Fig. 1b). According to a new classification, ROs are divided into five subclasses (Kweon et al. 2008) (older classification: IA, IB, IIA, IIB, and III) (Mason and Cammack 1992) based on the number of constituents and the nature of their redox centers (Fig. 2). Prominent examples for IIB RO systems are cumene dioxygenase (CDO) from *Pseudomonas fluorescens* IP01, biphenyl 2,3-dioxygenase (BPDO) from *Sphingobium yanoikuyae* B1 and carbazole 1,9a-dioxygenase (CARDO) from *Nocardioides aromaticivorans* IC177 containing a Rieske-type iron–sulfur cluster in the ferredoxin (Fd). Many of the characterized ROs are capable of oxidizing a broad range of substrates which goes well beyond the range of compounds that serves as growth substrates for the host bacterium (Boyd et al. 2001a; Gibson and Parales 2000; Parales and Resnick 2004). For instance, toluene dioxygenase (TDO) from *Pseudomonas putida* F1 is capable of oxidizing over 200 different substrates (Boyd et al. 2001b), and naphthalene dioxygenase (NDO) from *Pseudomonas* sp. NCIB 9816-4 has been reported to oxidize over 75 different substrates (Resnick et al. 1996). Moreover, the spectrum of reactions they are able to catalyze reactions range from *cis*-dihydroxylations, O- and N-dealkylations, desaturations, to the formation of chiral sulfoxides from sulfides (Boyd et al. 2001a; Hudlicky et al. 1999; Resnick et al. 1996). Thus, ROs have an enormous potential for manifold synthetically useful transformations.

2 Rieske non-Heme Iron Dioxygenases: Structural Insights and Catalytic Behavior

In contrast to many flavin-dependent monooxygenases, which directly receive the electrons from NAD(P)H, ROs depend on electron transfer from a reductase or the interplay of a reductase (FdR) and a ferredoxin (Fd). In the latter case, the cofactor-derived electrons are transferred via FdR and Fd to the terminal oxygenase (Oxy) component containing the active site (Fig. 1a) (Mason and Cammack 1992).

These complex redox machines require multistep electron tunneling architectures that can transfer electrons rapidly with only a small loss of free energy to the surface of the Oxy. Nature developed several strategies to transfer electrons over long distances along chains of closely spaced redox relays: iron–sulfur clusters, copper centers and hemes in respiratory enzymes (Leger et al. 2006; Leys and Scrutton 2004; Page et al. 2003). Biological electron-transfer proteins can be placed into several categories according to the complexity of the reaction and the structural architecture. The simplest ones are the ferredoxins. Fds are one-domain one-electron carrier proteins in which the electron resides transiently on an iron–sulfur cluster. These electron carriers are small proteins and act in diverse biochemical processes as universal electron carrier (Hall et al. 1971; Mathews and White 1993). In contrast to the majority of P450s, ROs usually display a strict specificity for their physiological redox partners. This dependency is based on the Rieske-type [2Fe–2S] center within the Fd, which is complexed by two conserved histidines and two conserved cysteines. This results in significantly different spectroscopic properties and a more positive midpoint reduction potential in contrast to plant-type [2Fe–2S] centers.

The multicomponent system of ROs comprises a terminal oxygenase where the substrate oxyfunctionalization takes place. Dependent on the RO class (Fig. 2), the Oxy either consists of two separate proteins (Fig. 1c), a large catalytic subunit (α) and a small subunit (β) in hetero-multimeric form ($\alpha_n\beta_n$), or lacks the β-subunit and exists in homo-multimeric form (α_n) (Chakraborty et al. 2012; Ferraro et al. 2005). The α-subunit contains the largely hydrophobic active site whereas the β-subunit is assumed to have a structural function. However, in distinct RO systems, the β-subunit was proposed to have an influence on substrate specificity (Hurtubise et al. 1998). ROs lacking a β-subunit display loops on the α-subunits that might enhance their stability and remove the need for a stabilizing small Oxy subunit (Ferraro et al. 2005). The α-subunit of the Oxy contains two metal centers, a Rieske-type [2Fe–2S] cluster that mediates the electron transfer to the mononuclear iron in the active site where O_2 binding and activation is thought to take place (Bassan et al. 2004). The active site iron is highly accessible with the majority of its surface available for the binding of both oxygen atoms (Karlsson et al. 2003). All crystal structures which have been solved of ROs to date, possess either an α_3 or $\alpha_3\beta_3$quaternary structure (Fig. 1a). In this complex, electron transfer proceeds between the Rieske cluster (blue) and the iron center (magenta) of adjacent subunits via a bridging aspartate (black) due to the smaller distance compared to intramolecular transfer (Fig. 1b) (Ferraro et al. 2005).

The first RO for which the crystal structure has been solved is the NDO from *Pseudomonas* sp. NCIB 9816-4 (Kauppi et al. 1998). From the crystal structure, the active site geometry has been enlightened, which is characteristic for ROs (Ohta et al. 2008). The non-heme iron in the hydrophobic active site is coordinated by two histidines (H208 and H213) and a bidentate aspartate (D362) which constitute the 2-His-1-carboxylate facial triad (Kauppi et al. 1998). Additionally, one or two water molecules are found to be bound in the resting state when no substrate is present (Ohta et al. 2008). On the other hand, the unique Rieske center is coordinated by two cysteine residues (C81 and C101) and two histidines (H83 and H104) which are located in the highly-conserved motif $CXHX_{17}CX_2H$ (Kauppi et al. 1998; Mason and Cammack 1992). In the $\alpha_3\beta_3$ hexamer, the Rieske cluster of one α-subunit is closer to the catalytic iron of the adjacent subunit (~12 Å) than to the Rieske cluster within the same subunit (~44 Å) (Ohta et al. 2008). The coupling of neighboring non-heme iron and the [2Fe–2S] center is facilitated by D205 that is within hydrogen-bonding distance of the metal-coordinating amino acids and bridges H104 (coordinating the Rieske center) to H208 of the mononuclear iron in the active site. Most likely, D205 is the main electron transfer pathway from the Rieske cluster to the catalytic iron as proved by a NDO variant (D205A) with completely abolished activity (Parales et al. 1999).

In contrast to P450s, far less is known about the catalytic mechanism of ROs to date (Barry and Challis 2013). Due to difficulties in obtaining spectroscopic information of the mononuclear non-heme iron in the presence of the Rieske cluster, the mononuclear ferrous site is more challenging to access spectroscopically (Ohta et al. 2008). Though the catalytic cycle of these enzymes still needs to be fully elucidated, mechanistic evidence suggests the formation of a hydroperoxide complex that undergoes an O–O bond cleavage and forms an Fe(IV)=O species upon reaction with the aromatic substrate (Barry and Challis 2013). Another pathway involving a discrete O–O bond cleavage followed by formation of Fe(V)=O(OH) species prior to substrate oxidation has also been suggested (Rivard et al. 2015). Both proposed pathways eventually lead to the formation of a ferric alkoxide complex, which upon reduction and protonation leads to the release of the product and the regeneration of the catalytic non-heme iron center (King-Smith et al. 2018; Rivard et al. 2015).

ROs are characterized by their ability to catalyze the asymmetric *cis*-dihydroxylation of C=C double bonds with high stereo- and regioselectivity. It has been shown that their broad substrate spectrum comprises more than 300 compounds (Joern et al. 2001). Moreover, their catalytic abilities are not only limited to dihydroxylation reactions, but also monohydroxylations, sulfoxidations, dealkylations, desaturations, oxidative cyclizations and epoxidations can be performed by these enzymes (Table 1). A mechanism involving radical intermediates has been proposed by Chakrabaty and coworkers for monohydroxylation reactions (Chakrabarty et al. 2007). It is proposed that the mechanism includes a high-valent Fe(V)-oxo-hydroxo intermediate. Moreover, results for the oxidation of cyclopropane-containing substrates indicate the formation of a Fe(V) species prior to the insertion reaction (Chakrabarty et al. 2007).

3 Challenges in the Application of Rieske's

Although ROs display a great substrate scope and catalyze reactions which are difficult to achieve by chemical means, their practical applicability is generally limited to the use of whole-cell biocatalysts due to several challenges originating from the nature of the enzymes. One reason is that, like many other oxidoreductases, their catalytic activities rely on reducing equivalents provided by NAD(P)H requiring the regeneration by electron transfer proteins (Dror and Fishman 2012; Matsui et al. 2014). The regeneration process is typically a limiting step in industrial applications as it reduces cost effectiveness (Torres Pazmiño et al. 2010). Moreover, their multi-component nature negatively influences their enzymatic activity in vitro due to non-covalently linkage of each of the components. Moreover, ROs have been reported to only display a low stability once they are isolated (Catterall and Williams 1971; Sauber et al. 1977). However, bioprocesses usually require a tolerance of the biocatalysts against high temperatures, extreme pHs, high substrate and product concentrations, oxidants, and organic co-solvents over a long period (Dror and Fishman 2012; Matsui et al. 2014). One approach to improve stability, catalytic activity, and optimum temperature as well as to prevent inhibition by high substrate loadings or metal ions, is to immobilize the free enzyme or the whole-cell biocatalysts. For instance, catechol 1,2-dioxygenase from *Stenotrophomonasmaltophilia* KB2 has been entrapped on calcium alginate gel and catechol 2,3-dioxygenase from the *Bacillus stearothermophilus* has been immobilized on a glyoxal agarose gel. Both strategies retained their structural rigidity and showed resistance toward inactivation at high substrate concentrations (Fernandez-Lafuente et al. 2000; Guzik et al. 2014b). In a comparative study by Guzik et al. (2014a) it has been shown for protocatechuate 3,4- dioxygenase, that different agents used for immobilization results in the improvement of a different

Table 1 Chemistry performed by Rieske non-heme iron oxygenases

	Reaction type	Reaction example	Selected references
1	Dearomatizing *cis*-dihydroxylation	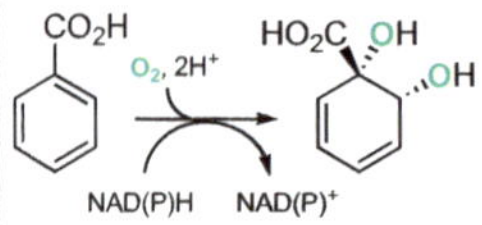	Rivard et al. (2015)
2	Aliphatic monooxygenation	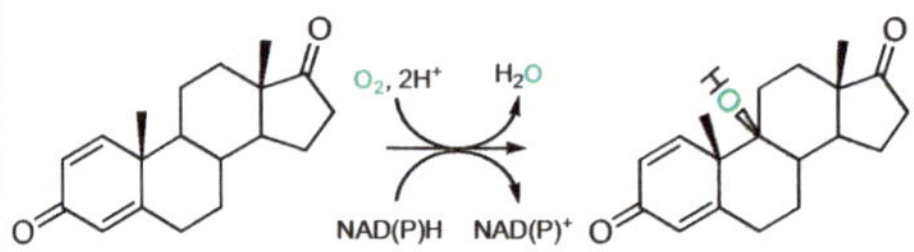	Capyk et al. (2009)
3	C–N-bond cleavage	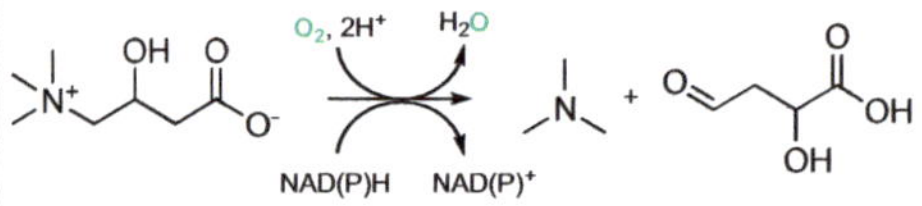	Kalnins et al. (2018)
4	O-demethylation	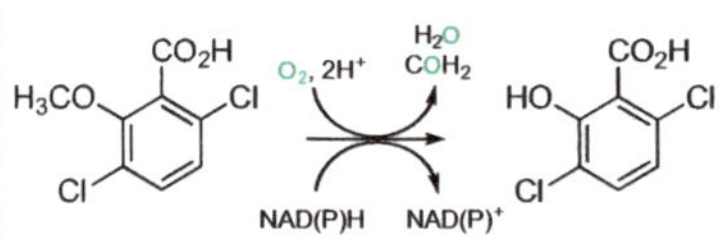	D'Ordine et al. (2009)
5	N-demethylation	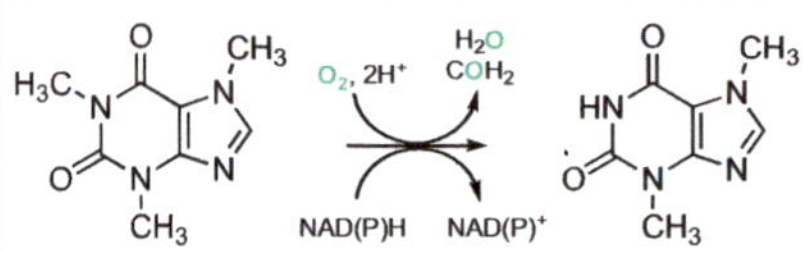	Summers et al. (2012)
6	Desaturation	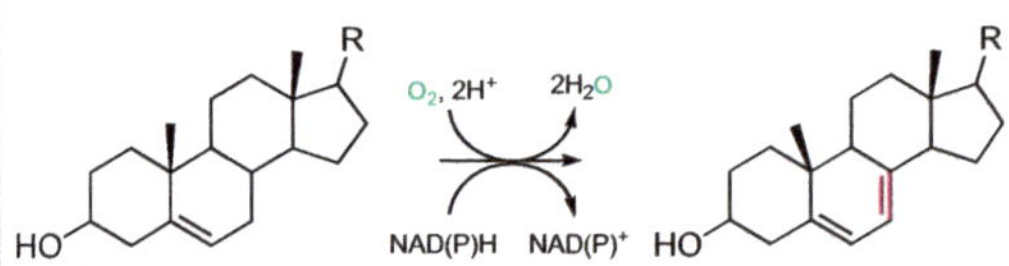	Yoshiyama-Yanagawa et al. (2011)
7	N-oxygenation	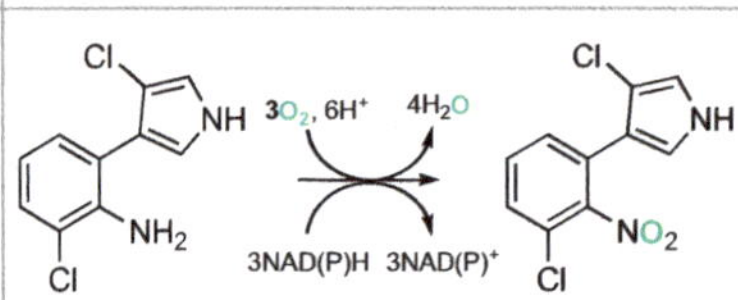	Lee and Zhao (2006)
8	C–C-bond formation	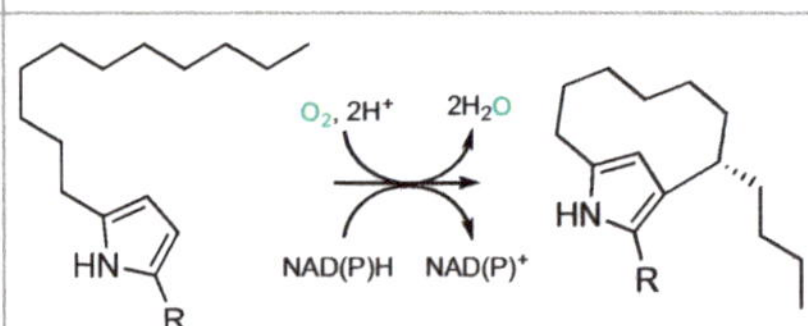	Sydor et al. (2011)

enzyme property. Moreover, the immobilization of the dioxygenase in both types of media resulted not only in an increased stability, but also in the decrease of optimum temperature by 5–10 °C. This feature is really desirable in industrial processes since it lowers the overall production cost (Matsui et al. 2014). For single component dioxygenases, many immobilization efforts have led to improved stability or enzymatic properties; however, these approaches have yet to be applied successfully for 2- or 3-component Rieske-type dioxygenases. Nonetheless, the immobilization of whole-cell biocatalysts remains a promising option to improve the properties of these enzymes which has been investigated for TDO. TDO producing *Pseudomonas putida* F1 cells have been entrapped in a mixture of recycled agave-fiber and polymer foam composites resulting in a lower rate of feedback inhibition by intermediary catechol produced in benzene degradation was observed (Robledo-Ortiz et al. 2011).

ROs usually display a strict selectivity for their physiological redox partners and only a few examples have been reported where this selectivity is less stringent (Armengaud et al. 2000; Armengaud and Timmis 1997). On the other hand, the specificity between the respective FdRs and their Fds is usually lower and both components are interchangeable. However, also those two components can display a high specificity for their physiological redox partners. This specificity not only influences their catalytic efficiency, but also hampers the synthetic applications of ROs. Next to the dependency of ROs on electron transfer from a Fd, many other enzymes like P450 monooxygenases and hydrogenases are also dependent on these electron carrier proteins. Bacterial and mitochondrial P450s rely on a shuttle of electrons from Fd and a FdR similar to the electron transfer chain of the RO system where the electrons are transferred via the Fd to the catalytic P450 heme domain (Munro et al. 2002). A similar electron transfer system based on Fd has been observed for hydrogenases from green algae. The [FeFe]-hydrogenases such as HydA1 are placed at the end of the photosynthetic electron transport chain and accept electrons from the photosynthetically reduced plant-type ferrodoxin-PetF (Thomas and Dirk 1993; Winkler et al. 2002, 2010). The electron transport chain of P450s and hydrogenases has been studied intensively in the last decades and is already well understood in terms of the underlying transfer mechanisms. However, in both cases the application of these enzymes proved to be challenging mainly due to their low catalytic activity, low stability, and the dependency on a complex electron transfer system. Especially in the case of P450s, many efforts had been made to surmount those limitations: cofactor regeneration systems (Kim et al. 2014; Lee et al. 2013), surrogate oxygen atom donors (Chen et al. 2012), direct chemical (Nazor et al. 2008), electrochemical (Udit et al. 2006) and light-driven reduction approaches (Jensen et al. 2011; Park et al. 2015; Tran et al. 2013). In particular, light-driven P450 reactions were established by using different mediators (Jensen et al. 2011). Also for the Fd-dependent hydrogenases, light-driven systems have been reported to simplify the complex natural electron transfer system by combining several photosensitizers with PetF and HydA1 (Adam et al. 2017). All these examples show that the electron transfer pathways of such multi-component enzymes can be successfully engineered to overcome the aforementioned limitations. Efficient and continuous supplementation of electrons to the Oxy is required to sustain their catalytic activity. In case of P450s, their reductase partners are synthesized as separate polypeptides. However, some self-sufficient P450s were found (De Mot and Parret 2002; Munro et al. 2007), where the P450 heme domain is fused to a

NADPH-P450 reductase. P450 BM3 from *Bacillus megaterium* is an entire class II P450 system in a single polypeptide (Munro et al. 2002). The arrangement in a fusion protein affords it the highest catalytic activity of any known P450 monooxygenase (Munro et al. 2002). Compared to P450-BM3, up till now no such prototype fusion protein exists in nature nor has been designed for ROs. Despite the high potential of such fusions to ensure an efficient and continuous supply of electrons, the design and generation of efficient RO fusions has not been accomplished yet.

One major drawback in achieving high expression levels and thus high catalytic efficiencies arises from the arrangement of the RO genes in an operon (Fig. 1c). Four genes encoding for the two subunits of the terminal Oxy, the Fd and the respective FdR have to be co-expressed together, which usually results in overall low expression levels of each protein and thus in generally low catalytic activities. A simplification of the complex RO electron transfer system is possible either by creating fusions to ensure the efficient supply of the terminal Oxy with electrons or by altering the acceptance of the Oxy towards non-physiological electron carriers. Such alterations would not only improve their potential for synthetic applications, but would also increase the mechanistic understanding of the underlying electron transfer processes in ROs.

4 Engineering the Substrate Scope and Catalytic Efficiencies

The application of enzymes in organic synthesis offers many advantages owing to their intrinsic high chemo-, regio- and stereoselectivity, however, biocatalytic syntheses routes might not always be efficient enough compared to chemical ones (Behrens et al. 2011; Hibbert et al. 2005). Protein engineering in general is a powerful tool to improve the performance of enzymes as catalysts in synthetic organic chemistry or biotechnology (Bornscheuer 2016; Bornscheuer et al. 2012). Two main strategies can be followed to alter enzyme properties for the desired application: (1) rational protein design and (2) directed evolution (Steiner and Schwab 2012). Both strategies or a combination thereof (semi-rational design) have been applied to alter the substrate scope, activity or selectivity of ROs (Dror and Fishman 2012). Especially NDO, TDO and BPDO have been subjected to various protein engineering studies in order to identify residues controlling selectivity and activity (Ferraro et al. 2017; Tan and Parales 2016). Table 2 gives an overview of selected protein engineering studies in which ROs have been improved or modified.

Table 2 Selection of protein engineering approaches to alter the substrate scope and/or the catalytic efficiencies of ROs

	Strain/Enzyme	Method	Result	Reference
1	*Ralstonia* sp. strain U2—Naphthalene dioxygenase (NDO)	Saturation mutagenesis DNA Shuffling	• Substrate specificity: F350T creates an enzyme with enhanced reaction rates toward the dinitrotoluene isomers • F350T/G407S higher activity	Keenan et al. (2005)
2	*Pseudomonas* sp. NCIB 9816-4- Naphthalene dioxygenase (NDO)	Site-directed mutagenesis	• Substrate specificity • Enantioselectivity – F352I: critical position	Yu et al. (2001)
3	*Pseudomonas* sp. strain NCIB 9816-4 Naphthalene dioxygenase (NDO)	Site-directed mutagenesis	• Substrate specificity • Regio- and stereospecificity • F224T successful binding-site of substrate	Seo et al. (2013)
4	*Burkholderia xenovorans* LB400 Biphenyl dioxygenase (BPDO)	Site-directed mutagenesis	• Substrate specificity • Regiospecificity • T335A or F336M	Kumar et al. (2011)
5	*Acidovorax* sp. Strain JS42- 2-Nitrotoluene 2,3-dioxygenase (2NTDO)	Site-directed mutagenesis	• Regiospecificity – A405G: not located or close to active site but still changes substrate pocket for 3NT	Mahan et al. (2015)
6	*Burkholderia xenovorans* LB400 – Biphenyl dioxygenase (BPDO)	Site-directed mutagenesis	• Higher catalytic efficiency and substrate specificity • T335A: significant changes	Mohammadi et al. (2011)
7	*Ralstonia* sp. PS12-Tetrachlorobenzene dioxygenase (TecA)	Site-directed mutagenesis	• Regioselectivity – F336I • Higher catalytic efficiency • L272	Pollmann et al. (2003)
8	*Pseudomonas* sp. NCIB 9816-4- Naphthalene dioxygenase (NDO)	Rational engineering (Molecular simulation)	• Substrate scope • F224 and L227 (bottleneck residues identified for mutation)	Escalante et al. (2017)
9	*Pseudomonas* sp. Cumene dioxygenase (CDO) Biphenyl dioxygenase (BPDO)	Site-directed mutagenesis	• Substrate specificity • Stereoselectivity – CDO M232A and BPDO M220A	Gally et al. (2015)

(continued)

Table 2 (continued)

	Strain/Enzyme	Method	Result	Reference
10	*Pseudomonas* sp. strain NCIB 9816-4 Naphthalene dioxygenase (NDO)	Semi-rational design	• Reaction specificity • Regio- and stereo-specificity • A206, V260 and H295	Halder et al. (2018)
11	*Mycobacterium neoaurum* ATCC 25795 (designed to Mn25795)—3-Ketosteroid 9-α-hydroxylase (KSH)	Site-directed mutagenesis	• Substrate specificity – V207T	Liu et al. (2018)
12	*Pseudomonas pseudoalcaligenes* JS45-Nitrobenzene dioxygenase (NBDO)	Directed evolution (Neutral drift libraries, Error-Prone PCR, DNA Shuffling) Site directed mutagenesis	• Catalytic efficiency • Regioselectivity – F222C/F251L/G253D	Bernath-Levin et al. (2014)
13	*Comamonas* sp. strain JS765—Nitrobenzene dioxygenase (NBDO)	Saturation Mutagenesis Site directed mutagenesis for double mutant	• Enantioselectivity • Catalytic efficiency – N258A/F293H	Shainsky et al. (2013)
14	*Rhodococcus* sp. strain DK17- O-Xylene dioxygenase	Site-directed mutagenesis	• Catalytic activity – L266F	Kim et al. (2013)
15	*Diaphorobacter* sp. strain DS2-3-Nitrotoluene dioxygenase (3NTDO)	Site-directed mutagenesis	• Catalytic efficiency – V350F, I204A, N258V	Kumari et al. (2017)

4.1 Rational Protein Design

Rational protein design is characterized by the introduction of site-specific mutations in order to alter or to induce the desired property (Reetz 2013). Usually, detailed knowledge about structural and mechanistic features in combination with in silico modeling is required to be able to predict the most promising residues for mutagenesis (Behrens et al. 2011). However, only for a small number of proteins the required structural and mechanistic data is available yet. Moreover, the in general very complex structure–function relationships in enzymes hamper the general application of rational design, which makes also the prediction of the mutation outcome on proteins difficult (Badenhorst and Bornscheuer 2018; Chica et al. 2005; Dror and Fishman 2012). Despite these significant limitations, rational protein design was successfully applied for the mutagenesis of ROs, thereby mainly targeting residues located in or close to the substrate binding pocket. The majority of mutations that beneficially affect enantioselectivity or substrate selectivity are usually located in or close to the active site and might include amino acid residues that are involved in substrate binding (Morley and Kazlauskas 2005). Especially for NDO, several residues in the active site have been identified and altered by site-directed mutagenesis in order to alter selectivity and product outcome (Table 2) (Parales 2003; Parales et al. 2000; Yu et al. 2001). It has been shown that especially one residue (F352) in NDO from *Pseudomonas* sp. strain NCIB 9816-4 mainly influences the regio- and stereoselectivity towards biphenyl and phenanthrene, thus offering the possibility to engineer dioxygenases by only introducing one mutation (Parales et al. 2000). This amino acid residue corresponds to F366 in tetrachlorobenzene dioxygenase from *Ralstonia* sp. PS12, which was also shown to control regioselectivity of the enzyme. The active site has been remodeled by site-directed mutagenesis revealing variant F366L which displayed significant changes in regioselectivity for 2-, 3-chloro-, 2,4-, 2,5- and 2,6-dichlorotoluene (Table 2, Entry 7) (Pollmann et al. 2003). Moreover, BPDOs from *Pseudomonas pseudoalcaligenes* KF707 and *Pseudomonas* sp. LB400 have been subjected to rational engineering in order to improve their ability to degrade polychlorinated biphenyls (PCBs) (Erickson and Mondello 1993; Suenaga et al. 2002). Amino acid mutations F227V, I335F, T376N, F377L and F377A resulted in enzyme variants with altered regioselectivities for dihydroxylation of various PCBs compared to the wild type enzyme for BPDO from *P. pseudoalcaligenes* (Suenaga et al. 2002). In contrast, the change of four amino acid residues (positions 335, 336, 338 and 341) by site-directed mutagenesis of BPDO from *Pseudomonas* sp. enhanced its capacity to degrade PCBs by extending the substrate range of BPDO (Erickson and Mondello 1993). Several studies have shown that ROs can be tailored towards the asymmetric dihydroxylation of olefins, which impressively demonstrates the power of protein engineering (Gally et al. 2015; Halder et al. 2018). In their first study, they have rationally engineered CDO and benzene dioxygenase (BPDO) from different *Pseudomonas* strains towards different substrates comprising mono-, *gem*-di-, *cis*-di and tri-substituted alkenes (Table 2, Entry 9) (Gally et al. 2015). The generation of single point mutations (CDO M232A and BPDO M220A) proved to be sufficient to display significant changes in the regio- and stereoselectivity of several aliphatic olefins. Especially the CDO variant M232A converted the olefins mostly with excellent stereoselectivity (>95%) and conversions (>90%) to the desired mono- or dihydroxylated products. Moreover, preparative scale biotransformations with monoterpenes yielded the hydroxylated products in mg amounts.

4.2 Directed Evolution

Due to the limited applicability of rational enzyme design and the limited sequence space that can be explored at a time, directed evolution methods have been applied to alter ROs and to identify residues that effect enzyme function (Ferraro et al. 2007). Using this random approach, mutations far from the active site that influence distinct enzyme properties like activity or thermostability can be identified (Morley and

Kazlauskas 2005). BPDO has been extensively studied by directed evolution in order to evolve the enzyme for bioremediation applications. It was found that BPDOs from different aromatics degrading strains show variations in the PCB congener selectivity patterns and activities (Kumamaru et al. 1998). The BPDOs from *P. pseudoalcaligenes* strain KF707 and *P. cepacia* LB400 differ in their substrate range despite their nearly identical amino acid sequence of the oxygenase α-subunits which vary in only 20 amino acid residues (96.4% identity). The construction of a chimeric BPDO from both strains led to the identification of a small number of amino acids at the C-terminus that are involved in substrate specificity and regioselectivity and thus for PCB recognition. The created KF707 variant was found to display an expanded range of biodegradable PCB congeners that could be generated (Kimura et al. 1997). DNA shuffling of the genes encoding for the α-subunits of BPDOs from *P. pseudoalcaligenes* KF707 and *P. cepacia* LB400 was then used to recombine as many differences in both enzymes as possible (Kumamaru et al. 1998). The screening of the generated variants led to the identification of several variants that showed enhanced degradation capabilities for PCB and related biphenyl compounds. Moreover, variants have been identified displaying a novel degradation capacity towards smaller aromatic substrates like benzene and toluene which are poor substrates for the wild-type enzyme.

For aniline dioxygenase (ADO), a random mutagenesis approach was applied in order to increase the activity against aniline derivatives. The variant V205A was used as template for one round of saturation mutagenesis on the active site residues followed by a round of error-prone PCR. Although the variant V205A exhibited an extended substrate specificity for 2-isopropylaniline, the change of valine to alanine at position 205 decreased the activity for aniline (8.4-fold) and 2,4-dimethylaniline (28-fold). The ep-PCR approach was successful to identify variants with improved activity towards the desired substrates (Ang et al. 2009). Similarly, the substrate scope of TDO towards 4-picoline was altered by a combined approach of ep-PCR and saturation mutagenesis (Sakamoto et al. 2001).

4.3 Semi-Rational Design

Although rational protein engineering is in general a powerful tool to precisely engineer certain enzyme properties, it suffers mostly from insufficient knowledge on structure–function relationships. Conversely, directed evolution approaches are mostly limited by the availability of fast and reliable high-throughput screening (HTS) methods for large mutant libraries (Behrens et al. 2011). Moreover, only a fraction of the possible protein variants is represented and biases resulting from the degeneracy of the genetic code further restrict the diversity and thus negatively impacting the quality of mutant libraries (Reetz et al. 2008). Thus, the focus in protein engineering has moved towards the design of smaller (focused) libraries with a high quality that eliminate the need for suitable HTS assays (Behrens et al. 2011; Davids et al. 2013). This methodology is called semi-rational design, which is characterized by the rational choice of positions to mutate and the reduced library size that results from the focus on specific amino acid residues that are targeted on the basis of prior structural or functional knowledge. The libraries which result from this strategy are more likely to yield an enzyme variant with the desired property; as multiple mutations can be introduced at positions where they might be most effective, e.g. in or near the active site when targeting enzyme properties like enantioselectivity, substrate specificity or new catalytic activities (Chica et al. 2005; Morley and Kazlauskas 2005).

Halder et al. (2018) investigated the semi-rational engineering towards the selective asymmetric dihydroxylation of differently substituted arene substrates (Table 2, Entry 10). The active site of NDO was analyzed towards the conversion of substituted arenes with varying side-chain length. In total, 12 first-shell-amino-acid residues up to 6 Å away from the catalytically iron ion were chosen to study their influence on reaction specificity as well as regio-and stereoselectivity. In total, 36 single variants were created by changing the respective residues either to the amino acids alanine, valine or isoleucine. Moreover, promising combinations enlarging the active site were identified by in silico docking experiments using (*R*)-limonene as substrate. This analysis revealed three more promising positions as potential starting point to reshape the active site cavity for non-planar, non-aromatic substrates. Thus, an additional set of 26 double variants comprising combinations of these positions have been created. The mutagenesis comprising the 12 first-shell amino acids of NDO revealed seven positions within the active site that showed a significant influence on product distribution and formation of the investigated substrate panel addressing reaction selectivity, regioselectivity, and stereoselectivity. Both studies show that ROs provide an ideal scaffold for the engineering towards both, the asymmetric *cis*-dihydroxylation and the regio- and stereospecific allylic monohydroxylation of various alkenes, but also targeting products difficult to access by Sharpless AD or Riley oxidation.

5 Dihydroxylations and Beyond: The Catalytic Diversity of RO's

Rieske non-heme iron oxygenases are most famous for their ability to selectively catalyze *cis*-dihydroxylations at aromatic rings (Table 1, entry 1). However, members of this enzyme family catalyze a wide variety of other transformations in diverse organisms ranging from bacteria, plants, insects and mammals (Barry and Challis 2013). Capyk and Eltis performed a phylogenetic analysis highlighting this important point by revealing that the bacterial arene-hydroxylating enzymes fall into a very small cluster within one of two main groups (Capyk and Eltis 2012). In the following section, we will highlight some recently discovered RO enzymes that are able to catalyze a range of transformations apart from dihydroxylations and thus illustrating the functional diversity within this family of enzymes.

KshAB is a RO involved in the cholesterol metabolism which has been characterized by Eltis and coworkers (Van der Geize et al. 2007). This enzyme catalyzes the monohydroxylation of 4-androstene-3,17-dione (AD) and 1,4-androstadiene-3,17-dione (ADD), however, both of the KshAB-catalyzed reactions are relatively slow and it was shown that the consumption of oxygen is faster in the presence of ADD than AD (Table 1, entry 2) (Capyk et al. 2009). These observations suggested the involvement of another crucial factor that may increase the catalytic efficiency of the enzymes in vivo or that a different intermediate in cholesterol catabolism is the true substrate for the enzyme. A sequence analysis revealed that KshAB only shows low sequence similarity ($\sim$11%) to other well described and typical ROs like phthalate dioxygenase from *Burkholderiacepacia*, 2-oxoquinoline 8-monoxygenase from *P. putida*, and Cardo13 from *Janthinobacterium* sp., suggesting that KshAB originates from a distinct subfamily. A deeper crytallographic analysis revealed that KshAB indeed contains a [2Fe−2S] cluster and a non-heme center, however, it lacks several other typical features of ROs (Capyk et al. 2009). Since the detailed nature of one of the non-protein ligands was not clear from this analysis, the authors suggested that a mixture of species with multiple occupancies is present. Inhibitors which could be potentially designed from docking studies in the substrate binding site of KshAB would be of particular interest, because they could both inhibit cholesterol catabolism and lead to the production of damaging reactive oxygen species in *M. tuberculosis* (Capyk et al. 2009).

CntA/CntB is a two-component RO which has been recently discovered in *Acinetobacter baumanii* (Zhu et al. 2014). It has been shown that CntA is capable of cleaving carnitine, γ-butyrobetaine, glycine betaine, and in one case, choline into trimethylamine (TMA) and malic semialdehyde (Table 1, entry 3). It has been found that TMA found in humans is exclusively of bacterial origin, whereby its derivative trimethylamine oxide has been associated with atherosclerosis and heart and renal failure. Tars and coworkers investigated and compared the substrate specificity of CntA oxygenases from different organisms and determined oxygen requirements for aerobic carnitine degradation by the TMA-producing *Providencia rettgeri* (Elssner et al. 1999; Kalnins et al. 2018). The isolated four CntA homologs did not display significant differences in their substrate scope, but all showed high activity toward carnitine/γ-butyrobetaine, medium activity toward glycine betaine, and very low activity toward choline. Moreover, TMA production in CntA-containing cell cultures also proceeded at low oxygen concentrations, indicating that CntA, although being an aerobic enzyme, could be responsible for carnitine degradation in the human digestive system.

Dicamba O-demethylase is able to catalyze the oxidative O-demethylation of the broadleaf herbicide dicamba as the first step of its degradation in plants (Table 1, entry 4) (Barry and Challis 2013). Thus, the gene encoding the O-demethylase could be a useful tool in the development of genetically modified crops that are resistant to dicamba (D'Ordine et al. 2009). Also, the dicamba O-demethylase shows like KshA, a low sequence similarity ($\sim$18%) compared to other known homologs. In 2009, the crystal structure of this enzyme has been solved as the first for a RO that catalyzes exocyclic monooxygenation reactions. Interestingly, dicamba O-demethylase possesses an asparagine residue in close proximity to the non-heme iron center. The amido group of asparagine could coordinate the ferrous iron in solution and is caused by a conformational change, which has also been similarly observed for the resting state of isopenicillin N synthase (IPNS), an enzyme belonging to the α-ketoglutarate-dependent non-heme iron oxygenase superfamily (although it does not require α-ketoglutarate as a cosubstrate) (Landman et al. 1997; Roach et al. 1997; Sami et al. 1997). In IPNS, substrate binding results in displacement of the Gln ligand (instead of Asn in dicamba O-demethylase) allowing the subsequent binding of dioxygen to the active site ferrous iron.

In addition to O-demethylation reactions, N-demethylation is also an unusual reaction observed for ROs. Subramanian and coworkers identified and characterized NdmA and NdmB, two ROs from *P. putida* CBB5 that are able to catalyze the oxidative N_1-demethylation and N_3-demethylation, respectively, of caffeine, as well as several methylated xanthine derivatives (Table 1, entry 5) (Summers et al. 2012). It has been proposed by the authors that both enzymes act sequentially in the demethylation of caffeine. Furthermore, NdmD was identified as the NADH-dependent reductase which shuttles the electrons to the [2Fe–2S] centers of NdmA and NdmB.

DAF-36/Nvd is a Rieske domain oxygenase that acts as cholesterol 7,8-dehydrogenase by converting cholesterol to 7-dehydrocholesterol in the first step of the ecdysteroid biosynthesis (Wollam et al. 2011; Yoshiyama-Yanagawa et al. 2011). The DAF-36/Nvd family of proteins is a conserved player in the cholesterol metabolism across the animal phyla such as roundworms, nematodes, arthropods as well as vertebrates and it has been shown that its deletion lead to *Drosophila* larval death (Yoshiyama-Yanagawa et al. 2011). The Ecdysteroid biosynthesis involves several chemical reactions catalyzed by ecdysteroidogenic enzymes including the Rieske oxygenase Nvd and the Halloween gene family of cytochrome P450s (Mykles 2011). In vitro studies suggests that DAF-36/Nvd acts as a cholesterol 7,8-dehydrogenase (Yoshiyama-Yanagawa et al. 2011) or cholesterol 7-desaturase (Wollam et al. 2011) which converts cholesterol to 7-dehydrocholesterol in an early ecdysteroid biosynthetic pathway (Table 1, entry 6). The conversion most probably proceeds via a monohydroxylated species that undergoes subsequent dehydration, although the direct desaturation is also possible. Although the biological relevance of this reaction is still not fully understood, recent studies suggest that, inter alia, Nvd function in molt regulation in arthropods (Sathapondecha et al. 2017).

PrnD is a RO that is able to catalyze the oxidation of an aniline derivative to the corresponding nitrobenzene derivative in the pyrrolnitrin biosynthesis (Table 1, entry 7) (Lee et al. 2005). Pyrrolnitrin is a broad-spectrum antifungal produced by several *Pseudomonas* and *Burkholderia* species (Mahenthiralingam et al. 2011; Pfender et al. 1993; Schmidt et al. 2009). PrnD is the first RO identified which is able to catalyze this type of reaction, although cytochrome P450 and non-heme diiron-dependent oxygenases have been found to catalyze this reaction in other biosynthetic pathways (Bernhardt and Urlacher 2014; Choi et al. 2008). Zhao and coworkers investigated the substrate scope of PrnD and found that other arylamines are converted to the corresponding nitroarenes, however, the catalytic efficiency was lower and the range of substrates limited (Lee and Zhao 2006; Lee et al. 2006). Moreover, the authors have investigated the catalytic mechanism of the oxidation of 4-aminobenzylamine to 4-nitrobenzylamine and were able to show that the reaction proceeds via a sequential oxidation of the amino group. On the basis of mutagenesis studies and the intermediates formed, the authors proposed that the conversion of the hydroxylamine to the nitrosoarene involves direct dehydrogenation rather than proceeding via a dihydroxylated amine intermediate.

Streptorubin B and metacycloprodigiosin belong to the family of antimalarial natural products produced by *Streptomyces coelicolor* M145 and *Streptomyces longisporusruber* DSM40667, respectively (Haynes et al. 2011; Hu et al. 2011; Papireddy et al. 2011). The gene redG encoding a protein with sequence similarity to a RO, is embedded within the gene cluster of genes responsible for the assembly of streptorubin B and undecylprodigiosin in *S. coelicolor* (Table 1, entry 8) (Cerdeno et al. 2001; Haynes et al. 2008; Stanley et al. 2006; Sydor et al. 2011). A sequence alignment of RedG with the α-subunit of NDO revealed that the N-terminal domain of RedG contains the His and Cys residues that bind the [2Fe−2S] cluster and the C-terminal domain contains the two His residues of the 2-His-1-carboxylate iron-binding triad in the non-heme center (Sydor et al. 2011). Moreover, it has been found that the "bridging" aspartate residue in NDO appears to be mutated to glutamate in RedG. Finally, it could be demonstrated that RedG catalyzes the oxidative carbocyclization of undecylprodigiosin to form streptorubin B. This unprecedented transformation of a RO has also been proven to be catalyzed byMcpG, an orthologue of RedG.

6 Application of ROs in Natural Product Synthesis

TDOs from different origins have been extensively used in chemo-enzymatic syntheses since their discovery by Gibson in 1968 (Gibson et al. 1968). Especially in the synthesis of many natural products such as (+)-lycoricidine (Hudlicky and Olivo 1992), entdeoxydihydrotsugicoline (Chang et al. 2016) and xylosmin, the RO-catalyzed dearomatization and desymmetrization has been applied. *Cis*-1,2-dihydroxyarenes are particularly interesting for natural product synthesis, since they offer a range of special features (Fig. 3).

One of these features is that the diols have a pro-enantiotopic plane of symmetry that enables the diastereoselective functionalization of the alkene (Hudlicky et al. 1992). Moreover, the cyclic backbone provides additional functionalities facilitating further reactions. In that way, a bromine or chlorine substituent lends itself as a partner in cross coupling reactions. On the other hand, the diene moiety is capable of undergoing cycloadditions or other selective olefin functionalizations like oxidative cleavage, aziridination or epoxidation (Hudlicky and Reed 2009). In the following section, the chemo-enzymatic synthesis of two selected natural products, namely ent-hydromorphone and (−)-patchoulenone will be described.

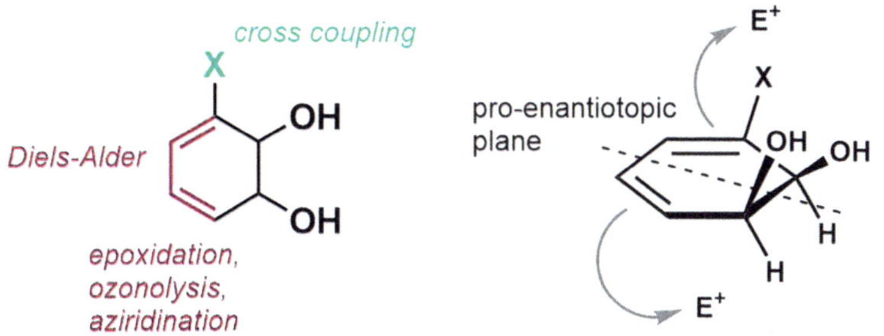

Fig. 3 The features of the dihydroxylated product that enables the synthetic versatility of *cis*-1,2-dihydroxyarenes

The synthesis of morphine and its congeners has been the subject of many synthesis efforts and many creative approaches having been published (Rinner and Hudlicky 2012). Hudlicky et al. developed a chemo-enzymatic synthesis route to ent-hydromorphonein in twelve steps starting from ß-bromoethylbenzene (Varghese and Hudlicky 2014). The key step involved the enzymatic dihydroxylation of the arene to the corresponding *cis*-dihydrodiol catalyzed by TDO, which was selectively reduced and protected to afford acetonide. The Mitsunobu coupling with the ring A fragment was followed by a Wittig olefination and MOM group deprotection. The phenol moiety was then oxidized to lead the Diels–Alder precursor, which underwent cycloaddition exclusively as the exocyclic diene to form the ring B of the morphinane. The synthesis was then completed by the intramolecular amination at C9. The authors hypothesized that the low yield of the reaction resulted from the [4 + 2] cycloaddition step due to the participation of only one diastereomer in the reaction. However, only a very small amount of product arising from the cycloadditon at the endocyclic diene was observed, although this motif is highly reactive. This could be explained by invoking steric factors, which prohibit the dienophile from coming into close proximity of the endocyclic diene.

The diene moiety of TDO products has been primarily subjected to Diels–Alder cycloaddition like applied in the synthesis of (−)-patchoulenone by Banwell et al. (1998, 2003). This compound shows potent antifungal activity against *Rhizoctoniasolani* and *Saprolegniaasterophora*, as well as in vitro activity against malarial parasite *Plasmodium falciparum* and has been first isolated in 1964 from *Cyperus rotundus* Linné (Thebtaranonth et al. 1995). The synthesis strategy developed by Banwell et al. starts with the protection of microbially oxidized toluene (up to 15 g/L) (Endoma et al. 2002) with *p*-methoxybenzaldehyde dimethyl acetal followed by a Diels–Alder cycloaddition to give a 4:1 product mixture in quantitative yield. In the following steps, the diastereomers were hydrolyzed, the product was *gem*-dimethylated, hydrogenated and the product was subjected an olefination. After a protecting group manipulation, an oxidation step and the addition of isopropenyllithium, the product underwent an anionic oxy-Cope rearrangement upon treatment with sodium hydride. The tricyclic core of patchoulene was established by a reductive cyclization reaction with samarium(II) iodide.

Debenzylation, Parikh–Doering oxidation, and dehydration completed the synthesis of the target natural product (−)-patchoulenone.

7 Concluding Remarks

Rieske non-heme iron oxygenases display extraordinary catalytic capabilities in terms of the vast array of substrates they can convert and their important biological functions. Although these enzymes are known since decades and many of their representatives have been characterized in detail, not every secret comprising their nature and catalytic features has been fully revealed up to know. Most of the considerable efforts in the field have focused on the dihydroxylation of aromatic compounds, however, limited data is available for the oxidation of unnatural substrates by this class of enzymes. Moreover, the exact nature of the catalytic cycle still needs to be fully elucidated. However, the study of structure–function relationships has facilitated the manipulation of their activity, either to probe their biological function or to promote a desirable phenotypic effect in the native organism. Thanks to their versatile reaction spectrum and broad substrate scope, ROs often have been considered as the non-heme analogues of cytochrome P450 monooxygenases. Although several of the reaction types described for P450s can also be performed by ROs, reactions like nitration, decarboxylation, aromatic coupling, C–C-bond cleavage, carbene transfer and sp^3 hydroxylation cannot be targeted with ROs to date (Li et al. 2007; Resnick et al. 1996; Zhang et al. 2014). However, methods of directed evolution and enzyme engineering could provide viable tools for the engineering towards novel types of reactions like the oxidation of unactivated C–H-bonds or to boost a desired activity towards higher efficiencies. Another feature that ROs share with P450s is their multicomponent nature and thus their dependency on a complex electron transport chain. In general, the applicability of enzymes depending on an electron transfer chain involving ferredoxins is still regarded as limited, and mostly restricted to the use of whole-cell biocatalysts. This is mainly due to the oxygen sensitivity of the [2Fe–2S] cluster, enzyme instability, and the expensive NAD(P)H co-substrate requirement. Especially in case of P450s, many efforts have been made to surmount these limitations by simplifying the electron transfer chain or by investigating light-driven systems to overcome the cofactor dependency (Dong et al. 2018; Lee et al. 2009, 2013; Park et al. 2015; Zhang and Hollmann 2018). This has led to an increased knowledge on the electron transfer pathways in oxidoreductases, however, the deeper understanding of the molecular interactions between Fds and their respective oxygenases remains to be elucidated.

Moreover, the uncoupling of NAD(P)H consumption and substrate oxidation in the presence of compounds lacking a productive binding mode in the active site lead to the formation of H_2O_2 which can inactivate the enzyme on the one hand, but also lead to a loss of reducing equivalents from the host cell and an increased oxygen demand as well as a lowered specific activity with regard to product formation (Lee

1999; van Beilen et al. 2003). For NDO, uncoupling has been described to constitute ~40–50% of the total O_2 consumption for benzene oxidation (Lee 1999). However, for P450s the reduction of uncoupling due to inappropriate positioning of the substrate in the enzyme active site has been overcome by protein engineering (van Beilen et al. 2003) and thus might be also applicable for ROs.

Moreover, the generation of functional fusion proteins between the redox partners and the oxygenase proved to be a successful approach to enhance product formation in P450-catalyzed reactions. Linking the single components for more efficient electron transport or higher expression levels due to a better coordination of transcription and translation has been successfully applied, resulting in an increased substrate oxidation (Scheps et al. 2013). Thus, the covalent linkage of the RO redox partners could be a successful strategy for improving reaction rates. It has been shown that the redox partner (physiological or artificial) and their fusion to the terminal oxygenase influences the type of reaction catalyzed by P450s (Zhang et al. 2014). This highlights the potential role of redox-partner protein–protein interactions in modulating the catalytic activity of a P450 and might also be applicable to other multi-component enzymes like ROs.

As Barry and Challis already outlined in 2013 (Barry and Challis 2013), a key challenge for the future applications of ROs is to translate fundamental work on these enzymes into synthetically useful catalysts. Whole-cell biocatalysts have found application in the production of homochiral starting materials for organic synthesis and in soil bioremediation via degradation of aromatic pollutants, however, the direct application of isolated enzymes still presents several substantial challenges as outlined above. However, many of the products targeted by RO catalysis are valuable building blocks for organic synthesis or might be applied in the flavor and fragrance industry. Moreover, the combination of RO-catalyzed oxyfunctionalization with further chemical transformations has opened up an emerging area in natural product synthesis. The use of tailored ROs will fuel their application in complexity-generating chemo-enzymatic and multi-enzyme reaction cascades.

Despite the significant hurdles to overcome, we do believe that by furthering our understanding of these fantastic enzymes, we will ultimately aid the development of efficient, inexpensive, and environment friendly biocatalysts capable of a wide variety of oxygenation and C–H activation reactions.

Acknowledgements This project has received funding from the European Union's Horizon 2020 research and innovation programme under the Marie Skłodowska-Curie grant agreement No 764920.

References

Adam D, Bosche L, Castaneda-Losada L, Winkler M, Apfel P, Happe T (2017) Sunlight-dependent hydrogen production by photosensitizer/hydrogenase systems. ChemSusChem 10:894–902

Ang EL, Obbard JP, Zhao HM (2009) Directed evolution of aniline dioxygenase for enhanced bioremediation of aromatic amines. Appl Microbiol Biotechnol 81:1063–1070
Armengaud J, Timmis KN (1997) Molecular characterization of Fdx1, a putidaredoxin-type [2Fe-2S] ferredoxin able to transfer electrons to the dioxin dioxygenase of Sphingomonas sp. RW1. Eur J Biochem 247:833–842
Armengaud J, Gaillard J, Timmis KN (2000) A second [2Fe-2S] ferredoxin from Sphingomonas sp strain RW1 can function as an electron donor for the dioxin dioxygenase. J Bacteriol 182:2238–2244
Axcell BC, Geary PJ (1975) Purification and some properties of a soluble benzene-oxidizing system from a strain of pseudomonas. Biochem J 146:173–183
Badenhorst CPS, Bornscheuer UT (2018) Getting momentum: from biocatalysis to advanced synthetic biology. Trends Biochem Sci 43:180–198
Banwell MG, Darmos P, McLeod MD, Hockless DCR (1998) From toluene to taxol (TM): chemoenzymatic and enantiodivergent routes to the AB-ring systems of taxoids and ent-taxoids. Synlett 8:897–899
Banwell MG, Hockless DCR, McLeod MD (2003) Chemoenzymatic total syntheses of the sesquiterpene (-)-patchoulenone. New J Chem 27:50–59
Barry SM, Challis GL (2013) Mechanism and catalytic diversity of Rieske non-heme iron-dependent oxygenases. ACS Catal 3:2362–2370
Bassan A, Blomberg MRA, Borowski T, Siegbahn PEM (2004) Oxygen activation by Rieske non-heme iron oxygenases, a theoretical insight. J Phys Chem B 108:13031–13041
Bataille CJR, Donohoe TJ (2011) Osmium-free direct syn-dihydroxylation of alkenes. Chem Soc Rev 40:114–128
Behrens GA, Hummel A, Padhi SK, Schatzle S, Bornscheuer UT (2011) Discovery and protein engineering of biocatalysts for organic synthesis. Adv Synth Catal 353:2191–2215
Bernath-Levin K, Shainsky J, Sigawi L, Fishman A (2014) Directed evolution of nitrobenzene dioxygenase for the synthesis of the antioxidant hydroxytyrosol. Appl Microbiol Biotechnol 98:4975–4985
Bernhardt R, Urlacher VB (2014) Cytochromes P450 as promising catalysts for biotechnological application: chances and limitations. Appl Microbiol Biotechnol 98:6185–6203
Bornscheuer UT (2016) Biocatalysis: successfully crossing boundaries. Angew Chem Int Ed 55:4372–4373
Bornscheuer UT, Huisman GW, Kazlauskas RJ, Lutz S, Moore JC, Robins K (2012) Engineering the third wave of biocatalysis. Nature 485:185–194
Boyd DR, Sheldrake GN (1998) The dioxygenase-catalysed formation of vicinal cis-diols. Nat Prod Rep 15:309–324
Boyd DR, Sharma ND, Allen CCR (2001a) Aromatic dioxygenases: molecular biocatalysis and applications. Curr Opin Biotechnol 12:564–573
Boyd DR, Sharma ND, Harrison JS, Kennedy MA, Allen CCR, Gibson DT (2001b) Regio- and stereo-selective dioxygenase-catalysed cis-dihydroxylation of fjord-region polycyclic arenes. J Chem Soc Perkin Trans 1:1264–1269
Capyk JK, Eltis LD (2012) Phylogenetic analysis reveals the surprising diversity of an oxygenase class. J Biol Inorg Chem 17:425–436
Capyk JK, D'Angelo I, Strynadka NC, Eltis LD (2009) Characterization of 3-ketosteroid 9 alpha-hydroxylase, a Rieske oxygenase in the cholesterol degradation pathway of *Mycobacterium tuberculosis*. J Biol Chem 284:9937–9946
Catterall FA, Williams PA (1971) Some properties of the naphthalene oxygenase from pseudomonas sp. NCIB 9816. Microbiology 67:117–124
Cerdeno AM, Bibb MJ, Challis GL (2001) Analysis of the prodiginine biosynthesis gene cluster of Streptomyces coelicolor A3(2): new mechanisms for chain initiation and termination in modular multienzymes. Chem Biol 8:817–829
Chakrabarty S, Austin RN, Deng DY, Groves JT, Lipscomb JD (2007) Radical intermediates in monooxygenase reactions of Rieske dioxygenases. J Am Chem Soc 129:3514–3515

Chakraborty J, Ghosal D, Dutta A, Dutta TK (2012) An insight into the origin and functional evolution of bacterial aromatic ring-hydroxylating oxygenases. J Biomol Struct Dyn 30:419–436

Chang DL, Zhang J, Witholt B, Li Z (2004) Chemical and enzymatic synthetic methods for asymmetric oxidation of the C-C double bond. Biocatal Biotransformation 22:113–130

Chang EL, Bolte B, Lan P, Willis AC, Banwell MG (2016) Chemoenzymatic total syntheses of the enantiomers of the protoilludanes 8-deoxydihydrotsugicoline and radudiol. J Org Chem 81:2078–2086

Chen MM, Coelho PS, Arnold FH (2012) Utilizing terminal oxidants to achieve P450-catalyzed oxidation of methane. Adv Synth Catal 354:964–968

Chica RA, Doucet N, Pelletier JN (2005) Semi-rational approaches to engineering enzyme activity: combining the benefits of directed evolution and rational design. Curr Opin Biotechnol 16:378–384

Choi YS, Zhang HJ, Brunzelle JS, Nair SK, Zhao HM (2008) In vitro reconstitution and crystal structure of p-aminobenzoate N-oxygenase (AurF) involved in aureothin biosynthesis. Proc Natl Acad Sci U S A 105:6858–6863

D'Ordine RL, Rydel TJ, Storek MJ, Sturman EJ, Moshiri F, Bartlett RK, Brown GR, Eilers RJ, Dart C, Qi Y, Flasinski S, Franklin SJ (2009) Dicamba monooxygenase: structural insights into a dynamic Rieske oxygenase that catalyzes an exocyclic monooxygenation. J Mol Biol 392:481–497

Davids T, Schmidt M, Bottcher D, Bornscheuer UT (2013) Strategies for the discovery and engineering of enzymes for biocatalysis. Curr Opin Chem Biol 17:215–220

De Mot R, Parret AHA (2002) A novel class of self-sufficient cytochrome P450 monooxygenases in prokaryotes. Trends Microbiol 10:502–508

Dong J, Fernandez-Fueyo E, Hollmann F, Paul C, Pasic M, Schmidt S, Wang Y, Younes S, Zhang W (2018) Biocatalytic oxidation reactions – a chemist's perspective. Angew Chem Int Ed 57:9238–9261

Dror A, Fishman A (2012) Engineering non-heme mono- and dioxygenases for biocatalysis. Comput Struct Biotechnol J J2:1–12

Elssner T, Preusser A, Wagner U, Kleber HP (1999) Metabolism of L(-)-carnitine by enterobacteriaceae under aerobic conditions. FEMS Microbiol Lett 174:295–301

Endoma MA, Bui VP, Hansen J, Hudlicky T (2002) Medium-scale preparation of useful metabolites of aromatic compounds via whole-cell fermentation with recombinant organisms. Org Process Res Dev 6:525–532

Erickson BD, Mondello FJ (1993) Enhanced biodegradation of polychlorinated-biphenyls after site-directed mutagenesis of a biphenyl dioxygenase gene. Appl Environ Microbiol 59:3858–3862

Escalante DE, Aukema KG, Wackett LP, Aksan A (2017) Simulation of the bottleneck controlling access into a Rieske active site: predicting substrates of naphthalene 1,2-dioxygenase. J Chem Inf Model 57:550–561

Fernandez-Lafuente R, Guisan JM, Ali S, Cowan D (2000) Immobilization of functionally unstable catechol-2,3-dioxygenase greatly improves operational stability. Enzym Microb Technol 26:568–573

Ferraro DJ, Gakhar L, Ramaswamy S (2005) Rieske business: structure-function of Rieske non-heme oxygenases. Biochem Biophys Res Commun 338:175–190

Ferraro DJ, Brown EN, Yu CL, Parales RE, Gibson DT, Ramaswamy S (2007) Structural investigations of the ferredoxin and terminal oxygenase components of the biphenyl 2,3-dioxygenase from Sphingobium yanoikuyae BI. BMC Struct Biol 7:10

Ferraro DJ, Okerlund A, Brown E, Ramaswamy S (2017) One enzyme, many reactions: structural basis for the various reactions catalyzed by naphthalene 1,2-dioxygenase. IUCrJ 4:648–656

Gally C, Nestl BM, Hauer B (2015) Engineering Rieske non-heme iron oxygenases for the asymmetric dihydroxylation of alkenes. Angew Chem Int Ed 54:12952–12956

Gibson DT, Parales RE (2000) Aromatic hydrocarbon dioxygenases in environmental biotechnology. Curr Opin Biotechnol 11:236–243

Gibson DT, Koch JR, Kallio RE (1968) Oxidative degradation of aromatic hydrocarbons by microorganisms. I. Enzymatic formation of catechol from benzene. Biochemistry 7:2653–2662

Guzik U, Hupert-Kocurek K, Krysiak M, Wojcieszynska D (2014a) Degradation potential of protocatechuate 3,4-dioxygenase from crude extract of Stenotrophomonas maltophilia strain KB2 immobilized in calcium alginate hydrogels and on glyoxyl agarose. Biomed Res Int 2014:1–8

Guzik U, Hupert-Kocurek K, Marchlewicz A, Wojcieszynska D (2014b) Enhancement of biodegradation potential of catechol 1,2-dioxygenase through its immobilization in calcium alginate gel. Electron J Biotechnol 17:83–88

Halder JM, Nestl BM, Hauer B (2018) Semirational engineering of the naphthalene dioxygenase from Pseudomonas sp NCIB 9816-4 towards selective asymmetric dihydroxylation. ChemCatChem 10:178–182

Hall DO, Cammack R, Rao KK (1971) Role for ferredoxins in the origin of life and biological evolution. Nature 233:136

Haynes SW, Sydor PK, Stanley AE, Song LJ, Challis GL (2008) Role and substrate specificity of the Streptomyces coelicolor RedH enzyme in undecylprodiginine biosynthesis. Chem Commun 16:1865–1867

Haynes SW, Sydor PK, Corre C, Song LJ, Challis GL (2011) Stereochemical elucidation of Streptorubin B. J Am Chem Soc 133:1793–1798

Hibbert EG, Baganz F, Hailes HC, Ward JM, Lye GJ, Woodley JM, Dalby PA (2005) Directed evolution of biocatalytic processes. Biomol Eng 22:11–19

Hu DX, Clift MD, Lazarski KE, Thomson RJ (2011) Enantioselective total synthesis and confirmation of the absolute and relative stereochemistry of streptorubin B. J Am Chem Soc 133:1799–1804

Hudlicky T, Olivo HF (1992) A short synthesis of (+)-lycoricidine. J Am Chem Soc 114:9694–9696

Hudlicky T, Reed JW (2009) Applications of biotransformations and biocatalysis to complexity generation in organic synthesis. Chem Soc Rev 38:3117–3132

Hudlicky T, Fan R, Luna H, Olivo H, Price J (1992) Enzymatic hydroxylation of arene and symmetry considerations in efficient synthetic design of oxygenated natural-products. Pure Appl Chem 64:1109–1113

Hudlicky T, Gonzalez D, Gibson DT (1999) Enzymatic dihydroxylation of aromatics in enantioselective synthesis: expanding asymmetric methodology. Aldrichim Acta 32:35–62

Hurtubise Y, Barriault D, Sylvestre M (1998) Involvement of the terminal oxygenase beta subunit in the biphenyl dioxygenase reactivity pattern toward chlorobiphenyls. J Bacteriol 180:5828–5835

Jensen K, Jensen PE, Moller BL (2011) Light-driven cytochrome P450 hydroxylations. ACS Chem Biol 6:533–539

Joern JM, Sakamoto T, Arisawa A, Arnold FH (2001) A versatile high throughput screen for dioxygenase activity using solid-phase digital imaging. J Biomol Screen 6:219–223

Kalnins G, Sevostjanovs E, Hartmane D, Grinberga S, Tars K (2018) CntA oxygenase substrate profile comparison and oxygen dependency of TMA production in *Providencia rettgeri*. J Basic Microbiol 58:52–59

Karlsson A, Parales JV, Parales RE, Gibson DT, Eklund H, Ramaswamy S (2003) Crystal structure of naphthalene dioxygenase: side-on binding of dioxygen to iron. Science 299:1039–1042

Kauppi B, Lee K, Carredano E, Parales RE, Gibson DT, Eklund H, Ramaswamy S (1998) Structure of an aromatic-ring-hydroxylating dioxygenase-naphthalene 1,2-dioxygenase. Structure 6:571–586

Keenan BG, Leungsakul T, Smets BF, Mori M, Henderson DE, Wood TK (2005) Protein engineering of the archetypal nitroarene dioxygenase of Ralstonia sp strain U2 for activity on

aminonitrotoluenes and dinitrotoluenes through alpha-subunit residues leucine 225, phenylalanine 350, and glycine 407. J Bacteriol 187:3302–3310
Kim D, Yoo M, Choi KY, Kang BS, Kim E (2013) Characterization and engineering of an o-xylene dioxygenase for biocatalytic applications. Bioresour Technol 145:123–127
Kim JH, Nam DH, Park CB (2014) Nanobiocatalytic assemblies for artificial photosynthesis. Curr Opin Biotechnol 28:1–9
Kimura N, Nishi A, Goto M, Furukawa K (1997) Functional analyses of a variety of chimeric dioxygenases constructed from two biphenyl dioxygenases that are similar structurally but different functionally. J Bacteriol 179:3936–3943
King-Smith E, Zwick CR 3rd, Renata H (2018) Applications of oxygenases in the chemoenzymatic total synthesis of complex natural products. Biochemistry 57:403–412
Kumamaru T, Suenaga H, Mitsuoka M, Watanabe T, Furukawa K (1998) Enhanced degradation of polychlorinated biphenyls by directed evolution of biphenyl dioxygenase. Nat Biotechnol 16:663–666
Kumar P, Mohammadi M, Viger JF, Barriault D, Gomez-Gil L, Eltis LD, Bolin JT, Sylvestre M (2011) Structural insight into the expanded PCB-degrading abilities of a biphenyl dioxygenase obtained by directed evolution. J Mol Biol 405:531–547
Kumari A, Singh D, Ramaswamy S, Ramanathan G (2017) Structural and functional studies of ferredoxin and oxygenase components of 3-nitrotoluene dioxygenase from Diaphorobacter sp strain DS2. PLoS One 12:1–16
Kweon O, Kim SJ, Baek S, Chae JC, Adjei MD, Baek DH, Kim YC, Cerniglia CE (2008) A new classification system for bacterial Rieske non-heme iron aromatic ring-hydroxylating oxygenases. BMC Biochem 9:11
Landman O, Borovok I, Aharonowitz Y, Cohen G (1997) The glutamine ligand in the ferrous iron active site of isopenicillin N synthase of Streptomyces jumonjinensis is not essential for catalysis. FEBS Lett 405:172–174
Lee K (1999) Benzene-induced uncoupling of naphthalene dioxygenase activity and enzyme inactivation by production of hydrogen peroxide. J Bacteriol 181:2719–2725
Lee J, Zhao HM (2006) Mechanistic studies on the conversion of arylamines into arylnitro compounds by aminopyrrolnitrin oxygenase: identification of intermediates and kinetic studies. Angew Chem Int Ed 45:622–625
Lee JK, Simurdiak M, Zhao HM (2005) Reconstitution and characterization of aminopyrrolnitrin oxygenase, a Rieske N-oxygenase that catalyzes unusual arylamine oxidation. J Biol Chem 280:36719–36728
Lee JK, Ang EL, Zhao HM (2006) Probing the substrate specificity of aminopyrrolnitrin oxygenase (PrnD) by mutational analysis. J Bacteriol 188:6179–6183
Lee SH, Nam DH, Kim JH, Baeg JO, Park CB (2009) Eosin Y-sensitized artificial photosynthesis by highly efficient visible-light-driven regeneration of nicotinamide cofactor. ChemBioChem 10:1621–1624
Lee SH, Kwon YC, Kim DM, Park CB (2013) Cytochrome P450-catalyzed O-dealkylation coupled with photochemical NADPH regeneration. Biotechnol Bioeng 110:383–390
Leger C, Lederer F, Guigliarelli B, Bertrand P (2006) Electron flow in multicenter enzymes: theory, applications, and consequences on the natural design of redox chains. J Am Chem Soc 128:180–187
Lewis JC, Coelho PS, Arnold FH (2011) Enzymatic functionalization of carbon-hydrogen bonds. Chem Soc Rev 40:2003–2021
Leys D, Scrutton NS (2004) Electrical circuitry in biology: emerging principles from protein structure. Curr Opin Struct Biol J 14:642–647
Li C, Zhang L, Zhang C, Hirao H, Wu W, Shaik S (2007) Which oxidant is really responsible for sulfur oxidation by cytochrome p450? Angew Chem Int Ed 46:8168–8170
Liu HH, Xu LQ, Yao K, Xiong LB, Tao XY, Liu M, Wang FQ, Wei DZ (2018) Engineered 3-ketosteroid 9 alpha-hydroxylases in *Mycobacterium neoaurum*: an efficient platform for production of steroid drugs. Appl Environ Microbiol 84(14):e02777–e02717

Mahan KM, Penrod JT, Ju KS, Al Kass N, Tan WA, Truong R, Parales JV, Parales RE (2015) Selection for growth on 3-nitrotoluene by 2-nitrotoluene-utilizing acidovorax sp strain JS42 identifies nitroarene dioxygenases with altered specificities. Appl Environ Microbiol 81:309–319

Mahenthiralingam E, Song LJ, Sass A, White J, Wilmot C, Marchbank A, Boaisha O, Paine J, Knight D, Challis GL (2011) Enacyloxins are products of an unusual hybrid modular polyketide synthase encoded by a cryptic Burkholderia ambifaria genomic island. Chem Biol 18:665–677

Mason JR, Cammack R (1992) The electron-transport proteins of hydroxylating bacterial dioxygenases. Annu Rev Microbiol 46:277–305

Mathews FS, White S (1993) Electron-transfer proteins enzymes. Curr Opin Struct Biol J 3:902–911

Matsui T, Dekishima Y, Ueda M (2014) Biotechnological production of chiral organic sulfoxides: current state and perspectives. Appl Microbiol Biotechnol 98:7699–7706

Mohammadi M, Viger JF, Kumar P, Barriault D, Bolin JT, Sylvestre M (2011) Retuning Rieske-type oxygenases to expand substrate range. J Biol Chem 286:27612–27621

Morley KL, Kazlauskas RJ (2005) Improving enzyme properties: when are closer mutations better? Trends Biotechnol 23:231–237

Munro AW, Leys DG, McLean KJ, Marshall KR, Ost TWB, Daff S, Miles CS, Chapman SK, Lysek DA, Moser CC, Page CC, Dutton PL (2002) P450 BM3: the very model of a modern flavocytochrome. Trends Biochem Sci 27:250–257

Munro AW, Girvan HM, McLean KJ (2007) Cytochrome P450–redox partner fusion enzymes. Biochim Biophys Acta Gen Subj 1770:345–359

Mykles DL (2011) Ecdysteroid metabolism in crustaceans. J Steroid Biochem 127:196–203

Nazor J, Dannenmann S, Adjei RO, Fordjour YB, Ghampson IT, Blanusa M, Roccatano D, Schwaneberg U (2008) Laboratory evolution of P450 BM3 for mediated electron transfer yielding an activity-improved and reductase-independent variant. Protein Eng Des Sel 21:29–35

Ohta T, Chakrabarty S, Lipscomb JD, Solomon EI (2008) Near-IR MCD of the nonheme ferrous active site in naphthalene 1,2-dioxygenase: correlation to crystallography and structural insight into the mechanism of Rieske dioxygenases. J Am Chem Soc 130:1601–1610

Page CC, Moser CC, Dutton PL (2003) Mechanism for electron transfer within and between proteins. Curr Opin Chem Biol 7:551–556

Papireddy K, Smilkstein M, Kelly JX, Shweta, Salem SM, Alhamadsheh M, Haynes SW, Challis GL, Reynolds KA (2011) Antimalarial activity of natural and synthetic prodiginines. J Med Chem 54:5296–5306

Parales RE (2003) The role of active-site residues in naphthalene dioxygenase. J Ind Microbiol Biotechnol 30:271–278

Parales RE, Resnick SM (2004) Aromatic hydrocarbon dioxygenases. In: Singh A, Ward OP (eds) Biodegradation and bioremediation. Springer, Heidelberg, pp 175–195

Parales RE, Parales JV, Gibson DT (1999) Aspartate 205 in the catalytic domain of naphthalene dioxygenase is essential for activity. J Bacteriol 181:1831–1837

Parales RE, Resnick SM, Yu CL, Boyd DR, Sharma ND, Gibson DT (2000) Regioselectivity and enantioselectivity of naphthalene dioxygenase during arene cis-dihydroxylation: control by phenylalanine 352 in the alpha subunit. J Bacteriol 182:5495–5504

Park JH, Lee SH, Cha GS, Choi DS, Nam DH, Lee JH, Lee JK, Yun CH, Jeong KJ, Park CB (2015) Cofactor-free light-driven whole-cell cytochrome P450 catalysis. Angew Chem Int Ed 54:969–973

Pfender WF, Kraus J, Loper JE (1993) A genomic region from pseudomonas-fluorescens Pf-5 required for pyrrolnitrin production and inhibition of pyrenophora-tritici-repentis in wheat-straw. Phytopathology 83:1223–1228

Pollmann K, Wray V, Hecht HJ, Pieper DH (2003) Rational engineering of the regioselectivity of TecA tetrachlorobenzene dioxygenase for the transformation of chlorinated toluenes. Microbiology 149:903–913

Punniyamurthy T, Velusamy S, Iqbal J (2005) Recent advances in transition metal catalyzed oxidation of organic substrates with molecular oxygen. Chem Rev 105:2329–2363

Reetz MT (2013) The importance of additive and non-additive mutational effects in protein engineering. Angew Chem Int Ed 52:2658–2666

Reetz MT, Kahakeaw D, Lohmer R (2008) Addressing the numbers problem in directed evolution. ChemBioChem 9:1797–1804

Resnick SM, Lee K, Gibson DT (1996) Diverse reactions catalyzed by naphthalene dioxygenase from Pseudomonas sp strain NCIB 9816. J Ind Microbiol Biotechnol 17:438–457

Rinner U, Hudlicky T (2012) Synthesis of morphine alkaloids and derivatives. Top Curr Chem 309:33–66

Rivard BS, Rogers MS, Marell DJ, Neibergall MB, Chakrabarty S, Cramer CJ, Lipscomb JD (2015) Rate-determining attack on substrate precedes rieske cluster oxidation during cis-dihydroxylation by benzoate dioxygenase. Biochemistry 54:4652–4664

Roach PL, Clifton IJ, Hensgens CMH, Shibata N, Schofield CJ, Hajdu J, Baldwin JE (1997) Structure of isopenicillin N synthase complexed with substrate and the mechanism of penicillin formation. Nature 387:827–830

Robledo-Ortiz JR, Ramirez-Arreola DE, Perez-Fonseca AA, Gomez C, Gonzalez-Reynoso O, Ramos-Quirarte J, Gonzalez-Nunez R (2011) Benzene, toluene, and o-xylene degradation by free and immobilized P. putida F1 of postconsumer agave-fiber/polymer foamed composites. Int Biodeterior Biodegrad 65:539–546

Sakamoto T, Joern JM, Arisawa A, Arnold FH (2001) Laboratory evolution of toluene dioxygenase to accept 4-picoline as a substrate. Appl Environ Microbiol 67:3882–3887

Sami M, Brown TJN, Roach PL, Schofield CJ, Jack E (1997) Glutamine-330 is not essential for activity in isopenicillin N synthase from *Aspergillus nidulans*. FEBS Lett 405:191–194

Sathapondecha P, Panyim S, Udomkit A (2017) An essential role of Rieske domain oxygenase Neverland in the molting cycle of black tiger shrimp, *Penaeus monodon*. Comp Biochem Physiol A Mol Integr Physiol 213:11–19

Sauber K, Frohner C, Rosenberg G, Eberspacher J, Lingens F (1977) Purification and properties of pyrazon dioxygenase from pyrazon-degrading bacteria. Eur J Biochem 74:89–97

Scheps D, Malca SH, Richter SM, Marisch K, Nestl BM, Hauer B (2013) Synthesis of omega-hydroxy dodecanoic acid based on an engineered CYP153A fusion construct. Microb Biotechnol 6:694–707

Schmidt S, Blom JF, Pernthaler J, Berg G, Baldwin A, Mahenthiralingam E, Eberl L (2009) Production of the antifungal compound pyrrolnitrin is quorum sensing-regulated in members of the *Burkholderia cepacia* complex. Environ Microbiol 11:1422–1437

Schmidt S, Scherkus C, Muschiol J, Menyes U, Winkler T, Hummel W, Groger H, Liese A, Herz HG, Bornscheuer UT (2015) An enzyme cascade synthesis of epsilon-caprolactone and its oligomers. Angew Chem Int Ed 54:2784–2787

Seo J, Ryu JY, Han J, Ahn JH, Sadowsky MJ, Hur HG, Chong Y (2013) Amino acid substitutions in naphthalene dioxygenase from Pseudomonas sp strain NCIB 9816-4 result in regio- and stereo-specific hydroxylation of flavanone and isoflavanone. Appl Microbiol Biotechnol 97:693–704

Shainsky J, Bernath-Levin K, Isaschar-Ovdat S, Glaser F, Fishman A (2013) Protein engineering of nirobenzene dioxygenase for enantioselective synthesis of chiral sulfoxides. Protein Eng Des Sel 26:335–345

Stanley AE, Walton LJ, Zerikly MK, Corre C, Challis GL (2006) Elucidation of the Streptomyces coelicolor pathway to 4-methoxy-2, 2′-bipyrrole-5-carboxaldehyde, an intermediate in prodiginine biosynthesis. Chem Commun 15:3981–3983

Steiner K, Schwab H (2012) Recent advances in rational approaches for enzyme engineering. Comput Struct Biotechnol J J2:1–12

Suenaga H, Watanabe T, Sato M, Ngadiman, Furukawa K (2002) Alteration of regiospecificity in biphenyl dioxygenase by active-site engineering. J Bacteriol 184:3682–3688

Summers RM, Louie TM, Yu CL, Gakhar L, Louie KC, Subramanian M (2012) Novel, highly specific N-demethylases enable bacteria to live on caffeine and related purine alkaloids. J Bacteriol 194:2041–2049

Sydor PK, Barry SM, Odulate OM, Barona-Gomez F, Haynes SW, Corre C, Song LJ, Challis GL (2011) Regio- and stereodivergent antibiotic oxidative carbocyclizations catalysed by Rieske oxygenase-like enzymes. Nat Chem 3:388–392

Tan WA, Parales RE (2016) Application of aromatic hydrocarbon dioxygenases. In: Davis UC (ed) Green biocatalysis. Wiley, Hoboken, NJ

Thebtaranonth C, Thebtaranonth Y, Wanauppathamkul S, Yuthavong Y (1995) Antimalarial sesquiterpenes from tubers of Cyperus-rotundus – structure of 10,12-peroxycalamenene, a sesquiterpene endoperoxide. Phytochemistry 40:125–128

Thomas H, Dirk NJ (1993) Isolation, characterization and N-terminal amino acid sequence of hydrogenase from the green alga *Chlamydomonas reinhardtii*. Eur J Biochem 214:475–481

Torres Pazmiño DE, Winkler M, Glieder A, Fraaije MW (2010) Monooxygenases as biocatalysts: classification, mechanistic aspects and biotechnological applications. J Biotechnol 146:9–24

Tran NH, Nguyen D, Dwaraknath S, Mahadevan S, Chavez G, Nguyen A, Dao T, Mullen S, Nguyen TA, Cheruzel LE (2013) An efficient light-driven P450 BM3 biocatalyst. J Am Chem Soc 135:14484–14487

Udit AK, Hill MG, Gray HB (2006) Electrochemistry of cytochrome P450BM3 in sodium dodecyl sulfate films. Langmuir 22:10854–10857

van Beilen JB, Duetz WA, Schmid A, Witholt B (2003) Practical issues in the application of oxygenases. Trends Biotechnol 21:170–177

Van der Geize R, Yam K, Heuser T, Wilbrink MH, Hara H, Anderton MC, Sim E, Dijkhuizen L, Davies JE, Mohn WW, Eltis LD (2007) A gene cluster encoding cholesterol catabolism in a soil actinomycete provides insight into *Mycobacterium tuberculosis* survival in macrophages. Proc Natl Acad Sci USA 104:1947–1952

Varghese V, Hudlicky T (2014) Short chemoenzymatic total synthesis of ent-hydromorphone: an oxidative dearomatization/intramolecular [4+2] cycloaddition/amination sequence. Angew Chem Int Ed 53:4355–4358

Winkler M, Heil B, Heil B, Happe T (2002) Isolation and molecular characterization of the [Fe]-hydrogenase from the unicellular green alga *Chlorella fusca*. Biochim Biophys Acta 1576:330–334

Winkler M, Hemschemeier A, Jacobs J, Stripp S, Happe T (2010) Multiple ferredoxin isoforms in *Chlamydomonas reinhardtii* – their role under stress conditions and biotechnological implications. Eur J Cell Biol 89:998–1004

Wollam J, Magomedova L, Magner DB, Shen YD, Rottiers V, Motola DL, Mangelsdorf DJ, Cummins CL, Antebi A (2011) The Rieske oxygenase DAF-36 functions as a cholesterol 7-desaturase in steroidogenic pathways governing longevity. Aging Cell 10:879–884

Yoshiyama-Yanagawa T, Enya S, Shimada-Niwa Y, Yaguchi S, Haramoto Y, Matsuya T, Shiomi K, Sasakura Y, Takahashi S, Asashima M, Kataoka H, Niwa R (2011) The conserved Rieske oxygenase DAF-36/Neverland is a novel cholesterol-metabolizing enzyme. J Biol Chem 286:25756–25762

Yu CL, Parales RE, Gibson DT (2001) Multiple mutations at the active site of naphthalene dioxygenase affect regioselectivity and enantioselectivity. J Ind Microbiol Biotechnol 27:94–103

Zaitsev AB, Adolfsson H (2006) Recent developments in asymmetric dihydroxylations. Synthesis-Stuttgart 37:1725–1756

Zhang WY, Hollmann F (2018) Nonconventional regeneration of redox enzymes – a practical approach for organic synthesis? Chem Commun 54:7281–7289

Zhang W, Liu Y, Yan JY, Cao SN, Bai FL, Yang Y, Huang SH, Yao LS, Anzai Y, Kato F, Podust LM, Sherman DH, Li SY (2014) New reactions and products resulting from alternative interactions between the P450 enzyme and redox partners. J Am Chem Soc 136:3640–3646

Zhu YJ, Jameson E, Crosatti M, Schafer H, Rajakumar K, Bugg TDH, Chen Y (2014) Carnitine metabolism to trimethylamine by an unusual Rieske-type oxygenase from human microbiota. Proc Natl Acad Sci U S A 111:4268–4273

Cross-linked Enzyme Aggregates: Current Developments and Applications

Rubia Noori, Mohammad Perwez, and Meryam Sardar

1 Introduction

Enzymes play important role in biotechnological industries due to their environment friendly nature and enhanced catalysis under mild conditions (Sheldon 2011), but their industrial application is confined because of their limited operational stability, recovery, storage, cost and reutilization (Sheldon and van Pelt 2013). Therefore, the stability of the free enzyme needs to be enhanced before it can be used for industrial applications. Immobilization is one of the approaches generally used to stabilize the enzyme (Perwez et al. 2017; Tischer and Wedekind 1999). The immobilization technique can be classified into two major categories, carrier bound and carrier free. Carrier bound immobilized enzymes reduces its cost by using the costly enzymes multiple times, furthermore, improving the stability and selectivity at the same time (Cao et al. 2003). Carrier bound immobilization can be achieved by linking an enzyme to the solid support either by physical methods or by covalent coupling (Tischer and Wedekind 1999; Husain 2010; Husain and Ulber 2011). Physical methods generally do not distort the conformation of enzymes, but the interaction between the enzyme and the support is not very strong which may lead to leaching of enzymes from the surface, the major demerit of physical method (Brena and Batista-Viera 2006; Kulshrestha and Husain 2006; Satar and Husain 2009). While in covalent coupling method chemical bond is formed between the enzymes and the support which might distort the structure, shape and changes aggregation state of the enzyme, thus reducing enzyme activity (Mateo et al. 2007b). Covalent coupling mainly occurs between the amino group of the lysine residue in enzyme with reactive group of the support (Mohamad et al. 2015). The use of support sometimes is costlier which increases the overall cost of production. Covalent binding does not allow leaching of the enzyme which is the major benefit obtained.

R. Noori · M. Perwez · M. Sardar (✉)
Department of Biosciences, Jamia Millia Islamia, New Delhi, India
e-mail: msardar@jmi.ac.in

Q. Husain, M. F. Ullah (eds.), *Biocatalysis*,
https://doi.org/10.1007/978-3-030-25023-2_5

Carrier free immobilization does not require support for immobilization rather it utilizes linking of enzymes with each other using cross linking agent such as glutaraldehyde which prevents loss of enzyme activity, reduces cost, prevents structural distortion of enzyme, purity of enzyme is achieved, reduces mass transfer limitations (Cui and Jia 2015). This method of immobilization is useful when greater yield and productivity is required.

CLEA is one of the carrier free immobilization methods which stabilizes the quaternary structure of enzymes due to the covalent linkage formed between enzymes using cross linking agent (Migneault et al. 2004). Enzymes are immobilized in CLEA by precipitating enzymes using inorganic salts, organic solvents and non-ionic polymers and then extensive cross linking occurs between lysine group of enzymes with cross linking agent (Talekar et al. 2013b). The greater advantage of forming CLEA is its stability towards temperature due to multipoint attachment of enzymes with each other and therefore it is resistant to heat denaturation. It also remains stable in organic solvents and resistant to proteolytic degradation in comparison to free enzymes (Sangeetha and Abraham 2008). Due to its operational stability, it can be reused which makes it cost effective. Therefore, in brief it can be stated that CLEA is a technology which uses two steps, purifies the enzyme using precipitation and immobilization of enzyme. So, crude enzyme extract can also be used to form CLEA as it purifies the enzyme before immobilization thus it is useful at the production site in industries.

2 Types of Immobilization

Confinement of enzymes via immobilization has been proven to be the oldest and most efficient method to produce biocatalyst which has the characteristics of maintaining enzyme catalytic activity and allowing them to be used repeatedly and continuously (Brena et al. 2013). Immobilization can be carried out using a support (called carrier-bound) or without the help of support (called carrier-free). A number of reviews, articles and books have appeared on different types of immobilization methods and their advantages and disadvantages. In this chapter we will discuss in detail only CLEA, a kind of carrier free method.

2.1 Carrier Bound Immobilization

Carrier-bound method can be achieved via physical or chemical interactions of enzyme with the support such as an ion exchange resin or silica. Different types of supports are described in the literature each having its own limitations (Mohamad et al. 2015), depending upon the enzyme and its application one has to select the support matrix. A foretime many efforts have been made to evolve carrier-bound immobilized enzymes in order to remove the cost pressure and promote enzyme reusability and their efficient separation (Sheldon et al. 2005).

In physical interaction, the enzyme is simply trapped or adsorbed via hydrophobic or ionic interactions on the support medium (Matsumoto and Ohashi 2003). Adsorption, encapsulation and entrapment are the basic form of enzyme immobilization (Cao 2006). For encapsulation/entrapment, polymeric matrices are used e.g., alginate, starch, pectin, carrageenan, collagen, polyacrylamide, gelatin, and the mixtures of polymers etc. (Haider and Husain 2007, 2008; Matto and Husain 2009; Cui and Jia 2015). Major advantages include maintaining the native state of enzymes, improved stability and most important, reusability (Matsumoto and Ohashi 2003). The biocatalyst thus formed has remarkable operational performances when compared to the free enzyme. Apart from the advantages, these methods are facing some drawbacks such as enzyme leakage from the matrix when the attachment is not strong enough, decreased biocatalyst activity and reaction productivity after the addition of noncatalytic carrier mass (Cao et al. 2003). Loss of enzyme activity occurs because of leaching of enzyme from the matrix (Sheldon et al. 2005).

While in chemical interactions, the enzyme is structurally modified. This interaction includes the formation of covalent bond between enzyme and the carrier molecule. Surface of the carrier binds the enzyme by chemical bond and changes its shape, rigidity and aggregation state. This type of attachment is irreversible in nature. Amino acids such as lysine, which is present on the surface of proteins, are involved in the covalent binding. They bind to a reactive group of the support and forms stable covalent bond (Velasco-Lozano et al. 2016). Generally, epoxides are used to react with the amino acid groups (Mateo et al. 2007a). This method tends to decrease enzyme leakage from the carrier. Although covalent bonding instates powerful link between enzyme and the matrix yet it has few drawbacks such as presence of the non-catalytic mass of the carrier and the high cost of production (Boller et al. 2002; Cao 2006). Therefore, there is an increasing need for the development of carrier-free immobilization methods with high catalytic activity and efficiency of enzymes.

2.2 *Carrier Free Immobilization*

Carrier-free immobilization or self-immobilization is simply linking of two enzyme molecules through bifunctional cross-linking agents or cross-linkers (Talekar et al. 2013b). Carrier free immobilized enzymes do not require additional mass of support medium (Cui and Jia 2015). They are produced either by cross linking the dissolved enzymes (CLE), by spray drying enzymes, by crystallized enzymes (CLEC), or aggregated enzymes (CLEA) (Velasco-Lozano et al. 2016). Different approaches to produce carrier-free immobilized enzymes are shown in Fig. 1.

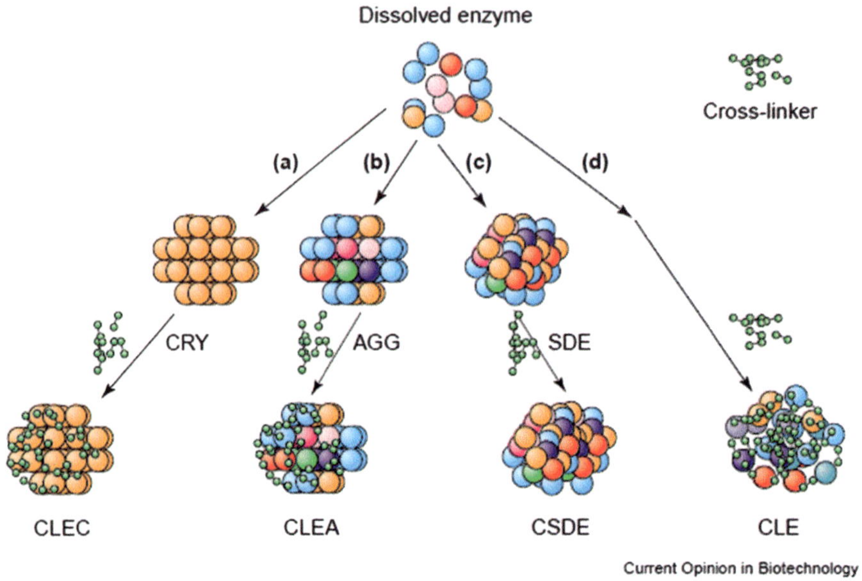

Fig. 1 The different approaches to produce carrier-free immobilised enzymes: (**a**) crystallization; (**b**) aggregation; (**c**) spray-drying; (**d**) direct cross-linking. AGG, aggregates; CRY, crystals; SDE, spray-dried enzyme. Reprinted with permission from (Cao et al. 2003)

2.2.1 Cross Linked Enzymes (CLE)

During 1960s, Quiocho and Richards reported the formation of cross-linked enzymes. The cross linking of insoluble proteins was done via glutaraldehyde which is a bifunctional cross-linking agent and these structures were known to retain their catalytic activity (Quiocho and Richards 1966). CLEs are obtained by adding an aqueous solution of the enzyme with that of glutaraldehyde (Sheldon 2011). The CLE size ranges from 1 to 100 μm (Velasco-Lozano et al. 2016). Although CLE showed increased thermal stability of enzymes, it still exhibited some drawbacks such as decreased mechanical stability and low reproducibility as well as retention activity (Roessl et al. 2010).

2.2.2 Cross Linked Spray Dried (CLSD)

In addition to all those methods implied to immobilize enzyme without any support (carrier-free), another strategy was designed in which spray-dried powders of enzymes were used (Amotz 1987). Hence, the technique gained the name as cross-linked spray dried (CLSD) enzymes. However, with this method, researchers observed reasonable or low activity retention by the enzyme (Cui and Jia 2015). The reason being, the enzyme was deactivated reversibly. So, this process was not utilised for further use (Cao et al. 1999).

2.2.3 Cross Linked Enzyme Crystals (CLEC)

Due to the drawbacks faced by CLSD, another methodology for cross linking was discovered known as cross linking of crystallized enzymes (CLECs) (Quiocho and Richards 1964). Enzymes which are highly compacted crystals can be bonded together by using cross linking agent such as glutaraldehyde and thus the CLEC formed are insoluble crystalline proteins. CLECs experienced a different fortune as it showed considerably high activity, enhanced mechanical strength, enhanced thermo stability and good stability against pH, organic solvents and proteolysis (Gogoi et al. 2010; Häring and Schreier 1999; Sheldon 2007a). As CLECs are stable under harsh conditions, they are conveniently used for biotransformation in non-aqueous media (Quiocho and Richards 1966). CLECs gained popularity as industrial biocatalysts in early 1990s and was commercialized by Altus Biologics (Margolin and Navia 2001). CLEC technology has been used in case of few enzymes like ribonuclease A, subtilisin, carboxypeptidase B, alcohol dehydrogenase and some lipases as CLECs exhibited properties like enhanced working stability, suitable particle size, high productivity, ease of recycling and high catalytic activity (Cao et al. 2003). These highly active immobilized enzymes vary in size from 1 to 100 μm. When the crystal size is decreased, the enzyme catalytic activity increases. A major drawback is that expensive and laborious method is followed to crystallize the enzyme in pure form (Brady et al. 2004). Requirement to build up a simpler and cost-effective method led researchers to develop new strategy which involved precipitation of enzymes called cross-linked enzyme aggregates (CLEAs).

2.2.4 Cross Linked Enzyme Aggregates (CLEA)

In 2000s, new method for carrier-free immobilization of enzymes, cross-linked enzyme aggregates (CLEAs), was developed and it was established as a new class of immobilized enzymes (Cao et al. 2000). Morphology of CLEA and its size decides the accessibility of substrate to the CLEA. The smaller size of CLEA allows more substrate binding than larger size of CLEA.CLEAs appears to be either fine grained particles with many cavities or course grained with fewer cavities (Schoevaart et al. 2004). The average size of CLEA particle is about 20 μm. The Scanning electron microscopy (SEM) image of CLEA of lipase is shown in Fig. 2. CLEAs are the result of precipitation of enzyme (not necessarily purified) from aqueous solution and cross-linking the resultant aggregates. The physical aggregates were held together by non-covalent binding. The cross-linking done via agent such as glutaraldehyde make the aggregates permanently insoluble after covalent binding and at the same time maintaining tertiary structure of enzyme as well as their catalytic activity (Cao et al. 2000; Musthapa et al., 2004; Sheldon 2007b). The idea was first demonstrated in 2000 by Cao et al. and gained momentum after the successful implication of this methodology on penicillin G amidase in the Sheldon's laboratories (Cao et al. 2000). CLEAs displayed 1000 times more activity than the

Fig. 2 SEM image of CLEA of lipase (*Thermomyces lanuginosa*). Reprinted with permission from (Gupta et al. 2009)

carrier-bound immobilization methods (Sheldon 2007a). This strategy proved to have much superior properties, to be described in the section below and also an ideal way to immobilize enzymes that can be used as biocatalyst for industrial purposes.

3 CLEA and Its Chemistry

CLEA preparation requires screening of water miscible organic solvents for greater activity recovery (Sheldon 2011). Organic solvents which are hydrophobic in nature are better precipitants than hydrophilic solvents. Cross linking step is performed usually with glutaraldehyde as it is easily available and is inexpensive (Migneault et al. 2004). Cross linking occurs between free amino group of lysine residue of amino acids and aldehydic group of monomer or polymer of glutaraldehyde (Vrsanska et al. 2017). The reaction involved is inter or intra molecular aldol condensation forming Schiff's base and addition of Michael 1, 4 type to α, β unsaturated aldehyde moiety (Migneault et al. 2004). The particle size, stability and the activity of CLEA is affected by the concentration of cross linker and enzyme, most importantly the ratio between the two, affects the properties of CLEA (Vrsanska et al. 2017). If the ratio decreases, then enough cross linker is not available to link enzymes which may lead to leaching of enzymes and if the ratio is too high then extensive cross-linking results in loss of flexibility of enzymes (Vrsanska et al. 2017). The lysine group of enzymes which is required for cross linking in such case will not be available which thus reduces the activity of CLEA. Therefore, for proper activity recovery of CLEA, the ratio of cross-linker and enzyme needs to be optimised. In cases where the enzyme lacks enough lysine residue, CLEA formed in such cases are unstable which results in leaching, so to increase stability extra lysine group is provided by co-immobilizing enzyme along with polyamines (López-Gallego et al. 2005). If the protein concentration is less in enzymatic preparation, then protein feeder is used like bovine serum albumin which enriches the enzyme with extra protein and thus promotes a stable leaching free CLEA (Shah et al. 2006). Different approaches have been made for preparing stable CLEAs of enzyme with low lysine residue using polymers containing numerous free amino groups for co-aggregation, e.g., poly-lysine or polyethyleneimine (Honda et al. 2006; López-Gallego et al. 2005; Yamaguchi et al. 2011). Certain proteins such as bovine serum albumin and hen egg white also promote stability of CLEAs as they are rich in lysine residue (Cabana et al. 2007; Karimpil et al. 2011).

Cross-linking with glutaraldehyde complicates conditions when enzyme is having lysine residue involved in catalytic activity. Also, glutaraldehyde sometimes non-specifically cross-links with the essential amino acid required for catalytic property of the enzyme (Mateo et al. 2004). Interesting alternatives have been found in place of glutaraldehyde. A macromolecular cross-linker dextran polyaldehyde was used to prepare β-mannanase CLEAs (Zhen et al. 2013). It displayed activity as high as sixteen times when compared to CLEAs prepared by glutaraldehyde (Netto et al. 2013; Talekar et al. 2012a). Another cross-linker, pectin, a high molecular weight, non-toxic, highly biocompatible and renewable natural polysaccharide, can also be used. The dialdehyde groups produced by pectin tends to react with the amino groups in lysine residues of enzymes to form Schiff's base (Guo et al. 2014; Wilson et al. 2006). Although CLEAs are novel and efficient method for enzyme immobilisation, there are few limitations to it. Low mechanical stability was observed when the cross-linking was not appropriate and as a result, leaching of enzyme from the reaction occurs. CLEA needs to be separated for reutilization from the reaction mixture via centrifugation or filtration which increases the size of CLEA due to aggregation, thus leading to internal mass transfer limitations (Talekar et al. 2012a).

4 Recent Developments in CLEA Formation

4.1 Magnetic-CLEAs (m-CLEAs)

Recent advances in the field of immobilization paved way for development of a new strategy that has been applied to immobilize few enzymes till now. Some researchers developed carrier-bound CLEA methodology as an updated version of CLEAs. It proved to be a favourable alternate to overcome the limitations showed by CLEAs.

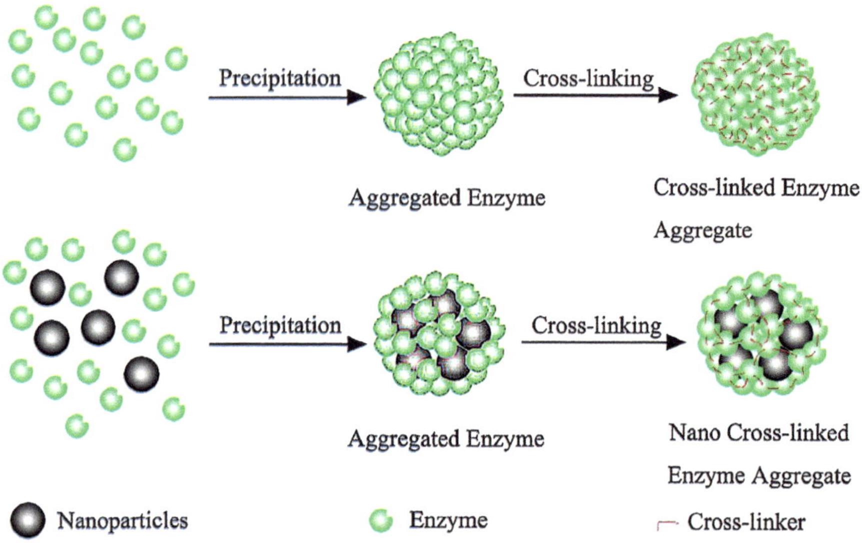

Fig. 3 Diagram depicting preparation of CLEA with or without Nanoparticles. (**a**) Formation of CLEAs without Nanoparticles; (**b**) Formation of CLEAs in the presence of Nanoparticles. Reprinted with permission from (Wang et al. 2017)

The cross-linking of enzyme was done on the surface of magnetic nanoparticles (MNPs). Nanoparticles have gained considerable attraction by researchers to use them as inert support in enzyme immobilisation (Husain 2016, 2017b; Lu et al. 2007). Due to the small size of the nanoparticles which ranges from 1 to 100 nm, it provides an additional benefit of larger surface area in a compact size which makes it a suitable candidate as a matrix for immobilizing enzymes.

The magnetic material as a support tends to have a few advantages which make them a good choice for immobilisation. Large surface to volume ratio of MNPs showed greater binding efficiency of enzyme and due to this, enzyme in large volume can be coated on their surface (Hu et al. 2009). These materials can be easily controlled and removed from the reaction mixture when it comes in contact with the external magnetic field and also provides stability upon repeated usage and improved the catalytic properties of CLEAs (Talekar et al. 2012a). The super paramagnetic behaviour provides ease of separation and selective recovery as well as their low toxicity makes them good candidate for support of cross-linked enzymes (Laurent et al. 2008; Netto et al. 2013; Verma et al. 2013).

To prepare magnetic cross-linked enzyme aggregates (m-CLEAs), the amino functionalized MNPs were added as an additive. The aggregates were formed by the precipitation of enzymes and cross-linking takes place between the enzyme aggregates and nanoparticles, shown in Fig. 3 (Hu et al. 2009). The added magnetite nanoparticles to the enzyme having low lysine residues achieve enough cross-linking of enzyme aggregates that will make the CLEAs mechanically stable and non-leachable. The magnet will easily retrieve the CLEAs from the reaction media, thus, eliminating the need for centrifugation and filtration (Takkinen et al. 1983).

The use of MNPs has allowed the enzyme to retain most of its original activity. MNPs along with 3-Aminopropyltriethoxysilane (APTES) help in providing efficient binding sites to enzyme as well as increases the stability of enzyme. Addition of APTES increases the concentration of amine groups on the surface of nanoparticles which helps to bind the enzyme aggregates (Talekar et al. 2012a). The metal binding with the amino acid residue of enzyme stabilizes its active conformation (Barbosa et al. 2014). The metals involved in MNPs can be iron, nickel, cobalt and their oxides like magnetite. Iron oxide nanoparticles have been used by Nadar et al. (2016) to synthesize m-CLEA of glucoamylase (Nadar and Rathod 2016). Recently, α-amylase m-CLEA have been prepared using magnetite nanoparticles which showed 100% recovery of catalytic activity with enhanced storage stability and reusability as well (Talekar et al. 2012a).

Magnetic-CLEAs prepared using alpha amylase proved to be the model enzyme for the preparation of m-CLEAs as the enzyme contains only 6% lysine residue out of the total amino acid content (Takkinen et al. 1983).

Cellulase CLEAs were prepared using iron oxide super paramagnetic nanoparticles which have been pre-treated with APTES. Covalent linkage via amine groups was established between cellulase aggregates and amine-functionalized MNPs (Jafari Khorshidi et al. 2016). Another industrially important enzyme, phenylalanine ammonia lyase (PAL), involved in bioconversion of L-phenylalanine to ammonia and trans-cinnamic acid, was used to prepare hybrid magnetic PAL-CLEAs utilising glutaraldehyde as cross-linker and magnetite nanoparticles as a co-aggregation unit. The effective result produced from the formation of m-CLEAs helped the researchers to enhance the catalytic properties of the enzymes which are having wide application in industries (Dong Cui et al. 2014). With the addition of MNPs, functional, operational and thermo-stability of the enzyme has been improved greatly and has reduced mass transfer problems (Cruz-Izquierdo et al. 2014).

4.2 Combi-CLEA: One Pot Multistep Reaction

Synthesis of chemicals requires large number of steps in a cascade reaction. Each step may deposit unstable intermediate or start unnecessary reaction and such multi-step reaction may result in low yield. So, to overcome the problem multistep reactions can be performed in a single step by immobilizing two or more enzymes together in CLEA, such immobilization strategy is called combi-CLEA. Combi-CLEA is useful in many aspects because it reduces cost of production at industrial level and produces better yield as compared to multi-step reactions. The combi-CLEA of porcine pancreas crude extract (containing lipase, α-amylase and phospholipase A2) have been reported (Dalal et al. 2007a) in which α-amylase in combi-CLEA showed decreased V_{max}/K_m but no change was observed for lipase and phospholipase A2.

Lipase and α-amylase retained activity for three cycles in combi-CLEA and the combi-CLEA was shown to be thermally stable. Similarly, combi-CLEA of cellulase and hemicellulase was prepared with commercially available Pectinex Ultra SP-L (Dalal et al. 2007b), the Propanol was selected as the best precipitant while glutaraldehyde was used as the cross linker at the concentration of 5 mM. No enzymatic activity was lost after combi-CLEA formation which showed its stability. There was no change in pH optima and temperature optima while significant changes were observed in thermal stability where cellulase showed maximum thermal stability at 70 °C while xylanase and cellulase showed thermal stability at 60 °C and 50 °C respectively and it could be reused upto four cycles. The combi-CLEA of glucose oxidase (GO) and horseradish peroxidase (HRP) were also prepared and utilised to increase the reaction rates for the two set of reactions involving hydrogen peroxide as the intermediates, this could be achieved due to the reduced distance between the GO and HRP in combi-CLEA. Glucose oxidase oxidises glucose to produce D-glucono-δ-lactone and hydrogen peroxide as the intermediate. Combi-CLEA is utilised to detect glucose as a colorimetric assay (Yang 2017). However, in another study, magnetic combi-CLEA of GO and HRP were prepared by precipitating and cross linking both the enzymes on the surface of amino functionalized MNPs (Zhou et al. 2016). The use ofMNPs in combi-CLEA formation is helpful in easy removal of enzymes from reaction mixture for its reusability. Ethanol was used as the precipitating agent and glutaraldehyde was used as cross linker. Magnetic combi-CLEA was efficient in removing direct black 38 (DB-38) dye by 92.28% while free enzyme could remove 46.82% and the same can be reused for 8 cycles (Zhou et al. 2016)

5 Optimizing Parameters of CLEAs

5.1 Additives Increases Enzyme Activity

Several approaches have been used for the optimization of CLEA, different parameters such as choice of precipitant; cross linking agent, additives, incubation time and pH play a major role in preparing stable CLEA. Enzyme activity is one of the factors which determine the characteristic properties of CLEA. CLEA prepared using enzymes which lack enough surface reactive amino groups are mechanically fragile, so such CLEA cannot be recovered efficiently (Sheldon 2007b). Researchers have been using additives such as polyamines, and BSA to overcome the loss of stability and activity due to less amine group and less protein content in the extract respectively used for preparing CLEA. Therefore, polyethyleneimine has been co-aggregated with enzyme to prepare CLEA of glutamyl acylase (López-Gallego et al. 2005). Vaidya et al. (2012) also used polyethyleneimine along with L-Aminoacylase which showed activity recovery of 74.9%. Apart from increasing the activity recovery, it also showed increased thermal and storage stability. BSA has also been used to compensate for the less protein content which was not enough for strong cross linking (Shah et al. 2006), CLEA of lipase and penicillin acylase formed in the presence of BSA retained 100% and 86% activity respectively while CLEA formed without BSA retained 0.4% and 50% respectively. Others also reported increase in activity of CLEA in presence of BSA, CLEA of amylase enzyme with BSA showed 22.33% recovery activity while without BSA the activity was reduced to 13.29% (Easa and Yusof 2006). The concentration of BSA plays an important role in maximum activity recovery, if the concentration remains low then insignificant activity recovery occurs and if the concentration is high, competition occurs between amino group of enzymes and BSA, thus reducing activity recovery due to insufficient enzyme cross linked. SDS, tween 80 and Triton X-100 has also been used as additives which increases the activity of lipases two to three-fold (Gupta et al. 2009). The use of additives helps the enzymes achieving active confirmation thus enhancing activity (Gupta et al. 2009; Topçular and Ayhan 2008). A simple polysaccharide starch was also used as an additive and co-precipitated with invertase enzyme using ammonium sulphate, the starch occupied the spaces in the precipitated enzyme. After the porous CLEA formed, starch was removed with the help of alpha amylase enzyme which leaves pores in the CLEA. Controlled pore size can be obtained by varying starch concentration till a limit which increases the enzyme activity but further increase in starch concentration decreases enzyme activity (Talekar et al. 2012b), thus the optimum pore size of CLEA is required for effective catalytic efficiency.

5.2 Precipitation: The Key Step in Immobilization

Precipitation of enzyme in CLEA formation has the greater impact on activity recovery as it helps in aggregating the enzymes into a configuration to make it stable and helps in yielding maximum activity recovery (Talekar et al. 2010). Precipitation is performed with salts, organic solvents, ionic and non-ionic polymers (Talekar et al. 2013b). The important parameter in precipitation is screening of precipitant, concentration, and time duration of incubation to allow for complete precipitation and cross linking. All these parameters change if enzyme and its source vary for CLEA preparation. So, for each enzyme all the parameters need to be optimized. Screening of precipitant is required to select the precipitant which keeps enzyme in most active conformation. Screening is performed by re-suspending the precipitated enzyme to measure activity. Enzyme may precipitate in an inactive conformation but when re-suspending retains its conformation and during cross linking, it again changes conformation and becomes inactive. Therefore, during screening of precipitant two or three precipitant should be selected which shows good activity recovery and then activity recovery in CLEA decides final selection of precipitant. In most of the cases, CLEA formation has been achieved with ammonium sulphate as salt for precipitation of enzymes (Cao et al. 2000; Rehman et al. 2016; Talekar et al. 2010; Vrsanska et al. 2017). During the precipitation step in CLEA formation the surface area of the enzyme decreases which increases the stability of enzymes. The concentration of precipitant determines the structure and morphology of CLEA formed, the decrease in the size of the tyrosinase CLEA was observed when the concentration of ammonium sulphate was increased and the CLEA formed were more ordered at higher concentration (Aytar and Bakir 2008). The loss in enzyme activity occurs during precipitation, the loss of activity can be minimized by the addition of sugars as the stabilising agent (Wang et al. 2011b). Most of the researchers have found that the activity recovery of CLEA of lipases varies from different microbial sources, this may be attributed to the fact that lipases bear different glycosylated groups on the surface, which during precipitation changes the conformation and the yield varies considerably (Prabhavathi Devi et al. 2009). Zhen et al. (2013) used all kinds of precipitants (neutral salts, organic solvents and non-ionic polymers) for preparing CLEA of β-mannanase. He observed that the activity recovery was best when organic solvents were used in comparison to polyethylene glycol (non-ionic polymer) and ammonium sulphate (neutral salts). Among organic solvents like methanol, ethanol, acetone and isopropanol, the solvent which is less hydrophobic showed lesser activity than the solvent which is more hydrophobic. This occurs because water molecule is essential for enzyme to remain in active conformation. The less hydrophobic molecule like methanol interacted more with the water molecule as a result water is not available for enzyme which is essential for its activity, thus changing the conformation of enzyme and reducing its activity. Therefore, methanol showed less activity while iso-propanol which is more hydrophobic showed best activity in organic solvents. The ratio of precipitants and enzyme also plays a very important role in preparing a much stable CLEA with high activity recovery. If the ratio is more, the active site of enzyme is damaged which causes loss in activity (Zhen et al. 2013). The combi-CLEA of neutrase, a commercial enzyme preparation, and papain was formed using ethanol as the precipitating agent (Chen et al. 2017). The relative enzyme activity of the combi-CLEA formed is higher than the free enzyme counterpart. Therefore, it is very difficult to generalize the use of solvents as precipitant for CLEA formation, which depends upon the enzyme, its source and other factors.

5.3 *Cross-linker: The Final Step in CLEA Preparation*

Cross linking is the crucial step in CLEA formation which joins enzymes to each other providing stability to enzymes. Generally, glutaraldehyde is used as the cross-linking agent which contains dialdehydic group helps facilitate bridging enzymes. But glutaraldehyde cannot be used for all enzymes due to its smaller size, high reactivity and instability at acidic pH etc. Dextran polyaldehyde was used for cross linking nitralases and it yielded activity recovery of 50%. When dextran polyaldehyde was used in combination with glutaraldehyde causes inactivation of enzyme, this occurs due to the smaller size of glutaraldehyde and its high reactivity as it penetrates and reacts with the amino group which is responsible for the catalytic activity of enzyme. Penicillin acylase enzyme also resulted into CLEA with 90% activity recovery with polyaldehyde dextran while 50% activity retains with glutaraldehyde. Dextran polyaldehyde do not react with the active sites of enzymes due to its larger size and the activity recovery improves (Mateo et al. 2004). Another commonly used cross-linker p-benzoquinone was used to prepare CLEA of lipases, and CLEA formed were stable in acidic pH while when CLEA formed using glutaraldehyde as the cross linker, schiff's base was formed and CLEA was found to be unstable in acidic pH which breaks down to give back aldehyde and amine. C–O and C–N bond was formed when p-benzoquinone is used as a cross linker and it forms a stable CLEA in acidic pH (Wang et al. 2011a). However, Yamaguchi et al. (2011) used glutaraldehyde and para-formaldehyde in the ratio of 1:16 to prepare CLEA of chymotrypsin, subtilisin, citrate synthase and laccases. Schiff's base formed was reduced with sodium cyanoborohydride ($NaCNBH_3$) to form irreversible cross linking. The yield of CLEA can be improved by two ways either by increasing the concentration of cross linker or by increasing the lysine residue on the enzyme surface. In the former case the conformation of the enzyme is hampered reducing the activity of the enzyme (Vrsanska et al. 2017). To increase the lysine group, poly lysine was added and it enhances the yield of the CLEA (Sheldon and van Pelt 2013; Yamaguchi et al. 2011). Laccase CLEA was prepared using chitosan as the cross linker due to the presence of amine group which reacts with the carboxylic group of non-essential amino acid. Chitosan as the cross linker is safe to use in comparison to glutaraldehyde which is toxic in nature (Arsenault et al. 2011).

Concentration of cross linker is a very important parameter which affects enzyme activity, particle size and stability. At a very low concentration of cross linker ineffective cross linking occurs which results into unstable CLEA and some of the enzyme remains in the solution reducing CLEA'S activity. At a very high concentration flexibility of CLEA is lost which causes rigidity that affects catalysis (Yu et al. 2006). Steric-hindrance also occurs due to which substrate cannot reach the active site of the enzyme leading to lower the catalysis and yield (Reshmi and Sugunan 2013). Therefore, optimization of cross linker concentration is important for a stable CLEA with appropriate size, morphology and enhanced activity recovery. Similarly, incubation time and temperature for cross linking is also important and needs to be optimized. Usually lower temperature is used for cross linking to maintain the enzyme activity, as at higher temperature enzyme may get denature. At higher incubation time extensive cross linking occurs which results in rigidity and finally inhibits catalysis. At lower incubation time proper cross linking does not occur and enzyme remains in solution reducing activity recovery (Majumder et al. 2008). Different ways of CLEA formation along with the precipitating agent cross linker and additives have been summarized in Table 1.

6 Applications of CLEAs

With all the characteristic features shown by CLEAs and its advantages over other immobilization techniques, it has gained popularity among researchers to develop more and more CLEAs and it has found wide range of applications in industries.

Table 1 Table summarizes different ways of CLEA formation

CLEA	Enzyme/Source	Precipitant	Cross-linker	Additives	References
CLEA	**Laccase/***Coriolopsis polyzona*	Polyethylene glycol	Glutaraldehyde	BSA	Cabana et al. (2007)
CLEA	**Feruloyl esterase/***Aspergillus awamori*	Ammonium sulphate	Glutaraldehyde	Triton X-100	Fazary et al. (2009)
CLEA	**Poly-3-hydroxybutyratede polymerase/***Streptomyces exfoliatus*	Acetone	Glutaraldehyde	BSA	Hormigo et al. (2012)
CLEA	**Hydrolase (BL28)/***Bacillus licheniformis*	Ammonium sulphate	Glutaraldehyde	–	Ju et al. (2013)
CLEA	**Subtilisin/***Bacillus subtilis*	Ammonium sulphate	Glutaraldehyde	Ammonium sulphate	Sangeetha and Abraham (2008)
CLEA	**Lipase/***Candida rugosa*	Ethanol	Glutaraldehyde	–	Kartal et al. (2011)
CLEA	**Laccase/***Trametes versicolor*	Acetonitrile	Glutaraldehyde	–	Seow and Yang (2017)
CLEA	**Lipase/***Burkholderia cepaci*	Acetone	Glutaraldehyde	BSA	Majumder et al. (2008)
CLEA	**Horseradish peroxidase/***Armoracia rusticana*	Acetone	Ethylene glycol-bis (succinic acid N-hydroxysuccinimide)	–	Bilal et al. (2017)
CLEA	**Acetyl xylan esterase/***Bacillus pumilus*	Ammonium sulphate	Glutaraldehyde	–	Montoro-García et al. (2010)
CLEA	**Lipase B/***Candida antarctica*	Ammonium sulphate	Glutaraldehyde	BSA	Torres et al. (2014)
CLEA	**Penicillin acylase**	Polyethylene gly-col-6000	Glutaraldehyde	–	Pchelintsev et al. (2009)
CLEA	**Chloroperoxidase/***Caldariomycesfumago*	Ammonium sulphate	Sodium borohydride	–	Roberge et al. (2009)
CLEA	**Invertase/***Saccharomyces cerevisiae*	Ammonium sulphate	Glutaraldehyde	–	Talekar et al. (2010)

CLEA	**α-amylase**/*Bacillus amyloliquefaciens*	Ammonium sulphate	Glutaraldehyde	–	Talekar et al. (2012c)
CLEA	**NADH-dependent nitrate reductase**/*Fusarium oxysporum*	Ammonium sulphate	Glutaraldehyde	–	Talekar et al. (2014)
CLEA	**Invertase**/*Saccharomyces cerevisiae*	Ammonium sulphate	Glutaraldehyde	–	Talekar et al. (2012b)
CLEA	**Lipase**/*Pencilluim notatum*	Ammonium sulphate	Ethylene glycol-bis (succinic acid N-hydroxysuccinimide)	–	Rehman et al. (2016)
CLEA	**α-amylase**	Ammonium sulphate	Dextran	–	Nadar et al. (2016)
CLEA	**Alkylsulfatase**/*Pseudomonas*	PEG4000	Glutaraldehyde	–	Li et al. (2016)
CLEA	**Horseradish peroxidase**/*Armoracia rusticana*	Acetone	Glutaraldehyde	Albumin	Šulek et al. (2011)
CLEA	**Laccase**/*Pleurotus ostreatus*	Ammonium sul-phate and tert-butanol	Glutaraldehyde	–	Kumar et al. (2012)
CLEA	**Lipase**/*E. coli*	BSA	p-benzoquinone	BSA	Wang et al. (2011a)
CLEA	**Papain**/*Carica papaya*	Ethanol	Glutaraldehyde	–	Wang et al. (2010)
CLEA	**Tyrosinase**/*Agaricus bisporus*	Ammonium sulphate	Glutaraldehyde	–	Aytar and Bakir (2008)
CLEA	**Penicillin G acylase**	Acetone	Glutaraldehyde	Trehalose	Wang et al. (2011b)
CLEA	**Penicillin G acylase**/*E. coli*	PEG 600	Glutaraldehyde	–	Wilson et al. (2004b)
CLEA	**Lipase**/*Alcaligenes* sp.	PEG 600	Glutaraldehyde	Dextran sulphate and polyethyleneimine	Wilson et al. (2006)
	Lipase/*Candida antarctica*	Ethylene glycol dimethyl ether	Glutaraldehyde	Dextran sulphate and polyethyleneimine	Wilson et al. (2006)

(continued)

Table 1 (continued)

CLEA	Enzyme/Source	Precipitant	Cross-linker	Additives	References
CLEA	**Lipozyme**/*Thermomyces lanuginosus*	PEG	Glutaraldehyde	D-sorbierite	Yang et al. (2012)
CLEA	**Laccase**/*Shewanella putrefaciens*	Ammonium sulphate	Glutaraldehyde	–	Sinirlioglu et al. (2013)
CLEA	**β-Mannanase**/*Bacillus licheniformis*	Isopropanol	dextran polyaldehyde	–	Zhen et al. (2013)
CLEA	**Lipase**/*Pseudomonas* sp.	Acetone	Glutaraldehyde	–	Zhao et al. (2008)
CLEA	**Laccase**/*Coriolopsis polyzona*	Ammonium sulphate	Chitosan	–	Arsenault et al. (2011)
m-CLEA	**Lipase**/*Aspergillus niger*	Dimethyl carbonate	Glutaraldehyde	–	Tudorache et al. (2013)
m-CLEA	**Glucoamylase**	Ammonium sulphate	Dialdehydic Pectin	–	Nadar and Rathod (2016)
m-CLEA	**Cellulase**/*Aspergillus niger*	Ammonium sulphate	Glutaraldehyde	–	Jafari Khorshidi et al. (2016)
m-CLEA	**Phenylalanine Ammonia Lyase**/*Rhodotorula glutinis*	Ammonium sulphate	Glutaraldehyde	–	dong Cui et al. (2014)
m-CLEA	**Lipase**/*Candida antarctica*	Ammonium sulphate	Glutaraldehyde	–	Cruz-Izquierdo et al. (2014)
m-CLEA	**α-amylase**/*Bacillus* sp.	Ammonium sulphate	Glutaraldehyde	–	Talekar et al. (2012a)
m-CLEA	**Xylanase**/*Bacillus gelatini*	Ammonium sulphate	Glutaraldehyde	–	Bhattacharya and Pletschke (2014)
m-CLEA	**Lipase**/*Candida rugosa*	Ammonium sulphate	Glutaraldehyde	–	Kim et al. (2010)
Combi-CLEA	**Amylosucrase**/*Deinococcus geothermalis*, **Maltooligosyltrehalose synthase** and **Maltooligosyltrehalosetrehalohydrolase**/*Brevibacterium helvolum*	Acetone	Glutaraldehyde	BSA	Jung et al. (2013)

Combi-CLEA	**Horseradish peroxidase** and **Glucose oxidase**/*Aspergillus niger*	Acetonitrile	Glutaraldehyde	–	Yang (2017)
Combi-CLEA	**Glucoamylase** and **Pullulanase**	Ammonium sulphate	Glutaraldehyde	–	Talekar et al. (2013a)
Combi-CLEA	**Peroxidase**/*Bjerkandera adusta* and **Glucose oxidase**/*Aspergillus niger*	Acetone	Glutaraldehyde	–	Taboada-Puig et al. (2011)
Combi-CLEA	Ultraflo L, Depol 740 L, Depol 670 L (**Feruloyl esterase**)	Ammonium sulphate or tert-butanol	Glutaraldehyde	–	Vafiadi et al. (2008)
Combi-CLEA	**Papain** and **Neutrase**	Ethanol	Glutaraldehyde	–	Chen et al. (2017)
Combi-CLEA	**α-L-arabinosidase** and **β-D-glucosidase**/*Vitis* sp.	Ammonium sulphate	Glutaraldehyde	–	Ahumada et al. (2015)
Combi-CLEA	**Pectinase** and **Cellulase**/ *Aspergillus niger*	Isopropanol	Glutaraldehyde	BSA	Dal Magro et al. (2016)
Combi-CLEA	**Pectinase**, **Xylanase** and **Cellulase**/*Aspergillus niger*	n-Propanol	Glutaraldehyde	–	Dalal et al. (2007b)

This section investigates the scope of the methodology pertaining to the potentiality of the CLEAs having high catalytic efficiency and activity. While in some cases, the activity of enzymes tends to increase as compared to their native form, from where they were derived. As already discussed above in details, this strategy is basically a single step process combining precipitation of enzyme and its cross-linking (Sheldon et al. 2005). The simplicity of this methodology is that it produces high-throughput result. CLEAs technology can be applied to almost any enzyme after optimizing best precipitant and cross-linker which may differ from one enzyme to another (Schoevaart et al. 2004; Wilson et al. 2004a).

CLEAs has been prepared for many enzymes till now, for e.g., hydrolases, proteases, amidases, lipases, esterases, nitrilases, glycosidases, oxidoreductases, peroxidases, and lyases. CLEAs has found applications in many industries like chemical, food processing, heavy industry, metal, construction, textile, medicine, pharmaceuticals, wastewater treatment etc. Some of the recent applications have been discussed in this section.

6.1 Biotransformation/Chemical Synthesis

CLEAs have been extensively used for biotransformation in different industries. CLEAs from a variety of proteases have been developed such as chymotrypsin, papain and alkaline protease using a source *Bacillus licheniformis* (Schoevaart et al. 2004). They have been used in laundry detergents, organic synthesis, e.g., in the resolution of (amino acid) esters, and amines (van Rantwijk and Sheldon 2004). Alcalase-CLEA which is involved in selective synthesis of α-carboxylic esters of N-protected amino acid and peptides have been reported (Nuijens et al. 2009a). These CLEA catalyses the α-hydrolysis of symmetrical aspartyl or glutamyl diesters to α-protected aspartic acid β-esters and glutamic acid γ-esters (Nuijens et al. 2009b). Aminoacylase CLEA from *Aspergillus* sp. was developed, it is an active and recyclable biocatalyst which catalyses the synthesis of (S)-aminoacids from N-acetyl amino acids via hydrolysis (Bode et al. 2003).

Lipase-CLEA has gained a lot of attention and it has been prepared from different sources like *Candida antarctica* B (CaLB CLEA) and *Candida rugosa*. CaLB CLEA was found to have high specific activity and it exhibits activities in supercritical carbon dioxide and ionic liquids and was also used to synthesize ethyl oleate from oleic acid and ethanol via esterification (Hobbs et al. 2006; Toral et al. 2007). Ethyl oleate finds variety of application in cosmetic and food additives industries, production of tailored triglycerides, in diesel fuel additives, in pharmaceutical industry as a solvent for lipophilic drugs such as esteroids and as a stationary phase in gaseous chromatography. On the other hand, *Candida rugosa* lipase CLEAs showed a two-fold increase in enantioselectivity in the kinetic resolution of racemic ibuprofen when esterification is done with 1-propanol (Yu et al. 2006). Similarly, lipase-CLEA has been prepared from *Thermomyces lanuginosa* with high stability. They can be used for hydrolysis of olive oil in isopropyl alcohol (Gupta et al. 2009). CLEA of lipase from *Pseudomonas* sp. (PSL-CLEA) with high activity and thermal stability were also reported. It was used successfully for the resolution of N-(2-ethyl-6-methylphenyl) alanine while at the same time maintaining an excellent enantioselectivity. PSL-CLEA was reused upto 10 cycles with minimal loss of enzyme activity (Zhao et al. 2008).

CLEAs from feruloyl esterase and acetylxylan esterase have been prepared and it plays essential role in the degradation of lignocellulose and is commercially important in biomass conversion processes (Vafiadi et al. 2008).

6.2 Polymers Synthesis

Synthetic polyamides found extensive use in industries due to their properties like high mechanical strength and good thermal resistance. But due to high melting temperature involved in its synthesis, the syntheses of polyamides are very difficult (Yamaguchi et al. 2018). Cutinase-CLEAs were developed which is involved in the synthesis of oligoamides having aliphatic–aromatic structures. It not only overcame the difficulties but also shown to have highest degree of polymerization via one-step or two-step synthesis of oligo (p-xylylene sebacamide) (Stavila et al. 2013).

6.3 Pharmaceuticals Chemicals

(−)-γ-Lactam have antiviral activity and play an important role in the synthesis of carbocyclic nucleosides. They inhibit the mode of action of viral reverse transcriptase. (+)-γ-Lactamase is the enzyme which facilitates the conversion of (+)-γ-lactam only and thus leaving (−)-γ-lactam in the racemic mixture. This (−)-γ-lactam can be recovered easily from the mixture. An efficient method has been reported in which CLEA-based enzyme microreactor (CEM) with low back pressure has been developed from cross-linked γ-lactamase. The developed CLEA retained 100% of its activity but after 10 h the activity got reduced to 52% (Hickey et al. 2009).

α-Cyanohydrins is the key molecule and has been extensively used in pharmaceuticals and agrochemical industries. Hydroxynitrile lyase has been obtained from *Davallia tyermannii* (DtHNL) which causes the enantioselective production of α-cyanohydrins. DtHNL-CLEAs were prepared with high increased stability and was used for the synthesis of (R)-mandelonitrile in a biphasic system as well as synthesis of (R)-hydroxypivaldehyde cyanohydrin (Lanfranchi et al. 2018).

Tyrosinase is an essential enzyme responsible for the synthesis of L-3,4-Dihydroxyphenylalanine (L-DOPA) from tyrosine and L-DOPA is used to treat Parkinson's disease. The CLEA of tyrosinase were prepared using ammonium sulphate as precipitating agent (Aytar and Bakir 2008). After 2 h of bioconversion process, 53% of the yield of L-DOPA was achieved with a productivity of 209 mg L^{-1} h^{-1}. After entrapping the tyrosinase-CLEAs in alginate beads, operational and mechanical stability was enhanced with a productivity of 57 mg L^{-1} h^{-1} and a long lifetime of >104 h (Xu et al. 2012).

Penicillin G amidase CLEA was developed with high activity in organic solvents and it is an excellent biocatalyst for the synthesis of cephalosporin and penicillin antibiotics (Wegman et al. 2001). Using *Rhodococcus erythropolis* source, amidase CLEA was successfully prepared. It gives rise to a pharmaceutical intermediate from prochiral diamide via enantioslective hydrolysis (Park et al. 2010). Lipase-CLEA- and alcalase-CLEA has been reported which are used in ester and peptide synthesis respectively in a single pot reaction method (Nuijens et al. 2011).

6.4 Nanoparticles Synthesis

Silver nanoparticles have been gaining popularity among researchers since few years and became the focus of much research. It has characteristics features like high catalytic efficiency, antimicrobial and optical properties. Synthesis of silver nanoparticles from conventional method is a tedious work that involves toxic solvents and additives which limits its applicability (Yamaguchi et al. 2018). Recently, nicotinamide adenine dinucleotide (NADH)-dependent nitrate reductase-CLEAs was reported for silver nanoparticles synthesis using silver nitrate. The CLEAs showed a catalytic efficiency of 80% even after many cycle of reusability with high stability (Talekar et al. 2016). CLEAs and combi-CLEAs of other enzymes can be exploited in future for the synthesis of different nanoparticles.

6.5 Food Industry

Aspergillus oryzae β-galactosidase CLEA was successfully prepared and used for the synthesis of galacto-oligosaccharides, which is a prebiotic food ingredients (Gaur et al. 2006). Similarly, levansucrase CLEA from *Bacillus subtilis* was prepared and proved to be an effective biocatalyst in synthesis of oligofructosides via trans-fructosylation. Oligofructosides have nutraceutical properties (Ortiz-Soto et al. 2009).

6.6 Elimination of Toxic Chemicals & Wastewater Treatment

Endocrine-disrupting chemicals (EDC) are found in wastewater. EDCs are toxic substances which are harmful to human body and disrupt the function of endocrine system and causes adverse effects in an organism. Enzymes of class oxidoreductases like oxidases and peroxidases are known to degrade EDC (Yamaguchi et al. 2018). Laccase-CLEA was reported which has a wide application in bioremediation of wastewater. The bioremediation process was performed in perfusion basket reactor and from urban wastewater it removed chemicals such as bisphenol A, 4-nonyl phenol, and triclosan (Cabana et al. 2007). Similarly, CLEA of tyrosinase was developed and showed similar property as laccase-CLEA. Tyrosinase is a copper-containing oxidase enzyme which causes removal of phenolic compounds from wastewater (Xu and Yang 2013). Peroxidase CLEA from *Roystonia regia* were reported with very high thermal stability and reusability and was used in the decolorization of wastewater containing azo dyes (Grateron et al. 2007). A combi-CLEA of peroxidase and glucose oxidase was designed which targeted the removal of EDC such as bisphenol A, nonylphenol, triclosan, 17 α–ethinylestradiol and 17 β-estradiol efficiently from the wastewater (Taboada-Puig et al. 2011).

6.7 Decolorization and Detoxification of Dyes

Application of enzymes in decolorization of dyes is widely known (Kulshrestha and Husain 2007; Husain and Husain 2012; Satar et al. 2012; Husain 2017a). Laccase is a well-known enzyme which has been exploited in textile, dyeing, printing and related industries. Magnetic-CLEAs were prepared using laccase enzyme. They exhibited good thermal and operational stabilities and showed catalytic degradation of about 61–96% of synthetic dyes such as Remazol brilliant blue R, malachite green, and Reactive Black 5 (Kumar et al. 2014). Horseradish peroxidase CLEA was prepared using *Armoracia rusticana* and their catalytic efficiency in bioremediation process was investigated. It was involved in 94.26% degradation of textile-based methyl orange dye, 91.73% of basic red 9, 84.35% of indigo, 81.47% of rhodamine B and 73.6% of rhodamine 6G. It was highly stable and retained 60% of its activity after seven cycles of methyl orange degradation (Bilal et al. 2017).

In another study, laccase CLEA from two different sources, *Fomes fomentarius* and *Trametes versicolor* was prepared and their decolourisation activity was studied on dyes like malachite green, bromothymol blue, and methyl red. 95 and 90% decolourisation potential for all dyes were shown by CLEA from *Trametes versicolor* and *Fomes fomentarius* respectively (Vrsanska et al. 2017).

6.8 Biodiesel Production

Biodiesel, a renewable carbon resource, is an interesting replacement to petroleum fuels as the non-renewable natural resources, fossil fuels, is depleting fastly in the present time. Oil crops are potential sources for biodiesel production. They are synthesized via direct use and blending, micro emulsions, thermal cracking (pyrolysis) and trans-esterification. Among various methods of preparation, trans-esterification of vegetable oils and animal fats are more commonly used (Mishra and Goswami 2018). Many researchers have taken interest in this field and successfully produced biodiesel from CLEAs of various enzymes. CLEA of lipase was prepared in three-dimensionally ordered macro porous silica. They showed high thermal as well as mechanical stability along with reusability. CLEA-lipase was found to be applicable in esterification and trans-esterification reactions. They exhibited trans-esterification of Jatropha oil with ethanol and resulted in the synthesis of fatty acid ethyl esters (FAEEs). FAEE is known to be the main components of biodiesel (Jiang et al. 2014). Similarly, lipase from *Penicillium expansum*, was used to develop lipase-CLEA. This paper reported high production of biodiesel from corn oil and microalgal oil using CLEA-lipase as a biocatalyst. It was found that the biodiesel was produced more in presence of ionic liquids as compared to organic solvent tert-butanol (Lai et al. 2012). Magnetic-CLEA of lipase has also been prepared using *Candida antarctica* and used as a robust biocatalyst to produce biodiesel from non-edible vegetable (such as unrefined soybean, jatropha, cameline) and waster frying oils (Cruz-Izquierdo et al. 2014). High conversion of vegetable oil such as olive oil via hydrolysis in isopropyl alcohol, was shown by Gupta et al. (2009), for the production of biodiesel. The lipase-CLEA was obtained by using *Thermomyces lanuginosus* as a source (Gupta et al. 2009). Similarily, lipase obtained from *Burkholderia cepacia* immobilised on iron oxideMNPs completely converted waste cooking oil to biodiesel (Karimi 2016). Using *Madhuca indica* oil, biodiesel was obtained in the form of fatty acid ethyl esters via trans-esterification of triglycerides from *Burkholderia cepacia* lipase CLEA (Kumari et al. 2007). The studies indicate that the CLEA can be efficiently used for biodiesel production.

7 Conclusion and Future Perspectives

Enzymes are promising industrial biocatalysts; however, two major drawbacks limit their applications; one is stability and second is their reusability. To overcome these limitations immobilization of enzymes has been investigated. A cross-linked enzyme aggregate (CLEAs) is a simple immobilization technique (a carrier free) to increase the stability and reusability of biocatalyst. The procedure for CLEAs preparation is simple, inexpensive, and does not require enzymes with high purity for the cross-linking. Basically, CLEA combines two steps purification and immobilization. CLEAs have improved performances in terms of activity, thermal and storage stability. They have a significantly enhanced shelf life and operational stability, are easy to recover and reuse, and are completely stable towards leaching in aqueous media. The main disadvantage of CLEAs is that they have poor mechanical stability, thus unsuitable for large reactors and difficult to separate from the reaction medium. In order to solve these problems, some strategies to improve CLEAs were developed. Addition of polymers and protein feeder during CLEA formation results in the improvement of enzyme recovery and stability. The use of magnetic support in CLEAs provides mechanical stability and also solves the separation problem. Cross-linking of two or more enzymes forms combi-CLEAs that are ideal for catalysing enzymatic cascade processes. Activity of CLEA depends on several factors such as precipitating agent, additive, cross-linker, cross-linking time, enzyme concentration, temperature, pH and agitation. Future studies should be aimed to improve CLEA's stability, performance in industrial processes and finding new additives and cross-linking agents so they can be widely applied in future in industrial biotransformations/bioconversions.

Glossary

°C Degree Celsius
APTES 3-aminopropyltriethoxysilane
BSA Bovine serum albumin
CaLBCLEA *Candida antarctica* B Cross-linked enzyme aggregates
CEM CLEA-based enzyme microreactor
CLE Cross linked enzymes
CLEAs Cross-linked enzyme aggregates
CLEC Cross linked enzyme crystals
DB-38 Direct Black-38
DtHNL-CLEAs *Davallia tyermannii* Hydroxynitrile lyase Cross-linked enzyme aggregates
EDC Endocrine disrupting chemicals
FAEEs Fatty acid ethyl esters
FAPEs Fatty acid propyl esters
GO Glucose oxidase
HRP Horseradish peroxidase
L-DOPA 3,4-Dihydroxyphenylalanine
m-CLEAs MagneticCross-linked enzyme aggregates
Mg/L/hr. Milligram/litre/hour
mM Millimolar
MNPs Magnetic nanoparticles
NAD Nicotinamide adenine dinucleotide
PAL-CLEAs Phenylalanine ammonia lyase- cross-linked enzyme aggregates
PSL-CLEA *Pseudomonassp.* Lipase Cross-linked enzyme aggregates
SDS Sodium dodecyl sulphate
SEM Scanning Electron Microscopy

References

Ahumada K, Urrutia P, Illanes A, Wilson L (2015) Production of combi-CLEAs of glycosidases utilized for aroma enhancement in wine. Food Bioprod Process 94:555–560

Amotz S (1987) Method for production of an immobilized enzyme preparation by means of a crosslinking agent. Google Patents

Arsenault A, Cabana H, Jones JP (2011) Laccase-based CLEAs: chitosan as a novel cross-linking agent. Enzyme Res 2011:1–10

Aytar BS, Bakir U (2008) Preparation of cross-linked tyrosinase aggregates. Process Biochem 43:125–131

Barbosa O, Ortiz C, Berenguer-Murcia A, Torres R, Rodrigues RC, Fernandez-Lafuente R (2014) Glutaraldehyde in bio-catalysts design: a useful crosslinker and a versatile tool in enzyme immobilization. RSC Adv 4:1583–1600

Bhattacharya A, Pletschke BI (2014) Magnetic cross-linked enzyme aggregates (CLEAs): a novel concept towards carrier free immobilization of lignocellulolytic enzymes. Enzym Microb Technol 61:17–27

Bilal M, Iqbal HM, Hu H, Wang W, Zhang X (2017) Development of horseradish peroxidase-based cross-linked enzyme aggregates and their environmental exploitation for bioremediation purposes. J Environ Manag 188:137–143

Bode ML, van Rantwijk F, Sheldon RA (2003) Crude aminoacylase from *Aspergillus* sp. is a mixture of hydrolases. Biotechnol Bioeng 84:710–713

Boller T, Meier C, Menzler S (2002) EUPERGIT oxirane acrylic beads: how to make enzymes fit for biocatalysis. Org Process Res Dev 6:509–519

Brady D, Steenkamp L, Skein E, Chaplin JA, Reddy S (2004) Optimisation of the enantioselective biocatalytic hydrolysis of naproxen ethyl ester using ChiroCLEC-CR. Enzym Microb Technol 34:283–291

Brena BM, Batista-Viera F (2006) Immobilization of enzymes. In: Guisan JM (ed) Immobilization of enzymes and cells, vol 22. Springer, Totowa, NJ, pp 15–30

Brena B, González-Pombo P, Batista-Viera F (2013) Immobilization of enzymes: a literature survey. In: Guisan JM (ed) Immobilization of enzymes and cells, vol 1051. Springer, Totowa, NJ, pp 15–31

Cabana H, Jones JP, Agathos SN (2007) Preparation and characterization of cross-linked laccase aggregates and their application to the elimination of endocrine disrupting chemicals. J Biotechnol 132:23–31

Cao L (2006) Carrier-bound immobilized enzymes: principles, application and design. Wiley, Weinheim

Cao L, Van Langen L, Janssen M, Sheldon R (1999) Preparation and properties of cross-linked aggregates of penicillin acylase and other enzymes. European Patent Application number EP1088887A1

Cao L, van Rantwijk F, Sheldon RA (2000) Cross-linked enzyme aggregates: a simple and effective method for the immobilization of penicillin acylase. Org Lett 2:1361–1364

Cao L, van Langen L, Sheldon RA (2003) Immobilised enzymes: carrier-bound or carrier-free? Curr Opin Biotechnol 14:387–394

Chen Z, Wang Y, Liu W, Wang J, Chen H (2017) A novel cross-linked enzyme aggregates (CLEAs) of papain and neutrase-production, partial characterization and application. Int J Biol Macromol 95:650–657

Cruz-Izquierdo A, Pico EA, Lopez C, Serra JL, Llama MJ (2014) Magnetic cross-linked enzyme aggregates (mCLEAs) of Candida antarctica lipase: an efficient and stable biocatalyst for biodiesel synthesis. PLoS One 9:e115202

Cui JD, Jia SR (2015) Optimization protocols and improved strategies of cross-linked enzyme aggregates technology: current development and future challenges. Crit Rev Biotechnol 35:15–28

Dal Magro L, Hertz PF, Fernandez-Lafuente R, Klein MP, Rodrigues RC (2016) Preparation and characterization of a Combi-CLEAs from pectinases and cellulases: a potential biocatalyst for grape juice clarification. RSC Adv 6:27242–27251

Dalal S, Kapoor M, Gupta MN (2007a) Preparation and characterization of combi-CLEAs catalyzing multiple non-cascade reactions. J Mol Catal B Enzym 44:128–132

Dalal S, Sharma A, Gupta MN (2007b) A multipurpose immobilized biocatalyst with pectinase, xylanase and cellulase activities. Chem Cent J 1:16

Dong Cui J, Li Cui L, Ping Zhang S, Fei Zhang Y, Guo Su Z, Hui Ma G (2014) Hybrid magnetic cross-linked enzyme aggregates of phenylalanine ammonia lyase from *Rhodotorula glutinis*. PLoS One 9:e97221

Easa MNB, Yusof FB (2006) Optimizing the preparation of Cross-Linked Enzyme Aggregates (Clea)-Amylase from supermeal worm (*Zophobas morio*). ARPN J Eng Appl Sci 10:9710–9716

Fazary AE, Ismadji S, Ju Y-H (2009) Biochemical studies on native and cross-linked aggregates of Aspergillus awamori feruloyl esterase. Int J Biol Macromol 44:240–248

Gaur R, Pant H, Jain R, Khare S (2006) Galacto-oligosaccharide synthesis by immobilized *Aspergillus oryzae* β-galactosidase. Food Chem 97:426–430

Gogoi P, Hazarika S, Dutta NN, Rao PG (2010) Kinetics and mechanism on laccase catalyzed synthesis of poly (allylamine)–catechin conjugate. Chem Eng J 163:86–92

Grateron C, Barbosa O, Rueda N, Ortiz-Lopez C, Torres R (2007) Azo dye decolorization by optimized cross linked enzyme aggregates (CLEAs) of a royal palm (*Roystonea regia*) peroxidase. J Biotechnol 2:S87

Guo J, Ge L, Li X, Mu C, Li D (2014) Periodate oxidation of xanthan gum and its crosslinking effects on gelatin-based edible films. Food Hydrocoll 39:243–250

Gupta P, Dutt K, Misra S, Raghuwanshi S, Saxena R (2009) Characterization of cross-linked immobilized lipase from thermophilic mould Thermomyces lanuginosa using glutaraldehyde. Bioresour Technol 100:4074–4076

Haider T, Husain Q (2007) An inexpensive and highly stable calcium alginate entrapped preparation of *Aspergillus oryzae* β galactosidase. Int J Biol Macromol 41(1):72–80

Haider T, Husain Q (2008) Concanavalin A layered calcium alginate-starch beads immobilized β galactosidase as a therapeutic agent for lactose intolerant patients. Int J Pharm 359(1–2):1–6

Haring D, Schreier P (1999) Cross-linked enzyme crystals. Curr Opin Chem Biol 3:35–38

Hickey AM, Ngamsom B, Wiles C, Greenway GM, Watts P, Littlechild JA (2009) A microreactor for the study of biotransformations by a cross-linked γ-lactamase enzyme. Biotechnol J 4:510–516

Hobbs HR, Kondor B, Stephenson P, Sheldon RA, Thomas NR, Poliakoff M (2006) Continuous kinetic resolution catalysed by cross-linked enzyme aggregates, 'CLEAs', in supercritical CO_2. Green Chem 8:816–821

Honda T, Miyazaki M, Nakamura H, Maeda H (2006) Facile preparation of an enzyme-immobilized microreactor using a cross-linking enzyme membrane on a microchannel surface. Adv Synth Catal 348:2163–2171

Hormigo D, García-Hidalgo J, Acebal C, de la Mata I, Arroyo M (2012) Preparation and characterization of cross-linked enzyme aggregates (CLEAs) of recombinant poly-3-hydroxybutyrate depolymerase from Streptomyces exfoliatus. Bioresour Technol 115:177–182

Hu B, Pan J, Yu H-L, Liu J-W, Xu J-H (2009) Immobilization of Serratia marcescens lipase onto amino-functionalized magnetic nanoparticles for repeated use in enzymatic synthesis of Diltiazem intermediate. Process Biochem 44:1019–1024

Husain Q (2010) β Galactosidases and their potential applications: a review. Crit Rev Biotechnol 30 (1):41–62

Husain Q (2016) Magnetic nanoparticles as a tool for the immobilization/stabilization of hydrolases and their applications: an overview. Biointerf Res Appl Chem 6(6):1585–1606

Husain Q (2017a) Nanomaterials as novel supports for the immobilization of amylolytic enzymes and their applications: a review. Biocatalysis 3:37–53

Husain Q (2017b) Nanomaterials immobilized cellulolytic enzymes and their industrial applications. JSM Biochem Mol Biol 4:1029

Husain Q, Husain M (2012) Peroxidases as a potential tool for the decolorization and removal of synthetic dyes from polluted water. In: Malik A, Grohmann E (eds) Environmental protection strategies for sustainable development. Springer, Berlin, pp 453–498

Husain Q, Ulber R (2011) Immobilized peroxidase as a valuable tool in the remediation of aromatic pollutants and xenobiotic compounds: a review. Crit Rev Environ Sci Technol 41(8):770–804

Jafari Khorshidi K, Lenjannezhadian H, Jamalan M, Zeinali M (2016) Preparation and characterization of nanomagnetic cross-linked cellulase aggregates for cellulose bioconversion. J Chem Technol Biotechnol 91:539–546

Jiang Y, Shi L, Huang Y, Gao J, Zhang X, Zhou L (2014) Preparation of robust biocatalyst based on cross-linked enzyme aggregates entrapped in three-dimensionally ordered macroporous silica. ACS Appl Mater Interfaces 6:2622–2628

Ju H, Jang E, Ryu BH, Kim TD (2013) Characterization and preparation of highly stable aggregates of a novel type of hydrolase (BL28) from *Bacillus licheniformis*. Bioresour Technol 128:81–86

Jung D-H, Jung J-H, Seo D-H, Ha S-J, Kweon D-K, Park C-S (2013) One-pot bioconversion of sucrose to trehalose using enzymatic sequential reactions in combined cross-linked enzyme aggregates. Bioresour Technol 130:801–804

Karimi M (2016) Immobilization of lipase onto mesoporous magnetic nanoparticles for enzymatic synthesis of biodiesel. Biocatal Agric Biotechnol 8:182–188

Karimpil JJ, Melo J, D'Souza S (2011) Hen egg white as a feeder protein for lipase immobilization. J Mol Catal B Enzym 71:113–118

Kartal F, Janssen MH, Hollmann F, Sheldon RA, Kılınc A (2011) Improved esterification activity of Candida rugosa lipase in organic solvent by immobilization as cross-linked enzyme aggregates (CLEAs). J Mol Catal B Enzym 71:85–89

Kim M et al (2010) Immobilization of cross-linked lipase aggregates onto magnetic beads for enzymatic degradation of polycaprolactone. J Basic Microbiol 50:218–226

Kulshrestha Y, Husain Q (2006) Adsorption of peroxidases on DEAE-cellulose directly from ammonium sulphate fractionated proteins of bitter gourd. Enzym Microb Technol 38:470–477

Kulshrestha Y, Husain Q (2007) Decolorization and degradation of acid dyes mediated by salt fractionated turnip (*Brassica rapa*) peroxidases. Toxicol Environ Chem 89:255–267

Kumar VV, Kumar MP, Thiruvenkadaravi K, Baskaralingam P, Kumar PS, Sivanesan S (2012) Preparation and characterization of porous cross linked laccase aggregates for the decolorization of triphenyl methane and reactive dyes. Bioresour Technol 119:28–34

Kumar VV, Sivanesan S, Cabana H (2014) Magnetic cross-linked laccase aggregates—bioremediation tool for decolorization of distinct classes of recalcitrant dyes. Sci Total Environ 487:830–839

Kumari V, Shah S, Gupta MN (2007) Preparation of biodiesel by lipase-catalyzed transesterification of high free fatty acid containing oil from *Madhuca indica*. Energy Fuel 21:368–372

Lai J-Q, Hu Z-L, Sheldon RA, Yang Z (2012) Catalytic performance of cross-linked enzyme aggregates of Penicillium expansum lipase and their use as catalyst for biodiesel production. Process Biochem 47:2058–2063

Lanfranchi E et al (2018) Production of hydroxynitrile lyase from Davallia tyermannii (DtHNL) in Komagataella phaffii and its immobilization as a CLEA to generate a robust biocatalyst. ChemBioChem 19:312–316

Laurent S, Forge D, Port M, Roch A, Robic C, Vander Elst L, Muller RN (2008) Magnetic iron oxide nanoparticles: synthesis, stabilization, vectorization, physicochemical characterizations, and biological applications. Chem Rev 108:2064–2110

Li S, Su Y, Liu Y, Sun L, Yu M, Wu Y (2016) Preparation and characterization of cross-linked enzyme aggregates (CLEAs) of recombinant thermostable alkylsulfatase (SdsAP) from Pseudomonas sp. S9. Process Biochem 51:2084–2089

Lopez-Gallego F, Betancor L, Hidalgo A, Alonso N, Fernandez-Lafuente R, Guisan JM (2005) Co-aggregation of enzymes and polyethyleneimine: a simple method to prepare stable and immobilized derivatives of glutaryl acylase. Biomacromolecules 6:1839–1842

Lu AH, Salabas EL, Schuth F (2007) Magnetic nanoparticles: synthesis, protection, functionalization, and application. Angew Chem Int Ed 46:1222–1244

Majumder AB, Mondal K, Singh TP, Gupta MN (2008) Designing cross-linked lipase aggregates for optimum performance as biocatalysts. Biocatal Biotransformation 26:235–242

Margolin AL, Navia MA (2001) Protein crystals as novel catalytic materials. Angew Chem Int Ed 40:2204–2222

Mateo C, Palomo JM, Van Langen LM, Van Rantwijk F, Sheldon RA (2004) A new, mild cross-linking methodology to prepare cross-linked enzyme aggregates. Biotechnol Bioeng 86:273–276

Mateo C et al (2007a) Advances in the design of new epoxy supports for enzyme immobilization–stabilization. Biochem Soc Trans 35:1593–1601

Mateo C, Palomo JM, Fernandez-Lorente G, Guisan JM, Fernandez-Lafuente R (2007b) Improvement of enzyme activity, stability and selectivity via immobilization techniques. Enzym Microb Technol 40:1451–1463

Matsumoto M, Ohashi K (2003) Effect of immobilization on thermostability of lipase from *Candida rugosa*. Biochem Eng J 14:75–77

Matto M, Husain Q (2009) A novel calcium alginate-starch hybrid support for both surface immobilization and entrapment of bitter gourd (*Momordica charantia*) peroxidase. J Mol Catal B Enzym 57(1–4):164–170

Migneault I, Dartiguenave C, Bertrand MJ, Waldron KC (2004) Glutaraldehyde: behavior in aqueous solution, reaction with proteins, and application to enzyme crosslinking. BioTechniques 37:790–802

Mishra VK, Goswami R (2018) A review of production, properties and advantages of biodiesel. Biofuels 9:273–289

Mohamad NR, Marzuki NHC, Buang NA, Huyop F, Wahab RA (2015) An overview of technologies for immobilization of enzymes and surface analysis techniques for immobilized enzymes. Biotechnol Biotechnol Equip 29:205–220

Montoro-García S, Gil-Ortiz F, Navarro-Fernandez J, Rubio V, García-Carmona F, Sanchez-Ferrer A (2010) Improved cross-linked enzyme aggregates for the production of desacetyl β-lactam antibiotics intermediates. Bioresour Technol 101:331–336

Musthapa S, Akhtar S, Khan AA, Husain Q (2004) An economical, simple and high yield procedure for the immobilization/stabilization of peroxidases from turnip roots. J Sci Ind Res 63:540–547

Nadar SS, Rathod VK (2016) Magnetic macromolecular cross linked enzyme aggregates (CLEAs) of glucoamylase. Enzym Microb Technol 83:78–87

Nadar SS, Muley AB, Ladole MR, Joshi PU (2016) Macromolecular cross-linked enzyme aggregates (M-CLEAs) of α-amylase. Int J Biol Macromol 84:69–78

Netto CG, Toma HE, Andrade LH (2013) Superparamagnetic nanoparticles as versatile carriers and supporting materials for enzymes. J Mol Catal B Enzym 85:71–92

Nuijens T, Cusan C, Kruijtzer JA, Rijkers DT, Liskamp RM, Quaedflieg PJ (2009a) Enzymatic synthesis of C-terminal arylamides of amino acids and peptides. J Org Chem 74:5145–5150

Nuijens T, Kruijtzer JA, Cusan C, Rijkers DT, Liskamp RM, Quaedflieg PJ (2009b) A versatile and selective chemo-enzymatic synthesis of β-protected aspartic and γ-protected glutamic acid derivatives. Tetrahedron Lett 50:2719–2721

Nuijens T, Cusan C, Schepers AC, Kruijtzer JA, Rijkers DT, Liskamp RM, Quaedflieg PJ (2011) Enzymatic synthesis of activated esters and their subsequent use in enzyme-based peptide synthesis. J Mol Catal B Enzym 71:79–84

Ortiz-Soto ME, Rudino-Pinera E, Rodriguez-Alegria ME, Munguia AL (2009) Evaluation of cross-linked aggregates from purified Bacillus subtilis levansucrase mutants for transfructosylation reactions. BMC Biotechnol 9:68

Park HJ, Uhm K-N, Kim H-K (2010) Biotransformation of amides to acids using a co-cross-linked enzyme aggregate of Rhodococcus erythropolis amidase. J Microbiol Biotechnol 20:325–331
Pchelintsev N, Youshko M, Svedas V (2009) Quantitative characteristic of the catalytic properties and microstructure of cross-linked enzyme aggregates of penicillin acylase. J Mol Catal B Enzym 56:202–207
Perwez M, Ahmad R, Sardar M (2017) A reusable multipurpose magnetic nanobiocatalyst for industrial applications. Int J Biol Macromol 103:16–24
Prabhavathi Devi B, Guo Z, Xu X (2009) Characterization of cross-linked lipase aggregates. J Am Oil Chem Soc 86:637–642
Quiocho FA, Richards FM (1964) Intermolecular cross linking of a protein in the crystalline state: carboxypeptidase. Proc Natl Acad Sci 52:833–839
Quiocho FA, Richards FM (1966) The enzymic behavior of carboxypeptidase-A in the solid state. Biochemistry 5:4062–4076
Rehman S, Bhatti HN, Bilal M, Asgher M (2016) Cross-linked enzyme aggregates (CLEAs) of Pencilluim notatum lipase enzyme with improved activity, stability and reusability characteristics. Int J Biol Macromol 91:1161–1169
Reshmi R, Sugunan S (2013) Improved biochemical characteristics of crosslinked β-glucosidase on nanoporous silica foams. J Mol Catal B Enzym 85:111–118
Roberge C, Amos D, Pollard D, Devine P (2009) Preparation and application of cross-linked aggregates of chloroperoxidase with enhanced hydrogen peroxide tolerance. J Mol Catal B Enzym 56:41–45
Roessl U, Nahálka J, Nidetzky B (2010) Carrier-free immobilized enzymes for biocatalysis. Biotechnol Lett 32:341–350
Sangeetha K, Abraham TE (2008) Preparation and characterization of cross-linked enzyme aggregates (CLEA) of subtilisin for controlled release applications. Int J Biol Macromol 43:314–319
Satar R, Husain Q (2009) Adsorption of peroxidase on Celite 545 directly from ammonium sulphate fractionated white radish (*Raphanus sativus*) proteins. Biotechnol J 4:408–416
Satar R, Jerah AAB, Husain Q (2012) Role of redox mediators in enhancing the decolorization and remediation of textile dyes by enzymes. In: Textiles: types, uses and production methods. Nova Science, New York
Schoevaart R et al (2004) Preparation, optimization, and structures of cross-linked enzyme aggregates (CLEAs). Biotechnol Bioeng 87:754–762
Seow N, Yang K-L (2017) Hollow cross-linked enzyme aggregates (h-CLEA) of laccase with high uniformity and activity. Colloids Surf B: Biointerfaces 151:88–94
Shah S, Sharma A, Gupta MN (2006) Preparation of cross-linked enzyme aggregates by using bovine serum albumin as a proteic feeder. Anal Biochem 351:207–213
Sheldon R (2007a) Cross-linked enzyme aggregates (CLEA® s): stable and recyclable biocatalysts. Biochem Soc Trans 35:1583–1587
Sheldon RA (2007b) Enzyme immobilization: the quest for optimum performance. Adv Synth Catal 349:1289–1307
Sheldon RA (2011) Characteristic features and biotechnological applications of cross-linked enzyme aggregates (CLEAs). Appl Microbiol Biotechnol 92:467–477
Sheldon RA, van Pelt S (2013) Enzyme immobilisation in biocatalysis: why, what and how. Chem Soc Rev 42:6223–6235
Sheldon R, Schoevaart R, Van Langen L (2005) Cross-linked enzyme aggregates (CLEAs): a novel and versatile method for enzyme immobilization (a review). Biocatal Biotransformation 23:141–147
Sinirlioglu ZA, Sinirlioglu D, Akbas F (2013) Preparation and characterization of stable cross-linked enzyme aggregates of novel laccase enzyme from Shewanella putrefaciens and using malachite green decolorization. Bioresour Technol 146:807–811
Stavila E, Alberda van Ekenstein G, Loos K (2013) Enzyme-catalyzed synthesis of aliphatic–aromatic oligoamides. Biomacromolecules 14:1600–1606

Šulek F, Fernandez DP, Knez Z, Habulin M, Sheldon RA (2011) Immobilization of horseradish peroxidase as crosslinked enzyme aggregates (CLEAs). Process Biochem 46:765–769
Taboada-Puig R, Junghanns C, Demarche P, Moreira M, Feijoo G, Lema J, Agathos SN (2011) Combined cross-linked enzyme aggregates from versatile peroxidase and glucose oxidase: production, partial characterization and application for the elimination of endocrine disruptors. Bioresour Technol 102:6593–6599
Takkinen K, Pettersson RF, Kalkkinen N, Palva I, Soderlund H, Kaariainen L (1983) Amino acid sequence of alpha-amylase from Bacillus amyloliquefaciens deduced from the nucleotide sequence of the cloned gene. J Biol Chem 258:1007–1013
Talekar S, Ghodake V, Kate A, Samant N, Kumar C, Gadagkar S (2010) Preparation and characterization of cross-linked enzyme aggregates of Saccharomyces cerevisiae invertase. Aust J Basic Appl Sci 4:4760–4765
Talekar S et al (2012a) Novel magnetic cross-linked enzyme aggregates (magnetic CLEAs) of alpha amylase. Bioresour Technol 123:542–547
Talekar S, Shah V, Patil S, Nimbalkar M (2012b) Porous cross linked enzyme aggregates (p-CLEAs) of Saccharomyces cerevisiae invertase. Cat Sci Technol 2:1575–1579
Talekar S, Waingade S, Gaikwad V, Patil S, Nagavekar N (2012c) Preparation and characterization of cross linked enzyme aggregates (CLEAs) of Bacillus amyloliquefaciens alpha amylase. J Biochem Technol 3:349–353
Talekar S et al (2013a) Carrier free co-immobilization of glucoamylase and pullulanase as combi-cross linked enzyme aggregates (combi-CLEAs). RSC Adv 3:2265–2271
Talekar S, Joshi A, Joshi G, Kamat P, Haripurkar R, Kambale S (2013b) Parameters in preparation and characterization of cross linked enzyme aggregates (CLEAs). RSC Adv 3:12485–12511
Talekar S et al (2014) Preparation of stable cross-linked enzyme aggregates (CLEAs) of NADH-dependent nitrate reductase and its use for silver nanoparticle synthesis from silver nitrate. Catal Commun 53:62–66
Talekar S, Joshi A, Chougle R, Nakhe A, Bhojwani R (2016) Immobilized enzyme mediated synthesis of silver nanoparticles using cross-linked enzyme aggregates (CLEAs) of NADH-dependent nitrate reductase. Nano Struct Nano Obj 6:23–33
Tischer W, Wedekind F (1999) Immobilized enzymes: methods and applications. In: Fessner W-D (ed) Biocatalysis-from discovery to application, vol 200. Springer, Berlin, pp 95–126
Topcular C, Ayhan H (2008) Carrier free cross-linked peroxidase aggregates: synthesis and characterization hacettepe. J Biol Chem 36:255–261
Toral AR, Antonia P, Hernandez FJ, Janssen MH, Schoevaart R, van Rantwijk F, Sheldon RA (2007) Cross-linked Candida antarctica lipase B is active in denaturing ionic liquids. Enzym Microb Technol 40:1095–1099
Torres MG, Foresti M, Ferreira M (2014) CLEAs of Candida antarctica lipase B (CALB) with a bovine serum albumin (BSA) cofeeder core: study of their catalytic activity. Biochem Eng J 90:36–43
Tudorache M, Nae A, Coman S, Parvulescu VI (2013) Strategy of cross-linked enzyme aggregates onto magnetic particles adapted to the green design of biocatalytic synthesis of glycerol carbonate. RSC Adv 3:4052–4058
Vafiadi C, Topakas E, Christakopoulos P (2008) Preparation of multipurpose cross-linked enzyme aggregates and their application to production of alkyl ferulates. J Mol Catal B Enzym 54:35–41
Vaidya BK, Kuwar SS, Golegaonkar SB, Nene SN (2012) Preparation of cross-linked enzyme aggregates of l-aminoacylase via co-aggregation with polyethyleneimine. J Mol Catal B Enzym 74:184–191
van Rantwijk F, Sheldon RA (2004) Enantioselective acylation of chiral amines catalysed by serine hydrolases. Tetrahedron 60:501–519
Velasco-Lozano S, Lopez-Gallego F, Mateos-Díaz JC, Favela-Torres E (2016) Cross-linked enzyme aggregates (CLEA) in enzyme improvement–a review. Biocatalysis 1:166–177

Verma ML, Barrow CJ, Puri M (2013) Nanobiotechnology as a novel paradigm for enzyme immobilisation and stabilisation with potential applications in biodiesel production. Appl Microbiol Biotechnol 97:23–39

Vrsanska M, Voberkova S, Jimenez Jimenez AM, Strmiska V, Adam V (2017) Preparation and optimisation of cross-linked enzyme aggregates using native isolate white rot fungi trametes versicolor and Fomes fomentarius for the decolourisation of synthetic dyes. Int J Environ Res Public Health 15:23

Wang M, Qi W, Yu Q, Su R, He Z (2010) Cross-linking enzyme aggregates in the macropores of silica gel: a practical and efficient method for enzyme stabilization. Biochem Eng J 52:168–174

Wang A et al (2011a) A facile technique to prepare cross-linked enzyme aggregates using p-benzoquinone as cross-linking agent Korean. J Chem Eng 28:1090–1095

Wang M, Qi W, Jia C, Ren Y, Su R, He Z (2011b) Enhancement of activity of cross-linked enzyme aggregates by a sugar-assisted precipitation strategy: technical development and molecular mechanism. J Biotechnol 156:30–38

Wang S, Zheng D, Yin L, Wang F (2017) Preparation, activity and structure of cross-linked enzyme aggregates (CLEAs) with nanoparticle. Enzym Microb Technol 107:22–31

Wegman MA, Janssen MH, van Rantwijk F, Sheldon RA (2001) Towards biocatalytic synthesis of β-lactam antibiotics. Adv Synth Catal 343:559–576

Wilson L et al (2004a) Cross-linked aggregates of multimeric enzymes: a simple and efficient methodology to stabilize their quaternary structure. Biomacromolecules 5:814–817

Wilson L, Illanes A, Pessela BC, Abian O, Fernandez-Lafuente R, Guisan JM (2004b) Encapsulation of crosslinked penicillin G acylase aggregates in lentikats: evaluation of a novel biocatalyst in organic media. Biotechnol Bioeng 86:558–562

Wilson L, Fernandez-Lorente G, Fernandez-Lafuente R, Illanes A, Guisan JM, Palomo JM (2006) CLEAs of lipases and poly-ionic polymers: a simple way of preparing stable biocatalysts with improved properties. Enzym Microb Technol 39:750–755

Xu D-Y, Yang Z (2013) Cross-linked tyrosinase aggregates for elimination of phenolic compounds from wastewater. Chemosphere 92:391–398

Xu D-Y, Chen J-Y, Yang Z (2012) Use of cross-linked tyrosinase aggregates as catalyst for synthesis of L-DOPA. Biochem Eng J 63:88–94

Yamaguchi H, Miyazaki M, Asanomi Y, Maeda H (2011) Poly-lysine supported cross-linked enzyme aggregates with efficient enzymatic activity and high operational stability. Cat Sci Technol 1:1256–1261

Yamaguchi H, Kiyota Y, Miyazaki M (2018) Techniques for preparation of cross-linked enzyme aggregates and their applications in bioconversions. Catalysts 8:174

Yang K-L (2017) Combined cross-linked enzyme aggregates of horseradish peroxidase and glucose oxidase for catalyzing cascade chemical reactions. Enzym Microb Technol 100:52–59

Yang XE, Zheng P, Ni Y, Sun Z (2012) Highly efficient biosynthesis of sucrose-6-acetate with cross-linked aggregates of Lipozyme TL 100 L. J Biotechnol 161:27–33

Yu H, Chen H, Wang X, Yang Y, Ching C (2006) Cross-linked enzyme aggregates (CLEAs) with controlled particles: application to Candida rugosa lipase. J Mol Catal B Enzym 43:124–127

Zhao L, Zheng L, Gao G, Jia F, Cao S (2008) Resolution of N-(2-ethyl-6-methylphenyl) alanine via cross-linked aggregates of Pseudomonas sp. lipase. J Mol Catal B Enzym 54:7–12

Zhen Q, Wang M, Qi W, Su R, He Z (2013) Preparation of β-mannanase CLEAs using macromolecular cross-linkers. Cat Sci Technol 3:1937–1941

Zhou L, Tang W, Jiang Y, Ma L, He Y, Gao J (2016) Magnetic combined cross-linked enzyme aggregates of horseradish peroxidase and glucose oxidase: an efficient biocatalyst for dye decolourization. RSC Adv 6:90061–90068

Immobilization of α-amylases and Their Analytical Applications

Om Prakash and Saumya Khare

Abbreviations

AFSMNPs	Amino-functionalized silica-coated magnetite nanoparticles
AAM	α-amylase
AgNP	Ag nanoparticles
AgNP/ESM	Ag-nanoparticle-decorated eggshell membrane
Ca^{+2}(Alg.St/PEI/GA)	Calcium (alginate.starch/polyethyleneimine/glutaraldehyde)
CLEA	Crosslinked enzyme aggregate
DEAE	Diethylaminoethylcellulose
EE	Encapsulated enzyme
ESM	Eggshell membrane
GNRs	Gold nanorods
HEMA-GMA	1-3 (hydroxyethylmethacrylate glycidyl methacrylate)
IPDI	Isophorone diisocyanate
LDH	Layered double-hydroxide
LOF	*Luffa operculata* fibre
M-CLEAs	Macromolecular cross-linked enzyme aggregates
MPIA	Magnetic particles immobilization α-amylase
MMIP	Magnetic Molecular Imprinted Polymer
NA	Naringin
NMOs	Nanostructured metal oxides
PPA	Porcine pancreatic α-amylase
PANI	Polyaniline
PEMA	3-D poly(ethylene-alt-maleic anhydride)
POMA	Poly(octadecene-alt-maleic anhydride)
PPyAgNp	Polypyrrole/silver nanocomposite

O. Prakash (✉) · S. Khare
Department of Biochemistry, Institute of Science, Banaras Hindu University, Varanasi, Uttar Pradesh, India

Q. Husain, M. F. Ullah (eds.), *Biocatalysis*,
https://doi.org/10.1007/978-3-030-25023-2_6

PUU	Polyurethane urea
PVA	Poly(vinyl alcohol)
PVA/PAA	Poly vinyl alcohol/poly acrylic acid

1 Introduction

Increasing world's population and industrialization leads to hasty use of toxic chemical catalyst for various industrial processes causing environmental deterioration. To circumvent these problems, chemical catalyst had been replaced with an environment friendly biological catalyst, which possesses high specificity and catalytic activity. This biocatalyst emerged several decades back to handle the non-systematic and incompetent processes in large-scale industries in a sound manner (Khare and Prakash 2017a). The alpha amylases are starch hydrolyzing enzymes having industrial relevance with major application in starch based food industries and detergents, including other sectors viz., textiles, papermaking, biofuel production, waste water treatment, pharmaceutics, and medicinal chemistry etc., (Prakash and Khare 2015a; Husain 2017a). Inspite of diverse industrial applications, the use of free enzyme is often limited. The major limiting factors includes cost intensive process of enzymes production, short lifespan, high sensitivity towards adverse reaction conditions with lack of reusability, operational and storage stability (DiCosimo et al. 2013; Khare and Prakash 2017b). Hence, the stability of enzymes as envisioned for industrial bioprocesses is of major concern considering its economical value. However, to meet these challenges, escalating knowledge about immobilization techniques is the key step that opens the new door for exploitation of enzymes for several analytical applications.

The term "*immobilized enzymes*" was considered for the first time in 1971, during Enzyme Engineering Conference (Hartmeier 1988). Immobilization is the state of art, which means physical confinement of enzyme molecules in a certain space with retention in the activity and ease to reuse continuously (Kragl 1996; Prakash and Khare 2015b). The first immobilized enzyme was "invertase" adsorbed to the surface of charcoal, used to hydrolyze sucrose (Nelson and Griffin 1916). Nonetheless, around the year 1966, the industrial applications of immobilized enzymes started for various purposes (Tosa et al. 1966). Enzyme immobilization often results into some loss of activity in contrast to their free counterparts, making the process highly expensive. However, it imparts enhanced stability and prolonged shelf life with ease to reuse which is worth a major prerequisite for their industrial application (Sheldon and Pereira 2017). This gives an idea to convert these requirements into a powerful toolbox, which is mainly concerned with improving the enzyme performance adopting suitable immobilization techniques. Moreover, the utility of the immobilized enzymes need a high quality of understanding in terms of technical and economical value (DiCosimo et al. 2013). In this context, advent of nanotechnology has added large avenue of interest in utilizing nanoscaffolds with outstanding properties, as efficient matrices for enzyme immobilization (Ansari and Husain

2012). Hence, the present chapter provides in depth knowledge about enzyme immobilization, various strategies, including its pros and cons. Further, the chapter highlights the role of nanotechnology in improving immobilization techniques with special emphasis on recent trends of immobilized alpha amylase on various support material and their promising applications for various industrial purposes.

2 Enzyme Immobilization

The characteristic features of immobilized enzyme depend on the physico-chemical properties of both the enzyme molecules as well as the support matrix (Fig. 1). The ideal support material must possess stability, inertness, physical strength, and potential to enhance the enzyme specificity with regenerability (Tischer and Wedekind 1999). Generally, the matrices used for enzyme immobilization includes organic, inorganic or the organic-inorganic hybrid materials. However, the support material should be insoluble and rigid. This reduces the chances of product inhibition, avoiding nonspecific adsorption and bacterial contamination (Singh 2009). The surface functional groups possess by the support material enable ease for attachment of enzyme under mild reaction conditions. This leads to augmented catalytic activity and stability of immobilized enzyme. Also, the large surface area, porosity and permeability of carrier material influences the enzyme loading, hence providing better accessibility for enzyme substrate reaction (Hettiarachchy et al. 2018). Beside this, the cost of immobilization is a key parameter to consider for industrial application. Hence, the choice of economically feasible matrices and subsequent selection of an immobilization method affects the trade-off between the cost and performance (Matto and Husain 2009; Agyei et al. 2015).

Another major challenge faced by the immobilization process is maintaining the structural integrity which directly affects the stability of the enzyme. However, the properties of the immobilized enzyme changes contrary to free enzyme, which largely depends on the inherent properties of enzyme molecules and the support material. The specific interaction between the latter provides discrete chemical, biochemical, mechanical and kinetic properties to the immobilized enzymes (Tischer and Wedekind 1999). The mass transfer effect constituting partition effect, binding steps and diffusion limitation also influences the performance of immobilized enzyme when enzyme and substrate reacts in the microenvironment as compared to their bulk environment. In addition, the optimization of reaction conditions during the immobilization process is a crucial step for effective enzyme immobilization. The parameters viz., temperature, pH, reaction time and buffer utilized influence the stability and catalytic activity of the enzyme (Dwevedi 2016). However, the enzyme immobilization can be achieved using physical and chemical approaches (Husain 2010; Veum and Hanefeld 2006). In general, immobilization involves binding of enzyme to suitable support material via adsorption or covalent interaction, by cross-linking enzyme aggregate or via entrapment/encapsulation method (Fig. 2). The subsequent section of the chapter further discusses the strategies adopted for enzyme immobilization.

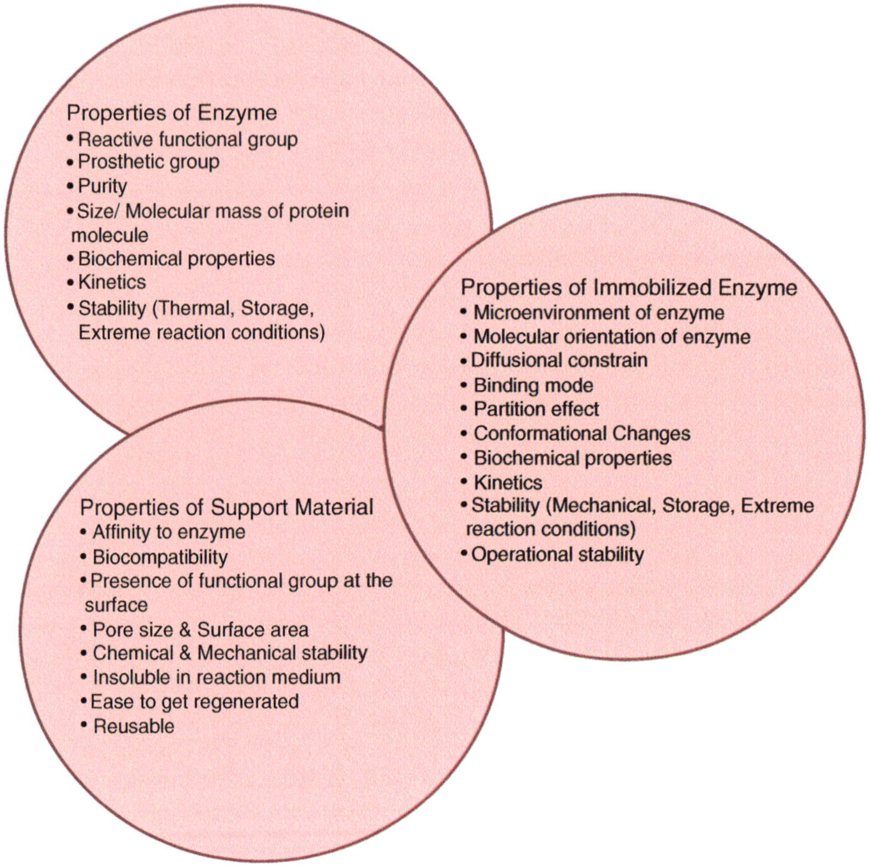

Fig. 1 The characteristic properties of immobilized enzyme governed by the physico-chemical properties of enzyme molecules and support matrix

3 Strategies Used for Enzyme Immobilization

3.1 Adsorption

The Physical adsorption of enzyme molecule to a carrier matrix is one of the oldest techniques, which mainly involves interactions such as electrostatic forces, hydrophobic, hydrogen bonds, or Van der Waal's forces (Michaelis and Ehrenreich 1908; Kulshrestha and Husain 2006; Jesionowski et al. 2014). The weak interactions involved in adsorption prevent the structural integrity of the enzyme molecules from being altered, thus retaining its activity (Hernandez and Fernandez-Lafuente 2011; Hwang and Gu 2013; Graebin et al. 2016). The efficient immobilization depends on physico-chemical properties of the support materials, such as surface area, size of the particle, pore size and surface functional groups. Moreover, the

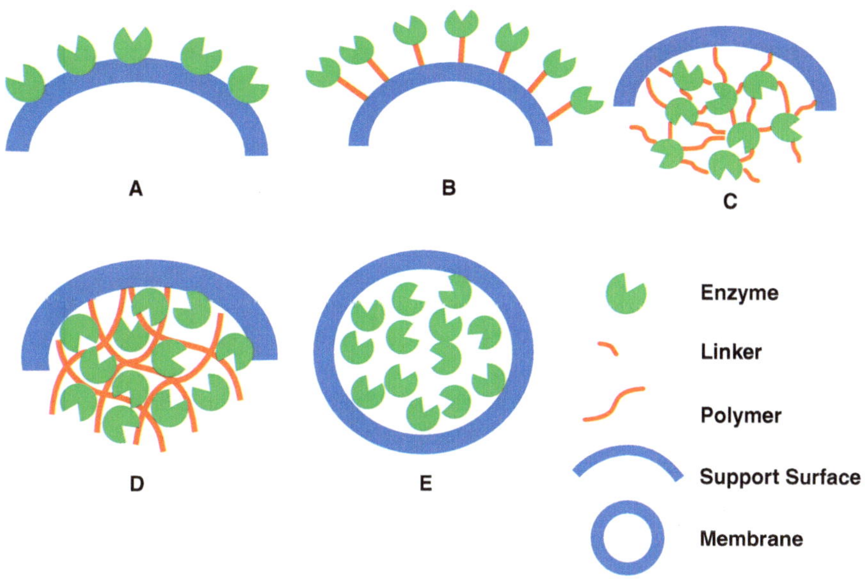

Fig. 2 Strategies used for enzyme immobilization: (**a**) Adsorption; (**b**) Covalent; (**c**) Cross-linking; (**d**) Entrapment; (**e**) Encapsulation

adsorption of enzyme molecules prerequisite higher affinity for support material which can be achieved by the presence of specific active functional groups on the surface of support matrices ensuring enzyme-carrier interactions (Jesionowski et al. 2014).

Adsorption is the simplest method of immobilization facilitating reuse of support material, providing economical feasibility for industrial applications (Es et al. 2015). Beside, weak interaction involved in adsorption may leads to enzyme leakage possessing lower operational stability with alteration in reaction conditions viz., temperatures, pH, ionic strength etc., or even in presence of substrate (Daneils and Farmer 1981; Salleh et al. 2006). Another, drawback of physical adsorption is non-specific interaction of biomolecules contrary to enzyme molecules, which may cause change in properties of immobilized enzyme (Norouzian 2003).

3.2 Entrapment/Encapsulation

Entrapment is confinement/caging of enzyme molecules, within an insoluble matrix viz., network of gel or fibre by covalent or non-covalent bonds (Singh 2009; Svec and Gemeiner 1996). The encapsulation of enzyme within a porous matrix prevents enzyme leakage with enhanced mechanical stability (Svec and Gemeiner 1996; Matto and Husain 2006). The entrapment involves occluding an enzyme within a polymer matrix (Sheldon and Pereira 2017). This is a simple technique of immobilization due to ease, cost effectiveness enabling free diffusion of low molecular weight biomolecules viz., substrate, proteins and reaction products. Therefore, the entrapment of higher molecular weight enzyme/protein is preferred due to large pore size of matrices. Entrapment also helps to retain the native structure of the enzyme molecule preventing direct interaction with the environment (Agyei et al. 2015; Franssen et al. 2013; Liu et al. 2018). Conversely, reducing the affect of adverse reaction conditions viz., gas bubbles, mechanical shear, hydrophobic solvents, change in pH and temperature whilst limit the mass transfer leading to lower enzyme loadings (Brady and Jordaan 2009; Graebin et al. 2016). There may also be some possibilities of enzyme damage due to chemical and heat used during the formation of gel (Edet et al. 2013). However, the problem of enzyme leakage can be overcome by controlling the polymerization conditions. This can be achieved by altering the pore size, surface functional group and network structure of polymeric material (Reetz and Jaeger 1998). Recently, entrapment of enzyme on nanostructured support materials viz., electrospun, nanofibers and pristine materials have attracted wider interest for various industrial applications (Datta et al. 2013).

3.3 Crosslinking

In last few years, cross-linked enzyme aggregates (CLEA) has emerged as a versatile carrier-free method of immobilization. This involves intermolecular cross-linking of enzyme molecule via surface functional groups of insoluble support matrice (Eldin et al. 2011). In this method, enzyme aggregates are prepared using precipitating agents e.g., organic solvents (acetone, ethanol, propanol, tert-butanol), salts (ammonium sulfate), or non-ionic polymers (polyethylene glycol) and subsequently binding to each other using glutaraldehyde as bifunctional crosslinking agent (Graebin et al. 2016; Sheldon and Van Pelt 2013; Talekar et al. 2012). The strong bond formation imparts stability, preventing enzyme leakage with lower desorption. Hence, CLEA owing higher enzymatic activity, stability, eliminates the use of additional carrier with ease of separation, which enables simple and cost effective production technique, has attained huge interest for industrial purposes (Sheldon 2011; Nadar et al. 2016). However, sometimes loss in the catalytic activity may occur during crosslinking due to distortion of native enzyme structure and alteration in active sites (Chakraborty et al. 2016). In addition, the process of crosslinking also faces the problem of diffusion limitation.

3.4 Covalent

The enzyme immobilization via covalent bonding can be achieved using different chemical bonding approaches viz., cross-linking, multifunctional reagents, or surface functionalized groups (Kumari and Kayastha 2011). Generally, the covalent binding of enzyme to support material can be achieved directly or indirectly. The direct crosslinking involves functional group of enzyme such as amino, carboxyl, sulfhydryl, hydroxyl, imidazole or phenolic groups, which play a key role in binding to the matrices (Eldin et al. 2011). However, the indirect attachment of enzyme molecule make use of spacer arm (viz., glutaraldehyde) and chemical reactions (viz., Schiff base, and imine bond formation etc.,) which imparts thermal stability (Husain and Ulber 2011; Datta et al. 2013). The spacer arm produces the distance between the enzyme molecule and the support material thereby minimizing the chances of denaturation (Chakraborty et al. 2016). Moreover, the covalent bonds are stronger which prevents enzyme leaching from the surface (Hwang and Gu 2013). Alternatively, the covalent interaction may lead to larger strain on the enzyme, along with the drastic change in structural conformation and catalytic properties. Conversely, this occurs due to harsh reaction conditions met by the immobilization process and presence of similar amino group at the active site of the enzyme, which may also interact with the support material (Dwevedi 2016).

4 Advantage Versus Disadvantage of Enzyme Immobilization

4.1 Advantages

Enzyme immobilization strategies have evolved to overcome the limitations of native enzyme. This provides enhanced operational and storage stability with peculiar properties like increased resistance against adverse reaction conditions viz., pH, temperature, inhibitors etc., Immobilized enzyme can perform better under mild physico-chemical conditions saving the use of energy (Hettiarachchy et al. 2018; Jesionowski et al. 2014; Meridor and Gedanken 2013). This provides ease of recovery with reusability of enzyme through series of cycle followed by separation of superior quality product, which are free of enzyme and reactant molecules (Marzadori et al. 1998; Mateo et al. 2007; Chakraborty et al. 2016; Husain 2016). Immobilization offers ease for termination of reaction with better control over the enzyme-substrate, which occurs when enzyme molecules is removed or added to the substrate and can facilitate more than one reaction at a time (Homaei et al. 2013; Liu et al. 2018). Immobilization provides larger surface area for enhanced enzyme loading with higher specific activity and enzyme immobilized efficiency (%). However, the immobilized enzymes are highly recommended for industrial applications as it avoids the use of reagents. Nevertheless, immobilization techniques are simple, ecofriendly, lacking tedious process of enzyme purification providing ease of recovery from the reaction mixture with reusability making the process economically viable and industrially relevant (Meridor and Gedanken 2013; Chakraborty et al. 2016; Hettiarachchy et al. 2018).

4.2 Disadvantages

Despite of various advantages, immobilization methods have certain limitations. It requires an additional step making the process tedious with intensive use of energy, time and capital (Liu et al. 2018). Depending upon the method adopted, it may lead to enzyme leakage due to large pore size of the matrices. However, the reduced product yield is achieved when enzyme attached to the support material is desorbed or enzyme active site is distorted. There also exist the possibilities of microbial contamination depending upon the support material (Hettiarachchy et al. 2018). The immobilization allows only limited movement of enzyme molecules because of physical/chemical interaction with support matrices (Es et al. 2015). Hence, it may also exhibit altered kinetics due to change in the orientation of the enzyme and diffusional resistance offered for the substrate molecule (Kress et al. 2002). Moreover, the immobilization strategies are often limiting if the substrate used is not soluble (Krishna 2011; Husain 2017b).

5 Nanomaterials as Versatile Support for Enzyme Immobilization

The recent progress in nanotechnology has witnessed a wealth of knowledge compounded with biotechnology with improved properties, tailored for interaction of enzyme to nanomaterials. During the pioneer phase of nanobiocatalysis, the most commonly used methods were adsorption and covalent linking which provides large surface area for enzyme immobilization (Kim et al. 2008). However, the extensive research in the arena of nanotechnology since 2014 has given great impetus for interaction of enzyme to the surface of nanomaterials viz., enzyme immobilization (Chen et al. 2017; Husain 2017c). This provided several nanomaterials such as nanoparticles, nanofibres, nanotubes, nanopores, nanorods, nanosheets, nanohybrids and nanocomposites etc., affordable for enzyme immobilization to be used efficiently in industrial bioprocesses (Husain 2017d).

The growing concern for using nanomaterials has aroused due to its unique properties which can be altered with ease, providing various advantages such as control over the size at the nanoscale, uniform size distribution same as the enzyme molecules, along with the optoelectronic properties and unique surface chemistry. The greater surface area to volume ratio and their surface functionalization via chemicals or bioconjugates involve synergistic effect of chemistry, material and biological sciences (Wang 2006; Kim et al. 2008; Biju 2014; Husain 2018a, b). The engineered/functionalized nanomaterials provide ease for fabrication with high enzyme loading, catalytic activity with prolonged reuse and improved biochemical properties viz., pH, thermal, operational and storage stability for specific application. The increased enzymatic activity upon immobilization on nanomaterials depends on the type of enzyme, carrier and immobilized reaction conditions. Moreover, it is vital to consider the alteration in structural and functional properties of enzyme molecule upon interaction with nanomaterials, which generally leads to conformational changes resulting in inhibition or augmentation of activity (Johnson et al. 2011; Chen et al. 2017).

Generally, the immobilized enzyme forms a stable monodispersed nanoparticle in aqueous medium possessing Brownian motion. Therefore, the mobility and diffusivity have to be smaller for nanoparticles with that of the free enzymes owing rather larger sizes as suggested by Stokes–Einstein equation. The difference in mobility plays determining role as directed towards transitional regions between the homogeneous catalysis with free enzyme, in contrast to heterogeneous catalysis upon enzyme immobilization. The Brownian motion is the crucial factor that may cause increased catalytic activity upon enzyme immobilization on nanoparticles (Jia et al. 2002; Wang 2006; Verma et al. 2013). Moreover, improved activity and stability of the immobilized enzyme on nanomatrices is devoted to low mass transfer resistance to substrates than those of macro-sized conventional matrices (Kim et al. 2006).

Nanomaterials, especially various types of nanoparticles possessing uniform size, high stability and dispersivity has attained immense interest as a novel support material for enzymatic catalysis. Among nanomaterials, magnetic nanoparticles are preferred due to ease of handling and separation from the reaction medium allowing retention of activity with increased reusability (Verma et al. 2013; Khan et al. 2018). Hence, it offers efficient operational stability avoiding the use of tedious process of separation such as centrifugation, leading to mechanical shearing which usually cause enzyme leakage with loss in the activity (Yiu and Keane 2012). Whilst, nanofibres provide enhanced flexibility in reactor designing due to ease of their preparation and handling (Nair et al. 2007). The small drop in pressure with elevated flow rate offered by the immobilized enzyme on nanofibre membrane is of great significance in bioreactors as compared to conventional methods (Verma et al. 2013). Hence, the enzyme immobilized on nanomaterials provide several unique advantages over other conventional matrices.

6 Immobilization of α-Amylase

In present scenario, the immobilization of α-amylase is the area of wider research interest for various technological processes in large-scale industries. The thermostable enzymes from various sources specifically from Bacillus sp. are of peculiar interest due to their outstanding expression system (Prakash and Jaiswal 2010). Besides, the inert environment and biocompatible nature of support material is crucial factor in immobilization, which alters the stability of enzyme providing reusability and economical feasibility. Generally, either organic or inorganic solid supports have been used for immobilization (Drevon 2002). The natural organic polymeric supports (agar, agarose, gelatin, alginate, chitosan and cellulose) along with synthetic organic polymers (amberlite, DEAE cellulose, polyesters beads, and activated organic supports) containing epoxide groups are the most widely studied materials (Dey et al. 2003; Jaiswal and Prakash 2011; Kumari and Kayastha 2011; Konsoula and Liakopoulou-Kyriakides 2006; Prakash and Jaiswal 2011; Datta et al. 2013; Singh and Kayastha 2014; Cakmakci et al. 2013). However, the major delinquent limiting the use of organic matrices are lack of stability towards microbial contamination, lower robustness to high temperatures and organic solvents, as well as disposable problems. Therefore, to circumvent these problems, inorganic supports such as clays, zirconia, alumina, and silica are used for α-amylases immobilization providing higher biocompatibility, inertness as well as chemical and physical stability (Reshmi et al. 2006, 2007; Sanjay and Sugunan 2005; Sohrabi et al. 2014).

Recently, the nano-scaled materials owing unique properties over traditional bulk materials viz., their miniature size, large surface area and high enzyme loading capacity has attracted huge attention for enzyme immobilization. Several organic or inorganic based nanomaterials including their nanohybrids and nanocomposites have attracted enormous interest of researchers and industrialist as promising support material for industrial application of immobilized enzymes. Table 1 represents the list of various types of nanomaterials used for immobilization of α-amylase. Further, the chapter gives an overview on α-amylase from various sources viz., microbes, plants and commercially available form, immobilized on different support materials emphasizing on nanomatrices for their specific applications. Table 2 summarizes the list of selective examples of immobilized α-amylase with their characteristic properties.

6.1 Microbes

The maltogenic α-amylase from *Bacillus stearothermophilus* (BsMa) have been immobilized onto porous polyurethane urea (PUU) microparticles via covalent linkages, which was achieved upon reaction of PVA (poly(vinyl alcohol) and IPDI (isophorone diisocyanate). The alteration in the concentration of PVA and IPDI played a crucial role in the synthesis of PUU microparticles producing free isocyanate groups affecting the surface area of PUU and immobilization efficiency. The free isocyanate group possessed by PUU microparticles favored immobilization of BsMa through amino or hydroxyl groups (Straksys et al. 2016).

Karam et al. (2016) used agro-industrial waste for isolation of amylase from *Aspergillus awamori* and immobilized it on unique support material viz., Ca^{+2}(Alg. St/PEI/GA) Calcium (alginate.starch/polyethyleneimine/glutaraldehyde) via covalent linkages. The thermodynamic studies provided insight about the stability of immobilized enzymes with energy of dissociation (Ed) value of 16.7 kJ mol^{-1} to those of soluble enzyme 8.71 kJ mol^{-1}. This suggested that additional energy was required to denature the immobilized enzyme due to conformational changes, which provided higher stability and rigidity. Moreover, the half-lives, D-values and Gibbs free energy (G°) for thermal energy of the immobilized form was increased compared to their free counterparts.

Table 1 List of various types of nanomaterials used for immobilization of α-amylase enzyme

Type of nanomaterial used	References
Nanoparticles	
Naringin functionalized magnetic nanoparticles	Defaei et al. (2018)
Magnetic nanoparticles	Eslamipour and Hejazi (2016)
Magnetic nanoparticle	Demir et al. (2012)
Magnetic Fe_2O_3 nanoparticles	Khan et al. (2012)
Silica nanoparticles	Soleimani et al. (2012)
PANI/Silver nanoparticles	Khan et al. (2013)
$CaCO_3$ nanoparticle	Demir et al. (2012)
Nanofibres	
PVA/PAA nanofibers	Basturk et al. (2013)
Nanotubes	
Halloysite nanotubes	Zhai et al. (2010)
Nanopores	
Nanopore zeolite	Talebi et al. (2016)
Nanosheet	
Graphene nanosheets	Singh et al. (2015)
Nanorod	
Gold nanorods	Homaei and Saberi (2015)
Nanocomposites	
PPy AgNp/Fe_3O_4-nanocomposite	Mohamed et al. (2018)
Nanohybrid	
Mg/Al-Layered double hydroxide nanohybrids	Bruna et al. (2015)

Recently, Mohamed et al. (2018) reported immobilization of α-amylase from *Trichoderma harzianum* on PPyAgNP/Fe_3O_4 nanocomposite using ionic interactions. The α-amylase with thiol/disulfide or amino groups among its chemical structure together with the synergistic effect of positive charges of PPy facilitated ionic binding while the thiol/disulfide group at the interface of enzyme was hypothesized for its attachment to AgNP. Moreover, the magnetite enabled ease of enzyme separation under magnetic field. The study suggested that outstanding properties of immobilized α-amylase on PPyAgNp/Fe_3O_4-nanocomposite could have promising application in biomedical.

6.2 Plants

Considering the industrial demand, the α-amylase from plant sources has also been explored for immobilization, in order to utilize its properties for specific bioprocess. The organic polymers possess high biocompatibility with network like structure and high hydrophilicity (Reddy et al. 2004; Isobe et al. 2011). Jaiswal et al. (2012) immobilized soybean α-amylase onto gelatin, a natural polymer, using glutaraldehyde

Table 2 List of selective examples of immobilized α-amylase with their characteristic properties

Source	Support material	Immobilization method	Immobilization (%)	Properties	References
Microbial					
Bacillus Stearothermophilus	Poly (urethane urea) microparticles	Covalent binding	68%	Shift in temperature optima from 60 to 70 °C of free and immobilized enzyme. Lower K_m value, storage stability for 28 days, reused for 7 cycles (retained 90% catalytic activity) at 40 °C.	Straksys et al. (2016)
Aspergillus awamori	Ca^{+2}(Alg.St/PEI/GA	Covalent linkages	–	Temperature optima: 50 °C and 55–60 °C, Ed value: 16.7 kJ mol^{-1} and 8.71 kJ mol^{-1} for free form contrary to immobilized enzyme. High K_m value, thermostable and reusable for 12 cycles (with 46% activity).	Karam et al. (2016)
Trichoderma harzianum	PPyAgNp/Fe_3O_4 nanocomposite	Ionic interaction	75%	Shift in pH from 6.0 to 6.5; temperature from 40 to 50 °C of soluble and immobilized enzyme. Thermostable with decreased K_m and recycled 10 times (80% retained activity).	Mohamed et al. (2018)
Plants					
Vigna radiata	Amberlite MB 150 & Chitosan beads	Covalent interaction	72% 69%	Free α-amylase showed optima pH of 5.6 while immobilized form at pH 7.0. Temperature optima of 65 °C for free and amberlite immobilized enzyme, while 75 °C for chitosan immobilized. Immobilized form showed higher K_m value and prolonged storage stability of 100 days with 10 cycles of reuse.	Tripathi et al. (2007)

Glycine max	Agarose & agar	Entrapment	75.3% 77%	Same pH optima of 7.0 for free and immobilized enzyme, lower temperature optima with increased K_m value and lower reusability upto 5 cycles.	Prakash and Jaiswal (2011)
T. aestivum	DEAE cellulose	Covalent interaction & ionic adsorption	86%	Shift in pH optima from 5.0 to 6.0, temperature optima 68–70 °C, K_m: 1.56–2.7 mg/mL, $t_{1/2}$: 80 min to $t_{1/2}$: 126 min for free and immobilized enzyme with enhanced storage stability of 60 days and recycled upto 10 cycles.	Singh and Kayastha (2014)
T. aestivum	Graphene sheet	Covalent linkage	85.16%	Broad range of stability for pH and temperature, improved thermal stability with storage stability for 60 days and reused for 10 cycles (73% activity). K_m: 1.56 mg/mL changed to 1.32 mg/mL for free form contrary to immobilized enzyme.	Singh et al. (2015)
Commercial					
Bacillus amyloliquefaciens	PANI/Ag nanocomposite	Covalent linkage	–	Shift in temperature optima from 50 to 60 °C, K_m: 16.30–23.90 mg/mL^{-1} for free and immobilized enzyme, reusable for 10 cycles.	Khan et al. (2013)
Bacillus subtilis	Gold nanorods	Ionic exchange & hydrophobic interactions	–	Temperature optima: 50–60 °C, pH: 6.0–7.0 for free and immobilized enzyme. Increased K_m value with prolonged storage stability of 2 months.	Homaei and Saberi (2015)
Aspergillus oryzae	Mg_3Al-layered double-hydroxide (LDH) matrix	Adsorption, electrostatic interaction	–	Stability of 5 weeks in dried form (90% initial activity).	Bruna et al. (2015)

(continued)

Table 2 (continued)

Source	Support material	Immobilization method	Immobilization (%)	Properties	References
Malt	Nano zinc oxide	Adsorption, Electrostatic interaction	80.5%	pH optima of 5.0 and 6.0 for free and immobilized enzyme. High K_m value, storage stability of 30 days (retained 70% activity) and lower reusability (4 times).	Antony et al. (2016)
Aspergillus oryzae	ESM & AgNP/ESM	Covalent interaction	–	Optimal pH 5 and temperature 55 °C was same for free and immobilized enzyme. Higher K_m for AgNP/ESM contrary to ESM immobilized enzyme with storage stability of 50 days.	Du et al. (2017)
Aspergillus oryzae	Titania/lignin hybrid	Covalent interaction, physical adsoption and ionic interaction	87%	Shift in pH optima from 5.5 to 6.5 and temperature optima from 50 to 60 °C for free form contrary to immobilized enzyme. Increased K_m value with storage stability of 30 days and reused for 10 cycles retained 80% activity.	Klapiszewski et al. (2018)
Bacillus subtilis	Naringin modified magnetic nanoparticles (MNP@SiO2/NA)	Ionic interaction	–	Shift in optimal pH 5.5–6.5, temperature 45–55 °C for free and immobilized enzyme, reused for 10 cycle retained 50% activity with storage stability for 6 weeks	Defaei et al. (2018)
Aspergillus oryzae	Magnetic nanoparticle	Adsorption	88.37%	Higher affinity for substrate with ease of recovery and reused for 5 cycles retaining 75% catalytic activity	Baskar et al. (2015)
Porcine pancreatic α-amylase	Modified magnetic Fe_3O_4 nanoparticles	Covalent	43.9%	Optimal shift in pH from 7.0 to 7.5 with optimal temperature 60 °C for free and immobilized enzyme. Increased K_m, storage stability 30 days retained 85.8% activity and reused for 6 times.	Guo et al. (2016)

Bacillus *sp*.	CLEA & Magnetic CLEA	Covalent	45% and 100%	Optima pH: 6.0, 7.0 and temperature optima: 45 °C, 50 and 60 °C for soluble and immobilized enzyme (CLEA and magnetic CLEA). Lower K_m value and storage stability for 18 days (retained 50 and 80% activity) recycled for 6 times.	Talekar et al. (2012)
Porcine pancreas	(PVA/PAA) nanofibers	Covalent linked	181.48 ± 3.60 mg/g	Shift in pH optima from 6.5 to 6.0; temperature optima from 30 to 50 °C for soluble and immobilized enzyme. Higher thermal stability and K_m value. Reusable upto 10 cycles retained 81.7% activity with storage stability of 30 days (82.9% activity).	Basturk et al. (2013)
Porcine pancreas	$CaCO_3$ nanoparticle	Covalent interaction	199.43 mg/g	Free and immobilized enzyme showed optima pH 6.5. Thermostable with low k_m value, recycled for 25 times and retained 71% activity after 40 days storage.	Demir et al. (2012)

as crosslinker under optimized reaction conditions. The enzyme was immobilized using covalent interaction and the process was optimized using response surface methodology, a statistical and mathematical tool (Jaiswal et al. 2012). The thermostable α-amylase from soybean seeds (*Glycine max*) was also immobilized on another organic polymer viz., agarose and agar matrices, using entrapment method with excellent capability for starch stain removal (Prakash and Jaiswal 2011). Immobilization of wheat α-amylase on DEAE-cellulose matrix via covalent interaction and ionic adsorption under optimal condition using response surface methodology has also been reported (Singh and Kayastha 2014). Immobilized enzyme involved covalent amide linkage between the enzyme and support material, which allowed controlled mobility of enzyme reducing the conformational flexibility with remarkable thermal stability (Singh and Kayastha 2014). The chitosan beads and Amberlite MB 150 showed potential as efficient matrices for immobilization of α-amylase from *Vigna radiate* and *Glycine max* using covalent interaction (Tripathi et al. 2007; Kumari and Kayastha 2011). Moreover, these polymeric matrices being economical with increased stability, appeared as suitable support material for industrial purposes.

Recently, Khare and Prakash (2017b) have purified thermostable α-amylase from radish seeds with potential of being utilized in various starch-based industries. Thus purified enzyme was covalently immobilized onto green synthesized silver nanoparticles using simple, sustainable and ecofriendly approach. The silver nanoparticles was functionalized using glutaraldehyde as a crosslinking agent, which facilitated attachment of the enzyme forming super nanobioconjugate structure with improved biochemical properties as compared to the soluble form. Silver nanoparticles owing unique opto-electronic properties with large surface area to volume ratio and biocompatible nature appeared as efficient matrice for immobilization, exhibiting augmented stability towards various chemicals, heat with remarkable storage stability and reusability (Khare 2018). Apart from this, graphene being a two-dimensional carbon nanomaterial with extraordinary electronic, thermal, and mechanical properties when combined with enormous surface area forms a suitable platform for immobilization of biomolecules (Sanchez et al. 2011). The functionalized graphene sheet exhibited a versatile matrice for α-amylase (*T. aestivum*) immobilization, using cysteamine as spacer arm and glutaraldehyde as crosslinking agent, which retained native biochemical properties (Singh et al. 2015).

6.3 Commercially Available

Polyaniline (PANI) a polymer owing high mechanical and chemical stability as well as resistance to microbial attack have provided new impetus in enzyme immobilization (Gospodinova and Terlemezyan 1998). The poyaniline (PANIs) in two forms, emeraldine salt and emeraldine base powder, have been explored for immobilization of diastase α-amylase via physical adsorption and covalent binding which showed enhanced storage stability as well as reusability (Ashly et al. 2011). The α-amylase from *Bacillus licheniformis* was reported to covalently immobilized onto another reactive polymers viz., hydrophilic 3-D poly(ethylene-alt-maleic anhydride)

(PEMA) copolymer which showed higher enzyme loading with increased catalytic activity contrary to hydrophobic poly(octadecene-alt-maleic anhydride) (POMA) (Cordeiro et al. 2011).

Poly vinyl alcohol (PVA), a biocompatible insoluble synthetic polymer possessing chemical and thermal stability with economical viability have wider applications in biomedical and enzyme immobilization. The electrospun PVA/PAA (Poly (vinyl alcohol)/poly (acrylic acid) nanofibers have been explored for α-amylase immobilization via covalent interaction. The PVA/PAA nanofibers was crosslinked via esterification and subsequently activated with the amine groups in the presence of 1,10-carbonyldiimidazole which facilitated ease for fabrication of enzyme on its surface with higher catalytic activity and stability (Basturk et al. 2013).

Another study discerns α-amylase (*Aspergillus oryzae*) immobilized on eggshell membrane (ESM) and Ag-nanoparticle-decorated eggshell membrane (AgNP/ESM) using glutaraldehyde as a cross-linking agent. The increased enzyme loading exhibited by AgNP/ESM (80.33 mg/cm^2) due to enhanced surface area, upon silver nanoparticles synthesis on ESM as compared to ESM alone (77.68 mg/cm^2). ESM, a byproduct of chicken eggs, is a natural biomaterial, which possess high specific surface area and intricate lattice network with highly crosslinked fibers. These are non-toxic, biocompatible, water-insoluble, infusible, and highly resistant to organic solvents and microbial contaminations. Hence, the ESM have emerged as unique insoluble support for enzyme immobilization (Du et al. 2017). Moreover, the naturally occurring *Luffa operculata* fibre (LOF) have also been used for α-amylase immobilization via adsorption which was stable over 3 weeks with remarkable operational stability (Morais et al. 2013).

To overcome the problems associated with organic polymeric material, inorganic materials viz., silica gels, alumina, and layered double hydroxides possessing thermal, mechanical stability, high resistance against microbial attacks and organic solvents are preferred for enzyme immobilization (Tischer and Wedekind 1999). The silica nanostructures possessing size in nm range, offered greater surface area, ordered structure, high stability to chemical and mechanical forces as well as resistance to enzyme contamination appeared as novel matrix. The α-amylase (termamylaze) obtained from Novozyme was adsorbed onto silica nanoparticles showed higher storage stability and shelf life of the product with potential for removal of starch soil from the cotton fabrics (Soleimani et al. 2012).

Bruna et al. (2015) for the first time reported α-amylase (AAM) (*Aspergillus oryzae*) immobilized on inorganic Mg_3Al-layered double-hydroxide (LDH) matrix via adsorption for the hydrolysis of starch. The enzyme adsorbed on LDH surface via electrostatic interaction involved multipoint attachment due to altered orientation and conformation of protein. The immobilized enzyme showed remarkable storage stability in dried form, which may have promising applications in industrial and pharmaceutical processes as a simple, ecofriendly, as well as cost effective biocatalyst. However, the critical limitations of physical adsorption method of immobilization involved desorption of enzyme from porous materials, which resulted in loss of activity with poor operational stability (Bruna et al. 2015). Apart from this, α-amylase (*Bacillus subtilis*) immobilized on alumina via adsorption showed improved thermal stability (Reshmi et al. 2006).

Recently, Klapiszewski et al. (2018) first time reported immobilization of α-amylase (*Aspergillus oryzae*) on a titania/lignin hybrid support. The studies depicted the synergistic properties of inorganic oxide and biopolymer, produces a composite material, which enables enzyme fabrication. The surface functionalized titanium dioxide via poly-L-lysine provided various functional group for attachment of activated lignin producing titania/lignin novel hybrid. The lignin molecule being biopolymer provided compatibility for enzyme immobilization mainly via covalent interaction and partially involved physical adsorption (Klapiszewski et al. 2018).

Nanostructured metal oxides (NMOs), a newly emerging domain of research have attracted wider interest for enzyme immobilization. NMO, owing effective surface area, proposed orientation, inert surface, better structural modification and non-toxic nature enables higher enzyme loading ability, reusability and biological activity (Tenne 2006; Khan et al. 2011; Soozanipour et al. 2015; Amirbandeh and Taheri-Kafrani 2016; Esmaeilnejad-Ahranjani et al. 2016; Shahrestani et al. 2016). Among NMO, magnetic nanoparticles has emerged as fascinating area for enzyme immobilization. The α-amylase (*Aspergillus oryzae* and *Bacillus subtilis*) immobilized on magnetic nanoparticles via simple adsorption and naringin (NA) functionalized magnetic nanoparticles, which involved ionic interaction, were recovered with ease from the reaction medium under magnetic field with higher reuse (Baskar et al. 2015; Defaei et al. 2018). The NA being a biocompatible flavonoid and hydrophilic in nature, has been used for modifying surface of magnetic nanoparticles which provided ease for enzyme immobilization (Defaei et al. 2018). The glutaraldehyde-activated amino-functionalized silica-coated magnetite nanoparticles alone (AFSMNPs) and covered with chitosan appeared as a suitable support for α-amylase (*Aspergillus Oryzae*) immobilization (Hosseinipour et al. 2015). These studies revealed magnetic nanoparticles as an efficient matrix for enzyme immobilization with improved biochemical properties, enhanced stability with ease of recovery and reusability.

The "Magnetic Molecular Imprinted Polymer" (MMIP), is another technique which involved covalent interaction with supports for bio-macromolecules immobilization due to its strong magnetism, low-toxicity and good reusability. Magnetic Particles Immobilized α-amylase (MPIA) was prepared via magnetic Fe_3O_4 nanoparticles synthesized by an improved co-precipitation method and subsequently covalent immobilized PPA (porcine pancreatic α-amylase) on carboxyl-terminated magnetic nanoparticles. The study showed that modification of magnetic Fe_3O_4 particles by silylation method provided good hydrophilicity. The surface of magnetic particles covered with large amount of amino groups and long chain carboxyl groups involved amination and carboxylation reaction. Moreover, amylase composed of many proteins with active center having large number of amino groups on its surface exhibited ease for attachment on the suitable magnetic $Fe_3O_4@SiO_2@NH_2$-COOH carriers by amide reaction (Guo et al. 2016).

The α-amylase (*Bacillus subtilis* and malt) immobilized onto solid support have been studied for gold nanorods (GNRs) and nano zinc oxide (Homaei and Saberi

2015; Antony et al. 2016). Gold nanorods used ionic exchange and hydrophobic interactions for stabilizing the enzyme molecule (Homaei and Saberi 2015) while zinc oxide having high isoelectric point was suitable for the adsorption of low IEP proteins (α-amylase) via electrostatic interaction (Antony et al. 2016). Beside, zinc oxide nanostructures provides several advantages over other nano metal oxides such as high biocompatibility, chemical stability, high specific surface area, environment friendly and cost effective (Rositza et al. 2012). Demir et al. (2012) immobilized α-amylase (porcine pancreas) onto glutaraldehyde activated silanized calcium carbonate nanoparticles via covalent interactions. This involved surface functionalization of $CaCO_3$ particles using silane-coupling agent (3-aminopropyl triethoxysilane) which provided functional group for attachment of enzyme via $-NH_2$ group using glutaraldehyde as crosslinker (Demir et al. 2012). Moreover, the polyaniline mediated synthesis of Ag nanocomposites was used to immobilize α-amylase (*Bacillus amyloliquefaciens*) via covalent binding using glutaraldehyde as a crosslinking agent. The free $-NH_2$ group of PANI involved reaction with the –CHO group of glutaraldehyde which facilitated attachment of enzyme. Thermoanalysis studies revealed increased thermal stability. Moreover, the Ag nanoparticles dispersed on PANI surface enabled high enzyme loading due to large surface area and showed improved properties with efficiency for starch hydrolysis (Khan et al. 2013).

Talekar et al. (2012) reported novel magnetic cross-linked enzyme aggregates of alpha amylase (*Bacillus sp.*) which involved chemical cross-linking of enzyme aggregates with amino functionalized magnetite nanoparticles and was easily recovered under the magnetic field. Addition of such amino functionalized magnetite nanoparticles into the solution of enzyme with low Lys residue content suggested formation of mechanically stable and non leachable CLEAs due to sufficient cross-linking of enzyme aggregate which retained 100% initial activity even after several reuse. The studies on macromolecular cross-linked enzyme aggregates (M-CLEAs) of α-amylase (commercial), have also been reported using precipitation method. The changes in secondary structures revealed increased structural and conformational rigidity due to cross-linkers, which imparts thermal stability. The study highlighted the potential of macromolecules (dextran and chitosan) as suitable biocompatible cross linkers for eco-friendly CLEAs preparation (Nadar et al. 2016).

7 Application of Immobilized α-amylase

The practical application of immobilized α-amylase have been well established in various sectors, with major use in food, pharmaceuticals, detergent industries and wastewater treatment plants. The α-amylase has attracted wider interest in food and fermentation industries particularly for the production of fermentable sugar (Reddy et al. 2003). The studies on α-amylase immobilized on gold nanorods showed enhanced stability towards extreme pH and temperature with prolonged storage stability, which was used as the potential candidate for large-scale production of high purity glucose syrups (Homaei and Saberi 2015). Bayramoglu et al. (2004)

reported α-amylase immobilized on epoxy support material i.e., poly HEMA-GMA-1-3 (hydroxyethylmethacrylate glycidyl methacrylate) membranes via covalent interaction and suggested improved thermal stability with prolonged storage compared to their free counterparts. The immobilized enzyme exhibited higher operational stability during the continuous process in enzyme reactor for 120 h, which revealed its potential use for the production of maltose and dextrin from starch. The another study on α-amylase immobilized on magnetic nanoparticles with superior properties contrary to native enzyme, indicated its efficient use for the hydrolysis of sweet potato starch and high amylose starch (Baskar et al. 2015; Guo et al. 2016). Similarly, the α-amylase immobilized on magnetic Fe_2O_3 and SnO_2 nanoparticles, and PANI assisted silver nanocomposite also showed potential for the hydrolysis of starch in batch process (Khan et al. 2012, 2013; Khan and Husain 2014).

The α-amylases have promising application as antistaling agent in bread and bakery. However, when using immobilized enzyme in food industries the food safety is the crucial factor. This mainly depends on the support matrices and the method of immobilization. Recently, encapsulated α-amylase into biocompatible beeswax using emulsion-congealing technique showed enhanced thermal and storage stability in contrast to their free counterparts. Further, the encapsulated enzyme (EE) was investigated for its antistaling property in gluten-free bread formulation by controlling qualitative properties of prepared batters and breads. The breads treated with EE revealed improved qualities viz., lower hardness and chewiness. This proved that beeswax might have protected the encapsulated enzyme during the batter preparation, fermentation, baking, and storage of bread. However, higher sensorial quality was exhibited by EE loaded bread compared to free enzyme treated breads (Haghighat-Kharazi et al. 2018).

The α-amylase from marine *Nocardiopsis sp.* was immobilized via ionotropic gelation technique using gellan gum. The entrapped enzyme in gellan gum beads was demonstrated for sustained release of enzyme over 7 h, which may have application in multiparticulate delivery system and digestive enzyme formulation including related biotechnological processes in pharmaceutics (Chakraborty et al. 2014). Apart from this, the α-amylase have ability to remove the tough starch stains from the cloths. The agarose and agar entrapped α-amylase also showed its potential use as additive in detergents, for the removal of starch stain from cloths. The washing performance of the detergents alone and synergistic effect of detergents with immobilized α-amylase revealed better washing performance (Prakash and Jaiswal 2011). The gelatin immobilized α-amylase with extraordinary thermal stability has also been studied to exhibit its role in detergents formulations (Jaiswal and Prakash 2011). The zirconia coated arylamine glass beads affixed inside a plastic beaker immobilized α-amylase with prolonged storage stability and reusability have showed potential for removal of starch stains as additive in detergents (Rani et al. 2007).

The radish α-amylase immobilized on green synthesized silver nanoparticles have showed enhanced stability towards various organic solvents and surfactants with remarkable reusability, appeared as potential candidate for removal of tough starch stains with superior washing performance as compared to detergents alone (Khare 2018). Similarly, α-amylase immobilized on silica nanoparticles also witnessed its efficient use for removal of starch stain (Soleimani et al. 2012). Hence, these studies suggested immobilized α-amylase as more promising candidate in detergent formulations providing improved washing performance and environmental feasibility. However, the Luffa operculata fibre (LOF) immobilized α-amylase as LOF-amy fragment with higher operational and storage stability have been used as kitchen grease trap component retaining 30% residual activity over a month. The study suggested the promising role of immobilized enzyme on biodegradable support material for wastewater treatment (Morais et al. 2013).

8 Summary

Immobilization is the strategy adopted to improve the properties of enzymes considering its activity, ease of recovery and reusability including operational as well as storage stability. These properties of immobilized enzymes are of utmost importance for industrial applications providing economical and environmental feasibility. In this context, the present chapter provides an insight about enzyme immobilization and various strategies considering its advantages as well as its shortcomings. The chapter gives an overview of industrial application of immobilized alpha amylases. It highlights the role of emerging technologies viz., the area of nanotechnology coupled with biotechnological advancement, which unfolds new dimension of research in the arena of enzyme immobilization. The excellent properties of nanomaterials have opened new opportunities for efficient use of enzyme with enhanced catalytic activity, stability and reusability. Moreover, the immobilized α-amylase have attracted wider interest considering its industrial relevance for various analytical application. Henceforth, the ongoing research in the field of nanotechnology suggests huge possibilities for enzyme immobilization, thereby making it more suitable for industrial processes. Moreover, there also exists need to develop cost effective and environment friendly nanomatrices with novel properties for enzyme immobilization; hence, leading the process industrially efficient.

References

Agyei D, Shanbhag BK, He L (2015) Enzyme engineering (immobilization) for food applications. In: Improving and tailoring enzymes for food quality and functionality. Woodhead, Sawston, pp 213–235

Amirbandeh M, Taheri-Kafrani A (2016) Immobilization of glucoamylase on triazinefunctionalized Fe_3O_4/graphene oxide nanocomposite: improved stability and reusability. Int J Biol Macromol 93:1183–1191

Ansari SA, Husain Q (2012) Potential applications of enzymes immobilized on/in nano materials: a review. Biotechnol Adv 30(3):512–523

Antony N, Balachandran S, Mohanan PV (2016) Immobilization of diastase a-amylase on nano zinc oxide. Food Chem 211:624–630

Ashly PC, Joseph MJ, Mohanan PV (2011) Activity of diastase α-amylase immobilized on polyanilines (PANIs). Food Chem 27(4):1808–1813

Baskar G, Banu NA, Leuca GH, Gayathri V, Jeyashree N (2015) Magnetic immobilization and characterization of α-amylase as nanobiocatalyst for hydrolysis of sweet potato starch. Biochem Eng J 102:18–23

Basturk E, Demir S, Danis O, Kahraman MV (2013) Covalent immobilization of a-amylase onto thermally crosslinked electrospun PVA/PAA nanofibrous hybrid membranes. J Appl Polym Sci 127(1):349–355

Bayramoglu G, Yilmaz M, Arica MY (2004) Immobilization of a thermostable α-amylase onto reactive membranes: kinetics characterization and application to continuous starch hydrolysis. Food Chem 84(4):591–599

Biju V (2014) Chemical modifications and bioconjugate reactions of nanomaterials for sensing, imaging, drug delivery and therapy. Chem Soc Rev 43(3):744–764

Brady D, Jordaan J (2009) Advances in enzyme immobilisation. Biotechnol Lett 31:1639–1650

Bruna F, Pereira MG, Polizeli MDLT, Valim JB (2015) Starch biocatalyst based on α-Amylase-Mg/Al-layered double hydroxide nanohybrids. ACS Appl Mat Interf 7(33):18832–18842

Cakmakci E, Danis O, Demir S, Mulazim Y, Kahraman MV (2013) Alpha-amylase immobilization on epoxy containing thiol-ene photocurable materials. J Microbiol Biotechnol 23:205–210

Chakraborty S, Jana S, Gandhi A, Sen KK, Zhiang W, Kokare C (2014) Gellan gum microspheres containing a novel α-amylase from marine *Nocardiopsis sp.* strain B2 for immobilization. Int J Biol Macromol 70:292–299

Chakraborty S, Rusli H, Nath A, Sikder J, Bhattacharjee C, Curcio S, Drioli E (2016) Immobilized biocatalytic process development and potential application in membrane separation: a review. Crit Rev Biotechnol 36(1):43–58

Chen M, Zeng G, Xu P, Lai C, Tang L (2017) How do enzymes 'meet' nanoparticles and nanomaterials? Trends Biochem Sci 1391:1–17

Cordeiro AL, Lenk T, Werner C (2011) Immobilization of bacillus licheniformis α-amylase onto reactive polymer films. J Biotechnol 154(4):216–221

Daneils J, Farmer DM (1981) UK Patent No. 2070002A

Datta S, Christena LR, Rajaram YRS (2013) Enzyme immobilization: an overview on techniques and support materials. 3 Biotech 3:1–9

Defaei M, Taheri-Kafrani A, Miroliaei M, Yaghmaei P (2018) Improvement of stability and reusability of α-amylase immobilized on naringin functionalized magnetic nanoparticles: a robust nanobiocatalyst. Int J Biol Macromol 113:354–360

Demir S, Gok SB, Kahraman MV (2012) α-Amylase immobilization on functionalized nano CaCO3 by covalent attachment. Starch-Starke 64(1):3–9

Dey G, Bhupinder S, Banerjee R (2003) Immobilization of alpha-amylase produced by Bacillus circulans GRS 313. Braz Arch Biol Technol 46(2):167–176

DiCosimo R, McAuliffe J, Poulose AJ, Bohlmann G (2013) Industrial use of immobilized enzymes. Chem Soc Rev 42(15):6437–6474

Drevon GF (2002) Enzyme immobilization into polymers and coatings. Doctoral dissertation, University of Pittsburgh

Du L, Huang M, Feng JX (2017) Immobilization of α-amylase on eggshell membrane and Ag-nanoparticle-decorated eggshell membrane for the biotransformation of starch. Starch-Starke 69:9–10

Dwevedi A (2016) Basics of enzyme immobilization. In: Enzyme immobilization. Springer, Cham, pp 21–44

Edet E, Ntekpe M, Omereji S (2013) Current trend in enzyme immobilization: a review. Int J Mod Biochem 2(1):31–49

Eldin M, Seuror E, Nasr M, Tieama H (2011) Affinity covalent immobilization of glucoamylase onto benzoquinone-activated alginate beads: II. Enzyme immobilization and characterization. Appl Biochem Biotechnol 164:45–57

Es I, Vieira JDG, Amaral AC (2015) Principles, techniques and applications of biocatalyst immobilization for industrial application. Appl Microbiol Biotechnol 99(5):2065–2082

Eslamipour F, Hejazi P (2016) Evaluating effective factors on the activity and loading of immobilized α-amylase onto magnetic nanoparticles using a response surface-desirability approach. RSC Adv 6(24):20187–20197
Esmaeilnejad-Ahranjani P, Kazemeini M, Singh G, Arpanaei A (2016) Study of molecular conformation and activity-related properties of lipase immobilized onto core–shell structured polyacrylic acid-coated magnetic silica nanocomposite particles. Langmuir 32(13):3242–3252
Franssen MC, Steunenberg P, Scott EL, Zuilhof H, Sanders JP (2013) Immobilised enzymes in biorenewables production. Chem Soc Rev 42(15):6491–6533
Gospodinova N, Terlemezyan L (1998) Conducting polymers prepared by oxidative polymerization: polyaniline. Prog Polym Sci 23(8):1443–1484
Graebin NG, Schoffer JDN, Andrades DD, Hertz PF, Ayub MA, Rodrigues RC (2016) Immobilization of glycoside hydrolase families GH1, GH13, and GH70: state of the art and perspectives. Molecules 21(1074):1–38
Guo H, Tang Y, Yu Y, Xue L, Qian JQ (2016) Covalent immobilization of α-amylase on magnetic particles as catalyst for hydrolysis of high-amylose starch. Int J Biol Macromol 87:537–544
Haghighat-Kharazi S, Milani JM, Kasaai MR, Khajeh K (2018) Microencapsulation of α-amylase in beeswax and its application in gluten-free bread as an anti-staling agent. LWT 92:73–79
Hartmeier W (1988) Immobilized biocatalysts: an introduction. Springer, Berlin
Hernandez K, Fernandez-Lafuente R (2011) Control of protein immobilization: coupling immobilization and site-directed mutagenesis to improve biocatalyst or biosensor performance. Enzym Microb Technol 48:107–122
Hettiarachchy NS, Feliz DJ, Edwards JS, Horax R (2018) The use of immobilized enzymes to improve functionality. In: Proteins in food processing, 2nd edn. Woodhead Publishing, Elsevier, Duxford, pp 569–597
Homaei A, Saberi D (2015) Immobilization of α-amylase on gold nanorods: an ideal system for starch processing. Process Biochem 50(9):1394–1399
Homaei AA, Sariri R, Vianello F, Stevanato R (2013) Enzyme immobilization: an update. J Chem Biol 6(4):185–205
Hosseinipour SL, Khiabani MS, Hamishehkar H, Salehi R (2015) Enhanced stability and catalytic activity of immobilized α-amylase on modified Fe_3O_4 nanoparticles for potential application in food industries. J Nanopart Res 17(9):1–13
Husain Q (2010) β Galactosidases and their potential applications: a review. Crit Rev Biotechnol 30 (1):41–62
Husain Q (2016) Magnetic nanoparticles as a tool for the immobilization/stabilization of hydrolases and their applications: an overview. Biointerf Res Appl Chem 6(6):1585–1606
Husain Q (2017a) Nanomaterials as novel support for the immobilization of amylolytic enzymes and their applications: a review. DeGruyter Biocatalysis 3(1):37–53
Husain Q (2017b) Nanomaterials immobilized cellulolytic enzymes and their industrial applications: a literature review. JSM Biochem Mol Biol 4(3):1029
Husain Q (2017c) High yield immobilization and stabilization of oxidoreductases using magnetic nanosupports and their potential applications: an update. Current Catal 6(3):168–187
Husain Q (2017d) Nanosupport bound lipases their stability and applications. Biointerf Res Appl Chem 7(6):2194–2216
Husain Q (2018a) Carbon nanotubes mediated immobilized glucose oxidase biosensors as an effective and sensitive analytical tool. Biointerf Res Appl Chem 8(1):3060–3074
Husain Q (2018b) Nanocarriers immobilized proteases and their industrial applications: an overview. J Nanosci Nanotechnol 18(1):486–499
Husain Q, Ulber R (2011) Immobilized peroxidase as a valuable tool in the remediation of aromatic pollutants and xenobiotic compounds: a review. Crit Rev Environ Sci Technol 41(8):770–804
Hwang ET, Gu MB (2013) Enzyme stabilization by nano/microsized hybrid materials. Eng Life Sci 13(1):49–61
Isobe N, Lee DS, Kwon YJ, Kimura S, Kuga S, Wada M, Kim UJ (2011) Immobilization of protein on cellulose hydrogel. Cellulose 18(5):1251–1256
Jaiswal N, Prakash O (2011) Immobilization of soybean α-amylase on gelatin and its application as a detergent additive. Asian J Biochem 6(4):337–346

Jaiswal N, Prakash O, Talat M, Hasan SH, Pandey RK (2012) α-Amylase immobilization on gelatin: optimization of process variables. J Genetic Eng Biotechnol 10(1):161–167
Jesionowski T, Zdarta J, Krajewska B (2014) Enzyme immobilization by adsorption: a review. Adsorp 20(5–6):801–821
Jia H, Zhu G, Vugrinovich B, Kataphinan W, Reneker DH, Wang P (2002) Enzyme-carrying polymeric nanofibers prepared via electrospinning for use as unique biocatalysts. Biotechnol Prog 18:1027–1032
Johnson PA, Park HJ, Driscoll AJ (2011) Enzyme nanoparticle fabrication: magnetic nanoparticle synthesis and enzyme immobilisation. Methods Mol Biol 679:183–191
Karam EA, Wahab WAA, Saleh SA, Hassan ME, Kansoh AL, Esawy MA (2016) Production, immobilization and thermodynamic studies of free and immobilized *Aspergillus awamori* amylase. Int J Biol Macromol 102:694–703
Khan MJ, Husain Q (2014) Influence of pH and temperature on the activity of SnO_2-bound alpha-amylase: a genotoxicity assessment of SnO_2 nanoparticles. Prep Biochem Biotechnol 44 (6):558–571
Khan MJ, Qayyum S, Husain Q, Alam F (2011) Effect of tin oxide nanoparticles binding on the structure and activity of α-amylase from *Bacillus amyloliquefaciens*. Nanotechnology 22 (45):455708
Khan MJ, Husain Q, Azam A (2012) Immobilization of porcine pancreatic α-amylase on magnetic Fe_2O_3 nanoparticles: applications to the hydrolysis of starch. Biotechnol Bioprocess Eng 17 (2):377–384
Khan MJ, Husain Q, Ansari SA (2013) Polyaniline-assisted silver nanoparticles: a novel support for the immobilization of α-amylase. Appl Microbiol Biotechnol 97(4):1513–1522
Khan M, Asmat S, Husain Q (2018) Synthesis, characterization, and applications of nanographene-armored enzymes. Method Enzymol 609:83–142
Khare S (2018) Physico-chemical characterization and immobilization studies of alpha amylase on green synthesized silver nanoparticles, Ph.D. Thesis. Banaras Hindu University, Varanasi, UP
Khare S, Prakash O (2017a) Current developments in biotechnology and bioengineering: production, isolation and purification of industrial products. J Clean Prod 158:380–381
Khare S, Prakash O (2017b) Purification and biochemical characterization of α-amylase from radish (Raphanus sativus L.) seeds using response surface methodology. Proc Natl Acad Sci India Sect B Biol Sci:1–10
Kim J, Jia H, Wang P (2006) Challenges in biocatalysis for enzyme based biofuel cells. Biotechnol Adv 24:296–308
Kim J, Grate JW, Wang P (2008) Nanobiocatalysis and its potential applications. Trends Biotechnol 26(11):639–646
Klapiszewski L, Zdarta J, Jesionowski T (2018) Titania/lignin hybrid materials as a novel support for α-amylase immobilization: a comprehensive study. Colloids Surf B: Biointerfaces 162:90–97
Konsoula Z, Liakopoulou-Kyriakides M (2006) Starch hydrolysis by the action of an entrapped in alginate capsules α-amylase from *Bacillus subtilis*. Process Biochem 41:343–349
Kragl U (1996) In: Enzymol I, Godfrey T, Wet S (eds) Immobilized enzymes and membrane reactor. Macmillan, London
Kress J, Zanaletti R, Amour A, Ladlow M, Frey JG, Bradley M (2002) Enzyme accessibility and solid supports: which molecular weight enzymes can be used on solid supports? An investigation using confocal Raman microscopy. Chem Eur J 8(16):3769–3772
Krishna PN (2011) Enzyme technology: pacemaker of biotechnology. PHI Learn, New Delhi, p 264
Kulshrestha Y, Husain Q (2006) Adsorption of peroxidases on DEAE-cellulose directly from ammonium sulphate fractionated proteins of bitter gourd. Enzym Microb Technol 38:470–477
Kumari A, Kayastha AM (2011) Immobilization of soybean (Glycine max) α-amylase onto chitosan and amberlite MB-150 beads: optimization and characterization. J Mol Catal B Enzym 69(1–2):8–14
Liu DM, Chen J, Shi YP (2018) Advances on methods and easy separated support materials for enzymes immobilization. TrAC Trends in Anal Chem 102:332–342

Marzadori C, Miletti S, Gessa C, Ciurli S (1998) Immobilization of jack bean urease on hydroxyapatite: urease immobilization on alkaline soils. Soil Biol Biochem 30:1485–1490

Mateo C, Palomo JM, Fernandez-Lafuente G, Guisan JM, Fernandez-Lafuente R (2007) Improvement of enzyme activity, stability and selectivity via immobilization techniques. Enzym Microb Technol 40:1451–1463

Matto M, Husain Q (2006) Entrapment of porous and stable Con A-peroxidase complex into hybrid calcium alginate-pectin gel. J Chem Technol Biotechnol 81:1316–1323

Matto M, Husain Q (2009) Decolorization of direct dyes by immobilized turnip peroxidase in batch and continuous processes. Ecotoxicol Environ Saf 72:963–971

Meridor D, Gedanken A (2013) Preparation of enzyme nanoparticles and studying the catalytic activity of the immobilized nanoparticles on polyethylene films. Ultrason Sonochem 20 (1):425–431

Michaelis L, Ehrenreich M (1908) The analysis of ferments by adsorption. Biochem J 10:283–290

Mohamed SA, Al-Harbi MH, Almulaiky YQ, Ibrahim IH, Salah HA, El-Badry MO, Abdel-Aty AM, Fahmy AS, El-Shishtawy RM (2018) Immobilization of *Trichoderma harzianum* α-amylase on PPyAgNp/Fe3O4-nanocomposite: chemical and physical properties. Artif cells Nanomed Biotechnol:1–6

Morais RR, Pascoal AM, Caramori SS, Lopes FM, Fernandes KF (2013) Immobilization of-amylase onto *Luffa operculata* fibers. Enzym Res 2013:803415

Nadar SS, Muley AB, Ladole MR, Joshi PU (2016) Macromolecular cross-linked enzyme aggregates (M-CLEAs) of α-amylase. Int J Biol Macromol 84:69–78

Nair S, Kim J, Crawford B, Kim SH (2007) Improving biocatalytic activity of enzyme-loaded nanofibers by dispersing entangled nanofiber structure. Biomacromolecules 8:1266–1270

Nelson JM, Griffin EG (1916) Adsorption of invertase. J Am Chem Soc 38(5):1109–1115

Norouzian D (2003) Enzyme immobilization: the state of art in biotechnology. Iran J Biotechnol 1 (4):197–206

Prakash O, Jaiswal N (2010) α-Amylase: an ideal representative of thermostable enzymes. Appl Biochem Biotechnol 160(8):2401–2414

Prakash O, Jaiswal N (2011) Immobilization of a thermostable-amylase on agarose and agar matrices and its application in starch stain removal. World Appl Sci J 13(3):572–577

Prakash O, Khare S (2015a) Industrial applications of novel alpha amylase. Prajna (Journal of Banaras Hindu University) Special Issue Sci Technol 60:1–5

Prakash O, Khare S (2015b) Applicability of nanomatrices immobilized α-amylase in biotechnology. HCTL Open Int J Technol Innovations Res 14:1–9

Rani P, Sharma M, Kumar V, Pundir CS (2007) Immobilization of amylase onto arylamine glass beads affixed inside a plastic beaker: kinetic properties and application. Indian J Biotechnol 6:230–233

Reddy NS, Annapoorna N, Sambasiva Rao KRS (2003) An overview of the microbial a-amylase family. Afric J Biotechnol 2(12):645–648

Reddy RCK, Srivastava PK, Dey PM, Kayastha AM (2004) Immobilization of pigeonpea (Cajanus cajan) urease on DEAE-cellulose paper strips for urea estimation. Biotechnol Appl Biochem 39:323–327

Reetz MT, Jaeger KE (1998) Overexpression, immobilization and biotechnological application of Pseudomonas lipases. Chem Phys Lipids 93(1):3–14

Reshmi R, Sanjay G, Sugunan S (2006) Enhanced activity and stability of α-amylase immobilized on alumina. Catal Commun 7(7):460–465

Reshmi R, Sanjay G, Sugunan S (2007) Immobilization of α-amylase on zirconia: a heterogeneous biocatalyst for starch hydrolysis. Catal Commun 8(3):393–399

Rositza Y, Linnea S, Volodymyr K, Ruth P, Anita LS Kajsa U (2012) ZnO materials and surface tailoring for biosensing. Front Biosci 4:254–278

Salleh AB, Abdul-Rahman RNZR, Basri M (2006) New lipases and proteases. Nova Publishers, New York, p 114

Sanchez VC, Jachak A, Hurt RH, Kane AB (2011) Biological interactions of graphene-family nanomaterials: an interdisciplinary review. Chem Res Toxicol 25(1):15–34

Sanjay G, Sugunan S (2005) Immobilization of α-Amylase onto K-10 montmorillonite: characterization and comparison of activity in a batch and a fixed-bed reactor. Clay Miner 40:499–510

Shahrestani H, Taheri-Kafrani A, Soozanipour A, Tavakoli O (2016) Enzymatic clarification of fruit juices using xylanase immobilized on 1, 3, 5-triazine-functionalized silica-encapsulated magnetic nanoparticles. Biochem Eng J 109:51–58

Sheldon RA (2011) Cross-linked enzyme aggregates as industrial biocatalysts. Org Process Res Develop 15:213–223

Sheldon RA, Pereira PC (2017) Biocatalysis engineering: the big picture. Chem Soc Rev 46 (10):2678–2691

Sheldon RA, Van Pelt S (2013) Enzyme immobilisation in biocatalysis: why, what and how. Chem Soc Rev 42:6223–6235

Singh BD (2009) Biotechnology expanding horizons. Kalyani, India

Singh K, Kayastha AM (2014) Optimal immobilization of α-amylase from wheat (Triticum aestivum) onto DEAE-cellulose using response surface methodology and its characterization. J Mol Catal B Enzym 104:75–81

Singh K, Srivastava G, Talat M, Srivastava ON, Kayastha AM (2015) α-Amylase immobilization onto functionalized graphene nanosheets as scaffolds: its characterization, kinetics and potential applications in starch based industries. Biochem Biophys Rep 3:18–25

Sohrabi N, Rasouli N, Torkzadeh M (2014) Enhanced stability and catalytic activity of immobilized α-amylase on modified Fe_3O_4 nanoparticles. Chem Eng J 240:426–433

Soleimani M, Khani A, Najafzadeh K (2012) α-Amylase immobilization on the silica nanoparticles for cleaning performance towards starch soils in laundry detergents. J Mol Catal B Enzym 74 (1–2):1–5

Soozanipour A, Taheri-Kafrani A, Isfahani AL (2015) Covalent attachment of xylanase on functionalized magnetic nanoparticles and determination of its activity and stability. Chem Eng J 270:235–243

Straksys A, Kochane T, Budriene S (2016) Catalytic properties of maltogenic α-amylase from Bacillus stearothermophilus immobilized onto poly (urethane urea) microparticles. Food Chem 211:294–299

Svec F, Gemeiner P (1996) Engineering aspects of carriers for immobilized biocatalysts. Biotechnol Genet Eng Rev 13(1):217–236

Talebi M, Vaezifar S, Jafary F, Fazilati M, Motamedi S (2016) Stability improvement of immobilized α-amylase using nano pore zeolite. Iran J Biotechnol 14(1):33–38

Talekar S, Ghodake V, Ghotage T, Rathod P, Deshmukh P, Nadar S, Mulla M, Ladole M (2012) Novel magnetic cross-linked enzyme aggregates (magnetic CLEAs) of alpha amylase. Bioresour Technol 123:542–547

Tenne R (2006) Inorganic nanotubes and fullerene-like nanoparticles. J Mater Res 21 (11):2726–2743

Tischer W, Wedekind F (1999) Immobilized enzymes: methods and applications. In: Biocatalysis-from discovery to application. Springer, Berlin, Heidelberg, pp 95–126

Tosa T, Mori T, Fuse N, Chibata I (1966) Studies on continuous enzyme reactions. I. Screening of carriers for preparation of water-insoluble aminoacylase. Enzymologia 31:214–224

Tripathi P, Kumari A, Rath P, Kayastha AM (2007) Immobilization of α-amylase from mung beans (Vigna radiata) on Amberlite MB 150 and chitosan beads: a comparative study. J Mol Catal B Enzym 49(1–4):69–74

Verma ML, Barrow CJ, Puri M (2013) Nanobiotechnology as a novel paradigm for enzyme immobilisation and stabilisation with potential applications in biodiesel production. Appl Microbiol Biotechnol 97(1):23–39

Veum L, Hanefeld U (2006) Carrier enabled catalytic reaction cascades. Chem Commun (8):825–831

Wang P (2006) Nanoscale biocatalyst systems. Curr Opin Biotechnol 17(6):574–579

Yiu HHP, Keane MA (2012) Enzyme-magnetic nanoparticle hybrids: new effective catalysts for the production of high value chemicals. J Chem Technol Biotechnol 87:583–594

Zhai R, Zhang B, Liu L, Xie Y, Zhang H, Liu J (2010) Immobilization of enzyme biocatalyst on natural halloysite nanotubes. Catal Commun 12(4):259–263

Immobilized Peroxidase Catalyzed Decolorization and Degradation of Industrially Important Dyes from Polluted Water

Qayyum Husain

Abbreviations

BGP	bitter gourd peroxidase
Con A	concanavalin A
CLEAs	crosslinked enzyme aggregates
CP	catalase peroxidase
CPP	cucurbita pepo peroxidase
CS	chitosan
DyP	dye decolorizing peroxidase
GA	glutaraldehyde
GMA	glycidyl methacrylate
GP	ginger peroxidase
HRP	horseradish peroxidase
LiP	lignin peroxidase
MnP	manganese peroxidase
MP-11	microperoxidase 11
PAN	polyacrylonitrile
PGP	pointed gourd peroxidase
rDyP	recombinant dye decolorizing peroxidase
SBP	soybean peroxidase
TP	turnip peroxidase
TMP	tomato peroxidase
UFM	ultrafiltration membrane
WRP	whiteradish peroxidase

Q. Husain (✉)
Department of Biochemistry, Faculty of Life Sciences, Aligarh Muslim University, Aligarh, India

Q. Husain, M. F. Ullah (eds.), *Biocatalysis*,
https://doi.org/10.1007/978-3-030-25023-2_7

1 Introduction

Synthetic dyes are used commonly in different industries ranging from food, textile, leather, painting, printing, plastic and pharmaceuticals. The majority of the dyes are recalcitrant, thus they can impart colour on various raw materials. Several dyes, dye precursors and aromatic amines derivatives obtained during degradation of dyes have been reported to be carcinogenic and mutagenic (Karim and Husain 2009; Zhu et al. 2012; Nguyen and Juang 2013). The majority of the textile dyes are harmful to the environment and potentially carcinogenic. The discharge of coloured compounds into the environment shared a small part of water pollution; however the coloured effluents create a serious human health and environmental problem (Husain 2006; Najafpoor and Davoudi 2017). The treatment of compounds present in wastewater coming out of industries is a great challenge for the last several decades; until now there is no single and cost effective method of their cleaning is available that can effectively remove dyes from textile and other industrial wastewater (Husain et al. 2009; Husain 2010).

Numerous chemical and physical methods or the combination of these have been applied for the remediation of pollutants from industrial effluents. In this century, chemical and other related industries focused a major shift both at the laboratory and industrial level from traditional chemical-based concepts to a more greener, sustainable and environmental-friendly catalytic alternative. Biocatalysts appeared more challenging remediation tools due to their various merits over the other classical methods. Enzymes based wastewater treatment has an advantage over the presently known physico-chemical methods because this method is environmental friendly and converts the pollutants into known compounds and it also does not create any sludge (Husain and Jan 2000; Husain and Qayyum 2013). These processes are energy efficient compared to traditionally used technology because it applied under moderate conditions, producing negligible amount of secondary by products (Lade et al. 2012; Silva et al. 2013).

The commercially available dyes are generally employed in distinct industries and greatly influenced the environment. Recently, the enzymatic approach has drawn great attention in the decolorization/degradation of industrially important dyes from wastewater as an alternative tool to conventional chemical, physical and biological treatments, which face serious limitations (Satar et al. 2012; Husain and Husain 2012; Bilal et al. 2017a). Although the biological methods, especially using enzymes, offer alternative and effective process for dye degradation and overcome the limitations of chemical and physical methods such as the instability, lack of reusability and high cost of free enzymes strictly, which limit their use in several scientific and technical applications. The activity of enzymes is rapidly decreasing in aqueous solutions against environmental changes due to very high susceptibility and unfavorable conformations. Enzymatic treatment appeared highly useful because of the action of enzymes on pollutants even when they are present in very dilute conditions and recalcitrant to the action of microbes involving in the degradation of dyes. The potential of a large group of oxidoreductive enzymes has been exploited in the decolorization and degradation of dyes (Chacko and Subramaniam 2011). Several dyes were found recalcitrant to the degradation/decolorization by the action of such enzymes. The addition of certain redox mediators enhanced the range of substrates and efficiency of degradation of recalcitrant compounds by peroxidases. Numerous redox mediators have been discussed in the literature, but very few of them are frequently used, e.g., 1-hydroxybenzotriazole (HOBT), veratryl alcohol (VA), violuric acid (VLA), 2-methoxy-phenothiazone (Husain and Husain 2008). Free enzymes cannot be applied at the large scale due to their limitations such as poor stability, difficulty to reuse them and non-applicability in continuous reactors. Therefore, the use of immobilized enzymes has significant advantages over soluble enzymes. In the near future, technology based on the enzymatic treatment of dyes present in the industrial effluents/wastewater will play a vital role. Treatment of wastewater on a large scale will also be possible by using reactors containing immobilized enzymes (Husain and Ulber 2011; Husain 2019).

The main objective of the present chapter is to review the literature based on the applications of distinct kinds of immobilized plant and microbial peroxidases engaged in the treatment of dyes present in wastewater/industrial effluents. The role of different immobilized peroxidases have been reviewed and described in detail. Diverse kinds of immobilized peroxidases have successfully been exploited in batch processes as well as in continuous reactors for the decolorization and degradation of commercially applied dyes. The toxicity and genotoxicity of the enzymatically targeted dyes products have also discussed in several cases.

2 Plant Peroxidases

Peroxidases are widely present in all vascular plants and are exploited for treatment of wide spectrum aromatic compound including distinct commercially used dyes (Akhtar et al. 2005a). These enzymes catalyze degradation of chemical structure of aromatic dyes either by precipitation or by opening the aromatic ring structure (Shaffiqu et al. 2002; Husain 2006). Plant peroxidases are stable catalysts that retain their biological activities over a broad range of pH and temperatures (Fatima et al. 2007; Fatima and Husain 2008). The performance of enzyme catalyzed decolorization and degradation processes depends on the structure of dyes and the operational parameters like concentration of dye, enzyme and H_2O_2, incubation time, pH and temperature. The recalcitrant and persistent dyes have successfully been treated by the action of peroxidases in the presence of redox mediators (Husain and Husain 2008; Satar et al. 2012). Thus, plant peroxidases are easily available, inexpensive, and ecofriendly biocatalysts for the bioremediation of polluted waters containing a wide range of synthetic dyes (Husain et al. 2009; Kalsoom et al. 2015).

2.1 Bitter Gourd Peroxidase

Bitter gourd (*Momordica charantia*) peroxidase (BGP) have been purified and characterized by various workers (Akhtar et al. 2005b; Fatima and Husain 2007). The Free BGP have used for the treatment of different phenolic compounds and synthetic dyes present in model wastewater (Akhtar et al. 2005c; Akhtar and Husain 2006). Table 1 demonstrates BGP immobilized preparations on various supports and their dyes/dye effluents decolorization applications. Concanavalin A (Con A)-Sephadex bound BGP has been used for the decolorization of reactive textile dyes (50–200 mg L^{-1}) in the buffer of pH 3.0–4.0 at 40 °C. The immobilized BGP retained about 50% dye decolorizing efficiency after 10 consecutive uses in batch processes. Mixtures of three, four and eight dyes were prepared and treated by both free and immobilized BGP. These mixtures were decolorized above 80% by immobilized BGP. Immobilized enzyme was capable of removing remarkably high concentration of color as well as total organic content (TOC) from individual dyes, mixture of dyes and dyeing effluent (Akhtar et al. 2005c). Calcium alginate-starch beads entrapped BGP was used for the degradation and decolorization of a textile industrial effluent in batch as well as in continuous reactor in the presence of a redox mediator, 1.0 mM HOBT. Above 70% of the effluent color was removed by immobilized BGP in a stirred batch process within 3 h. The bound BGP maintained 59% effluent decolorization activity after tenth repeated uses. The two-reactor system containing calcium alginate-starch entrapped BGP demonstrated about 50% decolorization efficiency even after 2 months of its continuous operation. The application of a two-reactor system containing immobilized enzyme and an adsorbent was highly useful in treating industrial effluents at large scale, and it had helped in producing pollutants free water (Matto et al. 2009).

BGP immobilized on the surface of Con A layered calcium alginate-starch beads was used along with few redox mediators for the decolorization of textile industrial

Table 1 BGP immobilized preparations on various supports and their dyes/dye effluents decolorization applications

Support material	Method of immobilization	Type of dye/effluent	Decolorization (%)	References
Con A Sephedex	Bioaffinity	Reactive textile dyes, their mixtures, Colored effluent	0–100	Akhtar et al. (2005c)
Calcium alginate-starch beads	Entrapment	Industrial effluents	70	Matto et al. (2009)
Con a layered calcium-alginate starch beads	Bioaffinity	Industrial effluents	90	Matto et al. (2009)
Calcium alginate-pectin beads	Entrapment	DB 1 and DR 17	51–85	Satar and Husain (2011)

effluent. The effluent was maximally decolorized upto 70% in the presence of 1.0 mM HOBT at pH 5.0 and 40 °C within 1 h. Immobilized BGP had efficiently removed above 90% color from the effluent after 3 h treatment in a batch process. The two-reactor system was capable of decolorizing 40% effluent even after 2 months of its continuous running (Matto and Husain 2009a). Calcium-alginate pectin entrapped BGP was used for decolorization and degradation of disperse dyes: disperse brown 1 (DB 1) and disperse red 17 (DR 17) in the presence of 0.2 mmol L^{-1} redox mediator, VLA. The dyes; DR 17 and DB 1were maximally decolorized to 85% and 51% at pH 3.0 and 60 °C within 2 h, respectively. The entrapped BGP (35 U) decolorized 98% DR 17 and 71% DB 1 in the batch processes in 90 min. UV-visible spectral analysis showed the loss of color and *Allium cepa* test exhibited removal of toxicity (Satar and Husain 2011).

2.2 *Horseradish Peroxidase*

Horseradish peroxidase (HRP) is one of the most popular peroxidase and has been extensively used in a number of commercial applications such as clinical, industrial and environmental (Veitch 2004; Alvarado and Torres 2009; Hamid and Rehman 2009). Table 2 depicts HRP immobilized on various supports and their dyes/dye effluents decolorization applications used acrylamide and alginate entrapped HRP for the decolorization of an azo dye. Acrylamide gel immobilized HRP showed better performance than the free HRP and alginate entrapped HRP. Alginate entrapped HRP exhibited poor dye decolorization compared to soluble enzyme. It was due to non-availability of the dye to the enzyme present within the polymeric network. Azo dyes are recalcitrant to the enzymatic degradation and persist for quite long during in the environment. An electroenzymatic process, an alternative to these processes, has been developed that include enzymatic catalysis and the electrochemical generation of H_2O_2. An electroenzymatic method along with immobilized HRP was developed to degrade orange II (azo dye) within a two-compartment packed-bed flow reactor. The orange II was partially decolorized due to its adsorption on the graphite felt. The overall application of the electroenzymatic led to a significantly higher degradation compared to the use of electrolysis alone. The obtained by-products were primarily consisted of an aromatic amine, sulfanilic acid and unknown compounds (Kim et al. 2005). In a further study, the similar group has investigated the use of electroenzymatic method for the degradation of orange II by using HRP bound to the inexpensive and stable inorganic beads and this preparation was employed in a continuous electrochemical reactor with in situ generation of H_2O_2. HRP was immobilized on Celite®R-646 as a porous support with 2% aqueous glutaraldehyde (GA), while the protein and activity yield were 3.6 mg protein and 5280 U g^{-1} celite, respectively. Orange II was degraded efficiently over 90% in a continuous operation for 36 h. In order to elucidate its breakdown pathway, the products were identified by GC/MS analysis. This study further strengthens

Table 2 HRP immobilized on various supports and their dyes/dye effluents decolorization applications

Support material	Method of immobilization	Type of dye/effluent	Decolorization (%)	References
Acrylamide/alginate	Entrapment	Acid dye		Mohan et al. (2005)
Celite (R) R-646	Covalent attachment	Orange II	90	Shim et al. (2007)
HMDA-GMA-g-PET	Covalent attachment	Azo dye	98	Arslan (2011)
Calcium alginate beads	Entrapment	AO7 & AB 25	75–84	Gholami-Borujeni et al. (2011)
Periodate activated dextran	Covalent attachment	CBB R-250	80–90	Altikatoglu and Celebi (2011)
CD-CS	Covalent attachment	Azo dye effluent	100	Karim et al. (2012)
PANI-g-PAN	Adsorption	Direct Blue 53, Direct Black 38	–	Bayramoglu et al. (2012)
Calcium alginate beads	Entrapment	AB25 & AO7	80	Gholami-Borujeni et al. (2013)
Polysulfone	Adsorption	Acid Blue 25 & Reactive Blue 19	70–100	Celebi et al. (2013)
Calcium alginate	Entrapment	Acid Blue 113 & tannery effluent	76	Preethi et al. (2013)
Polyurethane foam	Adsorption	Acid Red dye	61.2	Malani et al. (2013)
Fumed silica	Adsorption	AV109	55.1	Šekuljica et al. (2015)
Kaolin	Adsorption	AV109	87	Šekuljica et al. (2016a)
CLEA	Chemical aggregation	AV109	88.4–88.9	Šekuljica et al. (2016b, c)
CS, Al_2O_3 gel	Covalent attachment, adsorption, entrapment	CBB R-250, RBBR, Reactive Blue 52, Acridine Orange	78–100	Janović et al. (2017)
EPMA-g-PP	Covalent attachment	Basic Red 29	90	Kumar et al. (2016)
ZnO nanowires/SiO_2 composite/alginate	Covalent attachment	Acid Blue 113, Acid Black 10BX	90.3–95.9	Sun et al. (2017)
Calcium alginate beads	Entrapment	Reactived Blue 221 Reactive Blue 198	93 75	Farias et al. (2017)
CLEA	Chemical aggregation	MO Basic Red Indigo Rhodamin B Rhodamin 6G	94.9 91.73 84.35 84.47 73.6	Bilal et al. (2017c)
PAN beads		Acid Orange 20	98	Yincan et al. (2017)

applicability of an electroenzymatic process to degrade azo dyes from wastewater (Shim et al. 2007).

Gel/alginate entrapped HRP has been used in batch processes in the presence of H_2O_2 for the oxidation of direct yellow-12 dye. Immobilized HRP was quite effective in the reduction of this dye. The immobilized enzyme beads were reused 2–3 times for the removal of the dye with lower efficiency (Maddhinni and Vurimindi 2006). Bromophenol Blue (BPB) and methyl orange (MO) remediation potential of citraconic anhydride-modified HRP has been compared with its free form. Modified HRP showed a good decolorization over a wide range of dye concentration from 8 to 24 or 32 μmol L^{-1} at 300 μmol L^{-1} H_2O_2, it will be highly useful for industrial applications (Liu et al. 2006). Three fiber-forming polymeric materials (cellulose, chitosan (CS), and ethylene-vinyl alcohol copolymer) immobilized HRP was efficiently employed for the decolorization of dye solutions containing orange II, crystal violet (CV), astrogen red, and remazol brilliant blue R (RBBR) (Maki et al. 2006). Poly(ethylene terephthalate) (PET) fibers grafted with glycidyl methacrylate (GMA) using benzoyl peroxide (Bz_2O_2) as initiator and 1,6-diaminohexane (HMDA) was then covalently attached to GMA-PET fibers. HMDAGMA-*g*-PET fibers were activated by GA and it was used to immobilize HRP. Both free and immobilized enzyme preparations were applied in batch processes for the degradation of azo dye. About 98% of azo dye was removed by immobilized HRP in 45 min, while 79% dye was removed by free enzyme (Arslan 2011).

Calcium alginate gel beads immobilized HRP has been used for the decolorization textile of industrial effluent. The maximum dye removal was obtained at the ratio of H_2O_2/dye as 1.25. Optimum conditions for removal of acid orange 7 (AO7) and acid blue 25 (AB25) were: 90 min, reaction time; 0.8 U g^{-1}, enzyme concentration for 500 ADMI color, ratio of H_2O_2/dye, 1.25/1; temperature, 25–50 °C that cause 75% and 84% removal in AO7 and AB25 solution, respectively. There was a significant reduction in toxicity of dye solution after enzymatic treatment. The immobilized HRP catalyzed 80% color removal from polluted wastewater. Alginate entrapped enzyme was reused 10 times without any loss in its activity (Gholami-Borujeni et al. 2011, 2013). Preethi et al. (2013) used calcium alginate entrapped HRP (0.08 U) to degrade 3 mL of 30 mg L^{-1} C. I. acid blue 113 dye in the presence of 14 μL of H_2O_2 at pH 6.6 and 30 °C. The dye was maximally decolorized 76% within 45 min under the optimal experimental conditions. The immobilized HRP was recycled for 3 times for dye degradation. Kinetic parameters; showed lower K_m and higher V_{max} values, which demonstrated a higher affinity of HRP for the dye. The entrapped HRP was also employed for the treatment of real tannery dye-house wastewater.

HRP was attached to periodate-activated dextran and modified HRP has decolorized comassie brilliant blue (CBB) R-250 to 80–90% within 5 min and showed better performance as compared to free enzyme (Altikatoglu and Celebi 2011). Karim et al. (2012) coupled HRP to the β-cyclodextrin (CD)–CS complex and used this preparation for the removal of color from a textile effluent containing azo dyes. The maximum dye oxidation from the diluted textile effluent was observed in the presence of 0.6 mM H_2O_2 and 0.2 ml^{-1} at pH 8.0 and 40 °C for 2 h. In a batch process 100% color of the effluent was removed by immobilized HRP within 120 min. Moreover, β-CD–CS–HRP-containing bed-reactor was operated continuously and efficiently for complete removal of the dye from the effluent during 15 days of operation. The crosslinked HRP has shown its superiority over the un-crosslinked enzyme in the decolorization of dyes. The comet assay based genotoxicity test has demonstrated that the treated effluent was completely free from toxicity. Polyaniline (PANI) grafted polyacrylonitrile (PAN) film immobilized HRP was employed for the decolorization of two different benzidine-based dyes; direct blue 53 and direct black 38 in the presence of H_2O_2. These dyes were decolorized maximally at pH 6.0. The HRP-immobilized on PAN-g-PANI-3 film was highly effective in the removal of direct blue-53 from aqueous solution compared to direct black-38 (Bayramoglu et al. 2012). HRP immobilized covalently onto various polysulfone supports for the decolorization of textile diazo (acid black 1) and anthraquinone (reactive blue 19) dyes. The immobilized HRP was used 7-times in freshly prepared dye solutions for 63 days and it retained 70% of its dye decolorizing activity after 3-repeated uses (Celebi et al. 2013). Malani et al. (2013) have attempted to give a mechanistic insight into the interaction of sonochemical and enzymatic treatments using polyurethane foam immobilized HRP on the decolourization of acid red dye (an azo dye). The combined effect of enzymatic and sonochemical treatments have shown the highest decolourization of acid red (61.2%) in the presence of polyethylene glycol (PEG). HRP encapsulated in phospholipid-templated titania particles via biomimetic titanification was taken for the remediation of water contaminated with phenolic compounds and dye. The encapsulated HRP showed 79.72% removal efficiency for Direct Black-38 in the first treatment cycle. Moreover, the immobilized enzyme demonstrated superior dye removal efficiency and better reutilization compared to free HRP (Jiang et al. 2014).

Fumed silica adsorbed unpurified peroxidase (1.5 mg mL^{-1}) was used to optimize the decolorization of anthraquinone dye acid violet 109 (AV109) and the immobilized HRP was employed in a batch reactor. The obtained decolorization of dye was 55.1% under specified experimental conditions; pH 4.0, dye concentration 30 mg L^{-1}, H_2O_2 concentration 0.1 mM within 30 min. The immobilized HRP demonstrated remarkably high resistance against inhibitory influence of the dye and H_2O_2 than the free enzyme (Šekuljica et al. 2015). In a further study the same group has reported that the kaolin adsorbed HRP (0.1 U) decolorized 87% of the anthraquinone dye, AV109 under optimal experimental conditions at pH 5.0, 24 °C, dye concentration 40 mg L^{-1} and 0.2 mM of H_2O_2 within 40 min. The immobilized enzyme retained 35 ± 0.9% dye decolorization activity in a batch process after 7 repeated uses (Šekuljica et al. 2016a). Ten preparations of HRP-A isoenzyme were constructed by covalent binding of enzyme to CS and alginate, adsorption followed by cross-linking on Al_2O_3, and encapsulation in spherical calcium alginate beads via polyethylene glycol (PEG). The immobilized preparations; CS-HRP, Al-Gel-HRP and Al-HRP-Gel were investigated for the decolorization of dyes (50 to 175 mg L^{-1}). The impact of H_2O_2 concentration on dye decolorization was examined on the CS-HRP and the H_2O_2/dye ratio has played an important role in decolorization efficiency. The concentrations of H_2O_2 (0.22 to 4.4 mM) showed no difference in terms of impact on the biocatalyst decolorization efficiency. The high decolorization efficiency of the biocatalysts was validated by the removal of 25 and 100 mg L^{-1} anthraquinone (RBBR), triphenylmethane (CBB), acridine (acridine orange), and formazan metal complex dye (reactive blue 52). The decolorization was 53, 78, and 67% of their initial colors by Al-HRP-Gel, Al-Gel-HRP, and CS-HRP even after 7 repeated uses, respectively (Janović et al. 2017). Kumar et al. (2016) covalently immobilized HRP onto epoxy functionalized polypropylene (PP) films, fabricated *via* 60 Co-γ-radiation induced mutual irradiation grafting of 2,3-epoxypropyl methacrylate (EPMA) on to PP matrix. The immobilized HRP was applied for the treatment of textile basic red 29 dye wastewater. The immobilized enzyme successfully degraded, ~90% of the dye for over 20 days and was reused 5-times without any significant loss in its activity.

An effort has been done to develop a carrier-free technique for commercial HRP immobilization. The immobilized biocatalyst, HRP-CLEAs with 580 U g^{-1} of the activity was obtained under the following immobilization conditions: precipitation reagent 80% ammonium sulphate, cross-linking reagent 1% of GA and protein-fedder, bovine serum albumin (BSA) concentration 5 mg mL^{-1}. The obtained HRP-CLEAs showed great affinity towards anthraquinone dye, AV109 and it oxidized 88.4% of the dye in the buffer of pH 4, dye concentration 40 mg L^{-1} H_2O_2 concentration 1 mM and 0.1 U of HRP-CLEAs. The immobilized biocatalyst was used 5 and 8-oxidation cycles of the dye and pyrogallol and retained more than 80% of its original activity (Šekuljica et al. 2016b). In a further study these workers employed HRP-CLEAs for the oxidation of AV109 under experimental conditions of pH 4, dye concentration 30 mg L^{-1}, H_2O_2 concentration 0.1 mM. The dye was decolorized to 72.4% and 88.9% in a batch process and continuous packed bed reactor, respectively. CLEAs-HRP showed higher stability in the packed bed reactor (PBR) compared to batch reactor where the mechanical instability of aggregates under constant agitation has been noticed. The activity of CLEAs-HRP was maintained to 60 and 20% in PBR and batch process after 7 repeated applications, respectively. The reduction in the toxicity of the treated samples was evaluated by using brine shrimp, *Artemia salina assay* (Šekuljica et al. 2016c).

Bilal et al. (2016a) investigated the covalent immobilization of HRP on the calcium-alginate support via GA cross-linking and used this preparation for the detoxification and degradation of synthetic dyes. Moreover, the immobilized-HRP showed potential efficiency for the decolorization of dyes in sequential dye-decolorizing batch reactions. Cytotoxicity analysis using a plant bioassay and acute test demonstrated that the Ca-alginate immobilized-HRP was effectively used to clear the toxicity of treated dyes and retained a great potential for large-scale environmental remediation. In a most recent study, these workers have investigated the entrapment of HRP into CS beads and employed for the degradation of textile dyes. Moreover, the CTS-assisted entrapped-HRP was also used for the decolorization of four different textile dyes i.e. RBBR, reactive black 5 (RB5), Congo red (CR) and crystal violet (CV). The CTS-HRP showed considerable decolorization efficacy in six consecutive batch operations. The results of this work showed that

CTS-HRP is an attractive choice for use as industrial biocatalyst in large scale bioremediation of textile dyes and effluents (Bilal et al. 2017b).

A ZnO nanowires/macroporous SiO_2 composite was used as support to immobilize HRP by in-situ cross-linking method. The immobilized HRP showed high azo dye decolorization efficiency. The decolorization of acid blue 113 and acid black 10 BX was obtained as 95.4% and 90.3% under optimal experimental conditions, respectively. The immobilized HRP decolorized maximum dye as 50 mg L^{-1} within 35 min. The storage stability and reusability were greatly improved upon immobilization, from the decolorization of acid blue 113 it was found that 80.4% of initial efficiency retained after incubation at 4 °C for 60 days, and that 79.4% of decolorization potential retained after 12 repeated uses (Sun et al. 2017). A ZnO nanowires/macroporous SiO_2 composite was used as support to immobilize HRP simply by *in-situ* cross-linking method. As cross-linker was adsorbed on the surface of ZnO nanowires, the cross-linked HRP was quite different from the traditional cross-linking enzyme aggregates (CLEAs) on both structure and catalytic performance. Among three epoxy compounds diethylene glycol diglycidyl ether was found most suitable cross-linker, with which the loading capacity of HRP with pI of 5.3 obtained as high as 118.1 mg g^{-1} and specific activity was up to 14.9 U mg^{-1} support. The catalytic performance of immobilized HRP to decolorize anthraquinone dye was explored by using reactive blue 19 and AV109 as model substrates. The results revealed that the immobilized HRP demonstrated high decolorization efficiency and good reusability. The obtained decolorization was 94.3% and 95.9% of their original concentrations of the dyes, AV109 and reactive blue 19 within 30 min, respectively. A complete decolorization of these two dyes has been realized within 2–3 h by using this new biocatalysis system (Sun et al. 2018). HRP was cross-linked using a cross-linking agent, i.e., ethylene glycol-bis [succinic acid N-hydroxysuccinimide, (EG-NHS)], which is mild in nature as compared to GA. HRP-CLEAs were considered to investigate their bio-catalytic efficacy for bioremediation in a PBR system. A maximal of 94.26% degradation of textile-based methyl orange (MO) dye was recorded within a short time, following 91.73% degradation of basic red 9, 84.35% degradation of indigo, 81.47% degradation of rhodamin B, and 73.6% degradation of rhodamine 6G, respectively. HRP-CLEAs retained about 60% MO degradation efficiency after 7 reuses. There was a marked reduction in the toxicity after HRP-CLEAs treatment (Bilal et al. 2017c).

The synthetic effluents comprised of reactive blue 221 (RB 221) and reactive blue 198 (RB 198) were decolorized and treated by Ca-alginate beads immobilized HRP. A maximum cleaning of RB 221 and RB 198 at pH 5.5 and 30 °C, concentration of H_2O_2 of 43.75 μM, within 3 h was obtained as 93 and 75%. The HRP enzyme immobilized in Ca-alginate capsules showed a great potential for biotechnological applications, especially for the removal of reactive dyes (Farias et al. 2017). Yincan et al. (2017) immobilized HRP onto PAN based beads and used this preparation for the decolorization of acid orange 20 in synthetic solution. The obtained decolorization of dye was about 90% or 98% by free and immobilized HRP, respectively. Immobilized HRP decolorized 90% of the dye after 3 repeated uses. Bilal et al. (2017d) investigated the immobilization of HRP onto self-fabricated PVA-alginate beads using sodium nitrate as a cross-linker. The immobilized-HRP was used for MO degradation in a batch process. The degraded fragments of MO were monitored by ultra-performance liquid chromatography coupled with mass spectrometry and plausible degradation pathway for MO was proposed based on the identified intermediates. Recently, HRP was immobilized by copolymerization into cross-linked polyacrylamide gel and employed for the degradation and detoxification of MO. This immobilized HRP preparation could decolorize 90% of dye in a PBR. The acute toxicity assays showed that enzyme-based technology might be significantly useful for cleaning of dyes from wastewater at large-scale (Bilal et al. 2018).

2.3 *Other Plant Peroxidases*

Table 3 summarizes other plant peroxidases immobilized preparations on various supports and their applications in the decolorization of dyes/dye effluents.

2.3.1 Soybean Peroxidase

The efficacy of aminated soybean peroxidase (SBP) immobilized on the carbodiimide activated corncob powder was used to decolorize BPB. Immobilized SBP was used for the discoloration of BPB (0.02 mmol L^{-1}) from solution. After 30 min, 93 and 89% discoloration was achieved with the 10 mmol L^{-1} and 50 mmol L^{-1} derivatives, respectively. Moreover, these derivatives retained 60% of the catalytic activity even after 3 reuses. SBP was immobilized on a low-cost corncob powder support exhibited improved thermal stability (Galárraga et al. 2013). SBP was used as biocatalyst, both free in solution and immobilized on silica monoliths and TiO_2 as photocatalyst. The combined action of two catalysts catalyzed 2–4 times faster removal of all investigated orange dyes compared to the single systems. All dyes were completely removed by immobilized SBP within 2 h. As for carbamazepine, photocatalytic treatment prevails on the enzymatic degradation, but the synergistic effect of two catalysts led to a more efficient degradation; carbamazepine's complete disappearance was achieved within 60 min with combined system, while up to 2 h is required only with TiO_2 (Calza et al. 2016).

2.3.2 Turnip Peroxidase

An inexpensive immobilized turnip peroxidase (TP) has been employed for the decolorization of some direct dyes in batch and continuous reactors. Wood shaving was investigated as an inexpensive material for the preparation of bioaffinity support. Con A-wood shaving bound TP exhibited 67% activity. Both soluble and immobilized TP could effectively remove more than 50% color from dyes in the presence of metals/salt and 0.6 mM 1-HOBT within 1 h. The columns containing

Table 3 Plant peroxidases immobilized preparations on various supports and their applications in the decolorization of dyes/dye effluents

Enzyme	Support Material	Method of Immobilization	Type of dye/effluent	Decolorization (%)	References
SBP	Carbodiimde-corncob powder	Covalent attachment	BPB	89–93	Galárraga et al. (2013)
SBP	Silica mono-lith, TiO_2	Adsorption	Orange dyes	100	Calza et al. (2016)
TP	Con A-wood shaving	Bioaffinity	Direct Red 23 and mixtures of direct dyes	64 50	Matto and Husain (2009b)
TP	Calcium algi-nate beads	Entrapment	CR	94	Ahmedi et al. (2015)
CPP	Calcium algi-nate beads	Entrapment	DY106	69.71–75	Boucherit et al. (2012, 2013)
GP	Amino-SiO_2-TiO_2 NC	Adsorption	Acid Yellow 42	90	Ali et al. (2017)
GP	PCeGONC	Adsorption	Reactive Blue 4	99	Ali et al. (2018a)
GP	PZrSeI-CENC	Adsorption	MV6B	96	Ali et al. (2018b)
GP	ANGG-GP and AGG-GP	Entrapment	Industrial effluent	99	Ali and Husain (2018)
PGP	DEAE cellulose	Adsorption	DR 19 and mix-ture of direct dyes	88.2 77.4	Jamal and Goel (2014)
TMP	Con A-cellulose	Bioaffinity	Direct Red 23 and Direct Blue 80	–	Matto and Husain (2008)
AIP	CS beads	Entrapment	–	–	Pandey et al. (2017)
WRP	Celite	Adsorption	Reactive Red 120 and Reactive Blue 171	–	Satar and Husain (2009)

immobilized TP have decolorized 64% direct red 23 and 50% mixture of direct dyes at 4 and 3 months of operation, respectively. TOC analysis of treated dye or mixture of dyes revealed that these results were quite comparable to the loss of color from solutions. Thus, this study showed that the immobilized enzyme could be efficiently used for the removal of synthetic dyes from industrial effluents (Matto and Husain 2009b). Enzymatic discoloration of the diazo dye, CR, by immobilized has been done. Partially purified TP was immobilized by entrapment into spherical particles of calcium alginate and was evaluated for the discoloration of aqueous CR solution. The maximum 94% of CR (180 mg L^{-1}) decolorization was obtained in buffer of pH 2.0 at 40 °C in the presence of 10 mM H_2O_2 by alginate immobilized TP after 1 h incubation in a batch process. Immobilized TP retained 74% removal efficiency after 4 reuses (Ahmedi et al. 2015). TP was coupled with crosslinked chitosan beads via GA. The free and immobilized TP preparations were used to treat RBBR. During discoloration experiments with immobilized TP the two phenomena were noticed, (1) discoloration due to adsorption on the support (60.45%) and (2) dye degradation due to the enzyme action (27.50%). The free enzyme removed 62.86% of the color. The immobilized enzyme showed a potential of 61.17% for the removal of the dye color after 6 reuses (Chagas et al. 2014).

2.3.3 Cucurbita pepo (courgette) peroxidase (CPP)

Boucherit et al. (2012) have investigated the dye decolorization potential of the CPPs extracted from courgette and immobilized in calcium alginate beads. The maximum decolorization of direct yellow 106 (DY106) at pH 2, dye concentration 180 mg L^{-1} and H_2O_2 dose of 0.12 M was 69.71%. The degradation pathway and the metabolic products formed after the degradation were also predicted using UV-vis spectroscopy analysis. In a further study these workers entrapped all CPPs into calcium alginate beads and employed for the decolourization and degradation of DY106 in a buffer of pH 2, at 20 °C and H_2O_2 concentration of 100 mmol L^{-1}. The highest decolourization of DY106 was 75% within 15 min by immobilized CPP. The DY106 removal has been monitored by UV-Vis analysis and Fourier transform infrared (FT-IR) spectroscopy was also done on DY106 and enzymatic treatment precipitated byproduct (Boucherit et al. 2013).

2.3.4 Ginger Peroxidase (GP)

Ali et al. (2016) reported the maximum decolorization of direct blue 80 (25 mg L^{-1}) by 0.166 U mL^{-1} of GP. Direct blue 80 was also successfully removed in stirred batch process in the presence of 0.3 mM H_2O_2 and 0.6 mM HOBT. Amino-functionalized silica-coated titanium dioxide nanocomposite bound GP was taken for removal of acid yellow 42. The immobilized GP retained nearly 62% of its dye decolorization efficiency even after its sixth reuse. Immobilized peroxidase was highly efficient in the removal of 90% of Acid Yellow 42 dye in a stirred batch process within 90 min (Ali et al. 2017). Further, Ali et al. (2018a) investigated the adsorption of GP on the surface of polypyrrole (Ppy)-cellulose-graphene oxide nanocomposite (PCeGONC). The recovery of activity was 128% of the initial activity. PCeGONC bound GP demonstrated remarkably high decolorization of reactive blue 4 (RB 4) dye, 99% after 3 h treatment in a batch process while the free enzyme could decolorize only 88% of the same dye. The immobilized GP maintained 72% dye decolorization efficiency after 10 reuses in the batch processes. GC-MS analysis was performed to identify degraded product and degradation pathway has been proposed that showed enzymatic breakdown of RB 4 into low molecular weight compounds. Genotoxic assessment of GP treated RB 4 revealed significant reduction of its genotoxic potential. Furthermore, it also revealed higher affinity of HOBT, a redox mediator and RB 4 for PCeGONC bound enzyme as compared to the free enzyme. GP has been immobilized on Ppy-zirconium(IV) selenoiodate cation exchanger nanocomposite (PZrSeI-CENC) via adsorption. The immobilized GP has removed about 96% of the methyl violet 6B (MV 6B) from polluted water in 3 h while the native enzyme could remove only 81% of the dye under similar treatment conditions. Immobilized GP showed remarkably high reusability in stirred batch process and decolorized around 60% of MB 6B after tenth reuses. A degradation pathway was elucidated in compliance with GC-MS and it revealed the formation of low molecular products and such products were appeared less toxic as compared to parent compound (Ali et al. 2018b). In a most recent investigation, GP was entrapped into the hydrogels of guar gum (GG)-alginate/agarose bio-composites and these immobilized GP preparations were employed for the treatment of textile effluent. ANGG-GP and AGG-GP were highly stable against various physical and chemical denaturants during the decolorization of textile effluent. Both immobilized GP preparations were more efficient in the decolorization of textile effluent in batch processes compared to free enzyme. ANGG-GP and AGG-GP have successfully cleaned about 68% and 55% of the color from textile effluent after tenth repeated use in batch processes, respectively. Continuous PBRs containing ANGG-GP and AGG-GP were able to decolorize nearly 80% and 69% colour of the effluent even after 30 days of their continuous operation at room temperature (30 °C), respectively. Genotoxicity of textile effluent was markedly decreased after enzymatic treatment (Ali and Husain 2018).

2.3.5 Pointed Gourd Peroxidase

Pointed gourd (*Trichosanthes dioica*) peroxidase (PGP) activity has been reported in an earlier study. This is a glycosylated enzyme and binds specifically to Con A (Jamal et al. 2012). Further these workers have used DEAE cellulose adsorbed PGP for the treatment of water contaminated with synthetic dyes. Immobilized PGP has efficiently decolorize up to 88.2% and 77.4% of direct red 19 (DR19) and mixture of dyes (DR 19 and DB9), respectively in the stirred batch processes at 40 °C. Immobilized enzyme in the PBR was used for the continuous removal of DR19 and a dye mixture and they were successful in removing the color from DR 19 and dye mixture to 69.4% and 51.4% after 50 days of their continuous operation, respectively (Jamal and Goel 2014).

2.3.6 Tomato Peroxidase

Con A-cellulose bound tomato peroxidase (TMP) catalyzed degradation and decolorization of direct red 23 and direct blue 80 dyes in the presence of 6 redox mediator. HOBT emerged as a potential redox mediator for TMP catalyzed decolorization of direct dyes. These dyes were maximally decolorized at pH 6.0 and 40 °C by soluble and immobilized TMP. Immobilized TMP showed a lower K_m than the soluble enzyme for both dyes. Soluble and immobilized TMP demonstrated remarkably high affinity for direct red 23 than direct blue 80. The decolorization of dyes was monitored by using spectrophotometry (Matto and Husain 2008).

2.3.7 Azadirachta indica Peroxidase

CS beads entrapped *Azadirachta indica* peroxidase (AIP) has demonstrated a remarkable enhancement in dye decolorization ability as compared to free enzyme. It showed that the purified AIP has a lot of potential in the treatment of industrial effluents (Pandey et al. 2017).

2.3.8 Turkish Black Radish Peroxidase

Altinkaynak et al. (2017) described the preparation of enzyme-inorganic hybrid nanostructures with flower-like shape comprising of Turkish black radish peroxidase (TBRP) and Cu^{2+} metal ions using an enzyme immobilization technique. The peroxidase-Cu^{2+} hybrid nanoflowers showed reamarkably very high stabililization against the denaturation mediated by varying pH and provided excellent decolorization efficiency for Victoria blue dye above 90% within 1 h. The immobilized TBRP was efficiently reused and retained about 77% decolorization efficiency even after 10 repeated uses.

2.3.9 White Radish Peroxidase

Celite adsorbed white radish peroxidase (WRP) was exploited for the treatment of reactive dyes: reactive red 120 and reactive blue 171 in the presence of various redox mediators: HOBT, syringaldehyde, VLA and VA. The investigated dyes were decolorized to different levels by the WRP in the presence of these mediators. However, HOBT has emerged as one of most effective mediator for decolorization of the dyes by WRP and the dyes were maximally decolorized in the buffer of pH 5.0 and at 40 °C within 1 h. The immobilized WRP has decolorized the higher concentration of dyes in the presence of sodium azide, organic solvents and $HgCl_2$ compared to its free form. The loss of color was evaluated by using spectrophotometry and toxicity was examined by *Allium cepa* test. Immobilized peroxidase decolorized dyes in batch processes more efficiently. The dye decolorization by immobilized WRP was monitored in a continuous reactor and the column containing immobilized enzyme retained 73% reactive red 120 decolorization efficiency after 1 month of its continuous operation (Satar and Husain 2009).

2.3.10 *Ziziphus mauritiana* Peroxidases

Peroxidases from *Ziziphus mauritiana* were immobilized via simple adsorption and cross-linking by glutaraldehyde on the calcium pectate microsphere. The stability of immobilized peroxidases against heat and extreme conditions of pH was remarkably enhanced. The packed bed reactors containing adsorbed and cross-linked peroxidases were continuously used for the degradation and decolorization of direct red 23 for over 30 days; adsorbed peroxidases maintained 52.86% dye decolorization efficiency while cross-linked peroxidases maintained more than 77% dye decolorization potential after 30 days of their continuous operation. Gas chromatography coupled with mass spectrometry has demonstrated the presence of four major metabolites (Khan and Husain 2019).

2.4 Microbial Peroxidases

Table 4 illustrates microbial peroxidases immobilized on various supports and their applications in the decolorization of dyes/dye effluents. A catalase peroxidase (CP) from the newly isolated Bacillus SF was used to treat textile-bleaching effluents. The enzyme was stable in the buffer of high pH values and at elevated temperatures, but was more sensitive to inactivation by H_2O_2 than monofunctional catalases. The enzyme was immobilized on various alumina-based carriers of different shapes and the specific activity was increased with the porosity of the carrier. The shape of the carrier had an important influence on the release of O_2 formed during the catalase reaction from the packed-bed reactor and Novalox saddles were found to be the most suitable shape. Bleaching effluent was treated in a horizontal packed-bed reactor containing 10 kg of the immobilized CP at a textile-finishing company. The treated liquid (500 L) was reused within the company for dyeing fabrics with various dyes, resulting in acceptable colour differences of below Delta $E^* = 1.0$ for all dyes (Fruhwirth et al. 2002).

Simulated textile wastewater was degraded using a membraneless electrochemical reactor with immobilized peroxidase on the porous Celite. The optimal current density was 10 A m^{-2}, at which the highest amount of H_2O_2 could be produced. The decolorization efficiency of the simulated wastewater using the electrochemical and electroenzymatic method was 35% and 92%, respectively. Biodegradability, the ratio of 5-d biochemical oxygen demand (BOD) to chemical oxygen demand (COD), was enhanced about 1.88 times when using the electroenzymatic treatment rather than raw wastewater, which could not be achieved by the electrochemical treatment (Cho et al. 2009).

Table 4 Immobilization of microbial peroxidases on various supports and their applications in the decolorization of dyes/dye effluents

Enzyme	Support Material	Method of Immobilization	Type of dye/effluent	Decolorization (%)	References
MnP/ LiP	Amberlite IRA-400 resin	Adsorption	Kraft E1 effluent	50	Peralta-Zamora et al. (1998)
MnP	Membrane	Encapsulation	Dye effluent	85	Lopez et al. (2004)
MnP	TMOS-PTMS	Entrapment	Textile industrial effluent	98.8–99.2	Iqbal and Asgher (2013a, b)
MnP	Calcium alginate	Entrapment	Sandal-fix Red C4BLN, Sandal-fix Turq Blue GWF, Sandal-fix Foron Blue E2BLN, Sandal-fix Black CKF & Sandal-fix Golden Yellow CRL	82.1–95.7	Bilal and Asgher (2015a)
MnP	PVA-alginate	Entrapment	Sandel reactive dyes, textile wastewater	78.14–92.29 61–80	Bilal and Asgher (2015b)
MnP	Gelatin	Encapsulation	Sandal-fix Red C4BLN (Reactive Red 195 A)	90	Bilal et al. (2016a)
MnP	Agar-agar	Entrapment	Dyes	78.6–84.7	Bilal et al. (2016c)
MnP	CS beads	Covalent attachment	Textile wastewater	97.31	Bilal et al. (2016d)
MnP	Calcium alginate	Entrapment	Sandal reactive dyes	80–93	Shaheen et al. (2017)
MnP	Agarose beads	Entrapment	Colored effluents	98.4	Bilal et al. (2017e)
LiP	CS beads	Covalent attachment	Textile effluent	95.45	Sofia et al. (2016)
DyP1B	Calcium alginate ferro-magnetic beads	Entrapment	Reactive Blue 5	94	Wasak et al. (2018)

2.4.1 Manganese Peroxidase and Lignin Peroxidase

This part of the present review will be devoted to the applications of immobilized lignin-modifying enzymes; lignin peroxidase (LiP) and manganese peroxidase (MnP) for the degradation, decolorization, or detoxification of industrial dyes and dye-based industrial wastewaters (Husain 2006; Husain and Ulber 2011; Bilal et al. 2017a). Color removal from kraft effluent by LiP and HRP was compared. Immobilization of LiP Type III, lyophilized fungal culture and HRP on CNBr-Sepharose 4B improved the decolorization by factor of 2.9, 4.5 and 2.6 within 48 h, respectively. LiP Type I was only effective in the immobilized form for decolorization of dyes. The immobilized form all the studied systems exhibited an average value around of 30% polymer consumption and very little of depolymerization. LiP and lyophilized fungal culture have shown remarkably high potential in the treatment of kraft effluents (Ferrer et al. 1991). Amberlite IRA-400 resin immobilized LiP and MnP from *Phanerochaete chrysosporium* were found quite effective in removing color and phenolic species from kraft E1 effluent. The effluent was decolorized over 50% during 3 h of enzymatic treatment (Peralta-Zamora et al. 1998).

The application of enzymatic membrane reactor (EMR) containing MnP has been examined in a continuous mode for the decolorization of a dye (Lopez et al. 2002). In a further study the same group has constructed an EMR for the oxidation of azo dyes by MnP. The configuration consisted of a stirred tank reactor coupled with an ultrafiltration membrane (UFM). The membrane permitted the recovery of all enzyme activity while both parent dye and degradation products has been removed. The dye decolorization was more than 85% when continuous operation was done under optimal experimental conditions and minimal enzymatic deactivation was noticed even after 18 days (Lopez et al. 2004). Gelatin-immobilized MnP from WRF was employed for the treatment of some azo dyes in static and shaky conditions. Remarkably high concentration of the dyes was effectively removed by such treatment. The immobilized MnP maintained almost its full activity after two repeated uses in batch processes (Cheng et al. 2007). Enayatzamir et al. (2010) immobilized MnP from *P. chrysosporium* into Ca-alginate beads and the entrapped enzyme was applied for the decolorization of recalcitrant azo dyes; reactive black 5, direct violet 51, Bismark brown R and ponceau xylidine in batch systems. These dyes were significantly decolorized by immobilized MnP. LiP from *P. chrysosporium* BKM-F-1767 covalently immobilized on mesoporous silica was applied for the decolorization of acid orange II. The immobilized LiP retained 50% dye decolorization efficiency after 5 repeated uses (Hu et al. 2013).

MnPs from *Ganoderma lucidum* were immobilized in a sol-gel matrix of tetramethoxysilane (TMOS) and proplytrimethoxysilane (PTMS) and used this immobilized enzyme preparation to decolorize various textile effluents. The industrial effluents were decolorized to varying extents with a maximum of 99.2% after 4 h treatment. The maximally decolorized effluent was analyzed for formaldehyde and nitroamines and results showed that the toxicity parameters were below the permissible limits. The findings of the work have demonstrated that sol-gel entrapped MnP has a lot of scope in the remediation of industrial effluents (Iqbal and Asgher 2013a). Further, these workers have evaluated the entrapment of MnP isolated from *Trametes versicolor* IBL-04 into xerogel matrix composed of TMOS and PTMS. The immobilized MnP was employed for decolorization of textile industrial effluent in a packed bed reactor system (PBRS). The 98.8% decolorization of effluent was obtained after 5 repeated uses during 5 h. The kinetic properties, storage stability and reusability of entrapped MnP demonstrated its potential as a biocatalyst for bioremediation (Iqbal and Asgher 2013b). Bilal and Asgher (2015a) developed a promising and eco-friendly approach to entrap *Ganoderma lucidum* IBL-05 MnP into Ca-alginate beads. The immobilized MnP was subsequently used for enhanced decolorization and detoxification of textile reactive dyes. The enhanced catalytic potential of immobilized MnP led to 87.5%, 82.1%, 89.4%, 95.7% and 83% decolorization of sandal-fix red C4BLN, sandal-fix turq blue GWF, sandal-fix foron blue E2BLN, sandal-fix black CKF and sandal-fix golden yellow CRL dyes, respectively. The insolubilized MnP was reusable for 7 repeated cycles in dye color removal. Furthermore, immobilized MnP also caused a significant loss in BOD (94.61–95.47%), COD) (91.18–94.85% and TOC) (89.58–95%) of aqueous dye solutions. Ca-alginate bound MnP was catalytically more stable, reusable and worked over wider ranges of pH and temperature as compared to its free form. Results of cytotoxicity like hemolytic and brine shrimp lethality tests suggested that Ca-alginate immobilized MnP may effectively be used for detoxification of dyes and industrial effluents.

Ganoderma lucidum IBL-05 MnP was immobilized onto polyvinyl alcohol-alginate beads and was used for the decolorization and detoxification of new class of reactive dyes and textile wastewater. The immobilized MnP showed high decolorization efficiency for sandal reactive dyes (78.14–92.29%) and textile wastewater (61–80%). The immobilized MnP retained 64.9% of its initial activity even after 6 repeated use for sandal-fix foron Blue E2BLN dye. After treatment with immobilized enzyme the water quality assurance parameters; BOD, COD and TOC and cytotoxicity; haemolytic and brine shrimp lethality parameters were significantly reduced for both the dyes aqueous solution and textile wastewater. The findings of the work have shown that immobilized MnP can be efficiently employed for the remediation of textile dyes containing wastewater effluents (Bilal and Asgher 2015b).

Further these researchers have carried out the encapsulation of MnP into gelatin. Over 90% decolorization of sandal-fix red C4BLN (reactive red 195A) dye was achieved with immobilized MnP in 5 h. It has also maintained over 50% of its original activity after its sixth repeated use. The water quality parameters like pH, COD, TOC and cytotoxicity (brine shrimp and *Daphnia magna*) studies demonstrated the non-toxic nature of the product dye treatment (Bilal et al. 2016b). In a further study these workers have investigated the entrapment of purified MnP from *Ganoderma lucidum* IBL-05 agar-agar support. Moreover, the decolorization of three structurally different dyes was monitored in order to assess the degrading capability of the entrapped MnP. The decolorization efficiencies for all the tested dyes were 78.6–84.7% after 12 h. The studies concluded that the toxicity of dyes aqueous solutions was significantly reduced after treatment (Bilal et al. 2016c). MnP was crosslinked onto CS beads via GA and used for the degradation and detoxification of dyes from textile wastewater. The obtained removal of color was 97.31% and total loss of COD, TOC and BOD were 82.40%, 78.30% and 91.7%, respectively. The cytotoxicity of bio-treated effluents reduced significantly and 38.46%, 43.47% and 41.83% *Allium cepa* root length, root count and mitotic index were increased, respectively, whereas brine shrimp nauplii death reduced up to 63.64%. The mutagenicity was decreased about 73.44% and 75.43% for TA98 and TA100 strains, respectively. The CI-MnP retained 60% activity after 10 repeated decolorization batches. The CI-MnP showed excellent efficiency for the bioremediation of textile effluents and can be used for the remediation of toxic agents in wastewater (Bilal et al. 2016d). Decolorization efficiencies of sandal reactive dyes after treating with *G. lucidum* IBL-05 LiP entrapped in Ca-alginate beads was in the range of 80–93%. There was a remarkable reduction in BOD (66.44–98.22%), COD (81.34–98.82%) and TOC (80.21–97.77%) after treatment with entrapped LiP. The cytotoxicity values for heamolytic and brine shrimp lethality of dye solutions treated with Ca-alginate immobilized LiP reduced up to 2.10–5.06% and 5.43–9.23%, respectively. It has been concluded on the basis of reduced toxicity and cytotoxicity that Ca-alginate beads entrapped LiP has a lot of scope in the bioremediation of coloured industrial effluents (Shaheen et al. 2017). Bilal et al. (2017e) entrapped a novel MnP into agarose beads and used this preparation to treat different coloured effluents. The treated effluents were decolorized to different extents with a maximum of 98.4% decolorization after 6-repeated uses. LiP containing extracts from *Pleurotus ostreatus* and *Ganoderma lucidum* were immobilized onto carbon nanotubes (CNTs) and the immobilized extracts were successfully reused in the dye decolourization. CNTs bound enzyme showed very high catalytic efficiency and capability to reuse it in the dye decolourization process (Oliveira et al. 2016). The purified LiP from *Schizophyllum commune* IBL-06 was immobilized on CS beads activated with GA. The maximum textile dye decolorization 95.45% was obtained with immobilized LiP at 30 °C without hemolytic toxicity. CS beads-LiP retained over 70% activity after three repeated uses that gradually decreased to 35% after seventh reuse. The immobilized LiP exhibited significantly higher dye removal compared to free enzyme. The high heat stability, lower K_m and high V_{max} values of immobilized LiP demonstrated its suitability in different industrial and environmental uses (Sofia et al. 2016). Table 4 represents the immobilization of microbial peroxidases on various supports and their applications in the decolorization of dyes/dye effluents.

2.4.2 Dye Decolorizing Peroxidase

Shakeri and Shoda (2008) carried out the immobilization of a recombinant dye-decolorizing peroxidase (rDyP) from *Aspergillus oryzae* using silica-based mesoporous materials, FSM-16 and AlSBA-15. rDyP immobilized on FSM-16 at pH 4 decolorized eight sequential batches of an anthraquinone dye, RBBR in repeated-batch decolorization, but only two batches for RBBR occurred by rDyP immobilized on AlSBA-15 because of leaching rDyP from AlSBA-15. In a further study these workers used similar rDyP immobilized in on silica-based mesocellular foam (25 nm average pore size) for the degradation and decolorization of an anthraquinone dye, RBBR. The dye was maximally decolorized by this immobilized rDyP preparation even after 20 repeated uses in batch processes (Shakeri and Shoda 2010). Crude preparation of peroxidases from *Pleurotus eryngii* was stabilized in PEG and glycine solutions and immobilized on monofunctional and heterofunctional agarose solid supports. MANA-glyoxyl heterofunctional supports were demonstrated to have the greatest enhancement of decolorization, 1.3-fold and velocity of substrate consumption, five-fold. The application of crude enzymatic extracts to dye decolorization, illustrates a cost-effective alternative to purified enzymes (Vásquez et al. 2014). DyP type 1B (DyP1B) from *Pseudomonas fluorescens* Pf-5 entrapped into Ca-alginate ferromagnetic beads was successfully used to decolorize RB 5 at concentrations of 0.1, 0.05, and 0.01% (w v^{-1}) with efficiency rates of about 20, 29, and 45%, respectively. The immobilization of DyP1B in alginate beads together with Fe_3O_4 resulted into enhancement of its catalytic and dye decolorization efficiency (Wasak et al. 2018).

2.5 *Microperoxidase 11*

Silica gel immobilized microperoxidase 11 (MP-11) was quite successful in decolorizing a water insoluble dye; normally this dye is used in leather industry (Kadnikova and Kostic 2003). MP 11 was immobilized in hybrid periodic mesoporous organosilica materials and in a nano-crystalline metal organic [Cu (OOC–C6H4–C6H4–COO)Æ½ C6H12N2]n framework. The conversion of amplex-ultra red and methylene blue to their respective oxidation products occurred successfully in the presence of immobilized MP 11 (Pisklak et al. 2006).

2.6 *Coimmobilized Peroxidases*

Immobilization of more than one peroxidase on the same support is an efficient technology for using them in a very wide range of pH, temperature and H_2O_2 concentration. Jin et al. (2018) have investigated the coimmobilization of chloroperoxidase (CPO) and HRP on ZnO nanowires/macroporous SiO_2 composite support via an in situ cross-linking by an anionic bi-epoxy cross-linker. CPO and HRP were co-immobilized with loading capacity of 79.6 mg CPO g^{-1} support and 52.8 mg HRP g^{-1} support, and total specific activity up to 15.7 U mg^{-1} support. The co-immobilized enzyme also showed good stability after 60 days of storage and remarkably high reusability over 20 repeated uses. For the decolorization of azo dyes the co-immobilized CPO (60%)/HRP (40%) exhibited high catalytic activity over the broad ranges of pH, temperature and H_2O_2 concentration. This coimmobilized CPO and HRP preparation catalyzed complete decolorization of all tested azo dyes within 3 h.

2.7 Conclusion

Peroxidases immobilized by using various kinds of methods and support have successfully been employed in batch processes and continuous reactors for the decolorization and detoxification of dyes/dye effluents from wide spectrum groups (Fig. 1). Mohan et al. (2005) . The effluents collected from different industrial sites were also treated efficiently with the help of immobilized peroxidases from plant and microbial sources. It can be concluded that the immobilized peroxidase have a great future in the remediation of coloured effluents/wastewater coming out of many industrial sites (Fig. 2).

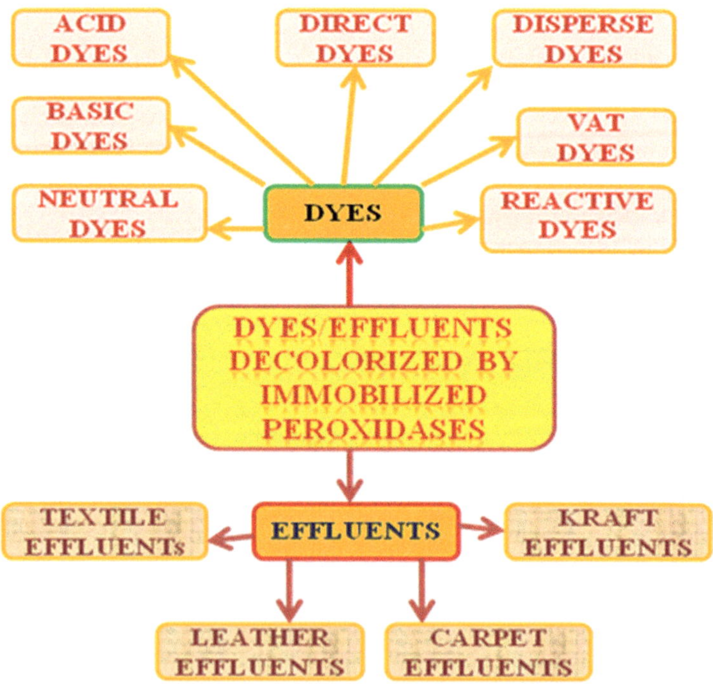

Fig. 1 Illustrates various dyes and dye effluents decolorized by immobilized peroxidases

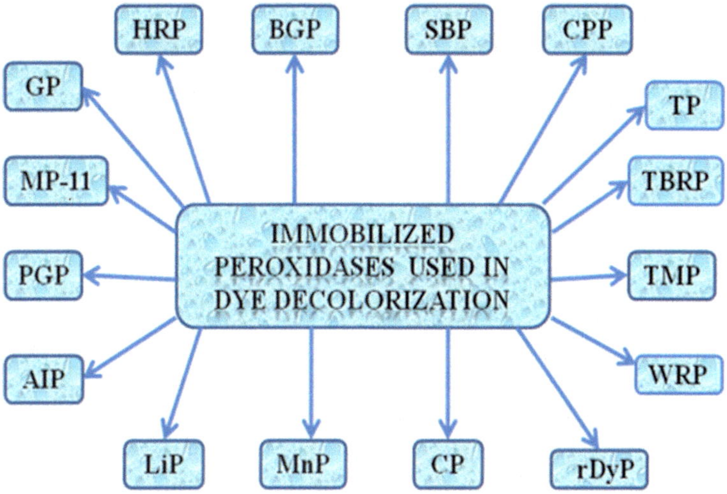

Fig. 2 Lists various kinds of immobilized peroxidases employed in the dye decolorization

References

Ahmedi A, Abouseoud M, Abdeltif A, Annabelle C (2015) Effect of diffusion on discoloration of Congo red by alginate entrapped turnip (*Brassica rapa*) peroxidase. Enzyme Res 2015:575618

Akhtar S, Husain Q (2006) Potential applications of immobilized bitter gourd (*Momordica charantia*) peroxidase in the removal of phenols from polluted water. Chemosphere 65:1228–1235

Akhtar S, Khan AA, Husain Q (2005a) Partially purified bitter gourd (*Momordica charantia*) peroxidase catalyzed decolorization of textile and other industrially important dyes. Bioresour Technol 96(16):1804–1811

Akhtar S, Khan AA, Husain Q (2005b) Simultaneous purification and immobilization of bitter gourd (*Momordica charantia*) peroxidases on bioaffinity support. J Chem Technol Biotechnol 80:198–205

Akhtar S, Khan AA, Husain Q (2005c) Potential of immobilized bitter gourd (*Momordica charantia*) peroxidases in the decolorization and removal of textile dyes from polluted wastewater and dyeing effluent. Chemosphere 60(3):291–301

Ali M, Husain Q (2018) Guar gum blended alginate/agarose biocomposite as a promising support for the entrapment of peroxidase: stability and reusability studies for the treatment of textile effluent. Int J Biol Macromol 116:463–471

Ali M, Ahmad M, Husain Q (2016) Redox mediated decolorization and detoxification of direct blue 80 by partially purified ginger (*Zingiber officinale*) peroxidase. Int J Environ Agr Res 2 (4):19–30

Ali M, Husain Q, Alam N, Ahmad M (2017) Enhanced catalytic activity and stability of ginger peroxidase immobilized on amino-functionalized silica-coated titanium dioxide nanocomposite: a cost-effective tool for bioremediation. Water Air Soil Poll 228:22

Ali M, Husain Q, Sultana S, Ahmad M (2018a) Immobilization of peroxidase on polypyrrole-cellulose-graphene oxide nanocomposite via non-covalent interactions for the degradation of reactive blue 4 dye. Chemoshpere 202:198–207

Ali M, Alam N, Ahmad M (2018b) Nano-peroxidase fabrication on cation exchanger nanocomposite: augmenting catalytic efficiency and stability for the decolorization and detoxification of Methyl Violet 6B dye. Sep Purif Technol 203:20–28

Altikatoglu M, Celebi M (2011) Enhanced stability and decolorization of coomassie brilliant blue R-250 by dextran aldehyde-modified horseradish peroxidase. Artif Cells Blood Substit Immobil Biotechnol 39(3):185–190

Altinkaynak C, Tavlasoglu S, Kalin R, Sadeghian N, Ozdemir H, Ocsoy I, Özdemir N (2017) A hierarchical assembly of flower-like hybrid Turkish black radish peroxidase-Cu2+ nanobiocatalyst and its effective use in dye decolorization. Chemosphere 182:122–128

Alvarado B, Torres E (2009) Recent patents in the use of peroxidases. Recent Pat Biotechnol 3 (2):88–102

Arslan M (2011) Immobilization horseradish peroxidase on amine-functionalized glycidyl methacrylate-*g*-poly(ethylene terephthalate) fibers for use in azo dye decolorization. Polym Bull 66:865

Bayramoglu G, Altintas B, Yakup Arica M (2012) Cross-linking of horseradish peroxidase adsorbed on polycationic films: utilization for direct dye degradation. Bioprocess Biosyst Eng 35(8):1355–1365

Bilal M, Asgher M (2015a) Dye decolorization and detoxification potential of Ca-alginate beads immobilized manganese peroxidase. BMC Biotechnol 15:111

Bilal M, Asgher M (2015b) Sandal reactive dyes decolorization and cytotoxicity reduction using manganese peroxidase immobilized onto polyvinyl alcohol-alginate beads. Chem Cent J 9:47

Bilal M, Iqbal HM, Hussain Shah SZ, Hu H, Wang W, Zhang X (2016a) Horseradish peroxidase-assisted approach to decolorize and detoxify dye pollutants in a packed bed bioreactor. J Environ Manag 183(Pt 3):836–842

Bilal M, Asgher M, Hu H, Zhang X (2016b) Kinetic characterization, thermo-stability and reactive red 195A dye detoxifying properties of manganese peroxidase-coupled gelatin hydrogel. Water Sci Technol 74(8):1809–1820

Bilal M, Asgher M, Shahid M, Bhatti HN (2016c) Characteristic features and dye degrading capability of agar-agar gel immobilized manganese peroxidase. Int J Biol Macromol 86:728–740

Bilal M, Asgher M, Iqbal M, Hu H, Zhang X (2016d) Chitosan beads immobilized manganese peroxidase catalytic potential for detoxification and decolorization of textile effluent. Int J Biol Macromol 89:181–189

Bilal M, Asgher M, Parra-Saldivar R, Hu H, Wang W, Zhang X, Iqbal HM (2017a) Immobilized ligninolytic enzymes: an innovative and environmental responsive technology to tackle dye-based industrial pollutants-A review. Sci Total Environ 576:646–659

Bilal M, Iqbal HM, Hu H, Wang W, Zhang X (2017b) Enhanced bio-catalytic performance and dye degradation potential of chitosan-encapsulated horseradish peroxidase in a packed bed reactor system. Sci Total Environ 575:1352–1360

Bilal M, Iqbal HM, Hu H, Wang W, Zhang X (2017c) Development of horseradish peroxidase-based cross-linked enzyme aggregates and their environmental exploitation for bioremediation purposes. J Environ Manag 188:137–143

Bilal M, Rasheed T, Iqbal HMN, Hu H, Wang W, Zhang X (2017d) Novel characteristics of horseradish peroxidase immobilized onto the polyvinyl alcohol-alginate beads and its methyl orange degradation potential. Int J Biol Macromol 105(Pt 1):328–335

Bilal M, Asgher M, Iqbal HMN, Hu H, Wang W, Zhang X (2017e) Biocatalytic performance and dye-based industrial pollutants degradation potential of agarose-immobilized MnP using a packed bed reactor system. Int J Biol Macromol 102:582–590

Bilal M, Rasheed T, Iqbal HMN, Hu H, Wang W, Zhang X (2018) Horseradish peroxidase immobilization by copolymerization into cross-linked polyacrylamide gel and its dye degradation and detoxification potential. Int J Biol Macromol 113:983–990

Boucherit N, Abouseoud M, Adour L (2012) Degradation of disperse dye from textile effluent by free and immobilized *Cucurbita pepo* peroxidase. EPJ Web Conferences 29:00008

Boucherit N, Abouseoud M, Adour L (2013) Degradation of direct azo dye by *Cucurbita pepo* free and immobilized peroxidase. J Environ Sci 25(6):1235–1244

Calza P, Zacchigna D, Laurenti E (2016) Degradation of orange dyes and carbamazepine by soybean peroxidase immobilized on silica monoliths and titanium dioxide. Environ Sci Poll Res Int 23(23):23742–23749

Celebi M, Kaya MA, Altikatoglu M, Yildirim H (2013) Enzymatic decolorization of anthraquinone and diazo dyes using horseradish peroxidase enzyme immobilized onto various polysulfone supports. Appl Biochem Biotechnol 171(3):716–730

Chacko JT, Subramaniam K (2011) Enzymatic degradation of azo dyes-A review. Int J Environ Sci 1(6):1249–1260

Chagas PMB, Torres JA, Nogueira FGE, dos Santos CD, Correa AD (2014) Catalytic stability of turnip peroxidase in free and immobilized form on chitosan beads. Int J Curr Microbiol Appl Sci 3(11):576–595

Cheng X, Jia R, Li P, Zhu Q, Tu S, Tang W (2007) Studies on the properties and co-immobilization of manganese peroxidase. Chin J Biotechnol 23:90–96

Cho SH, Shim J, Moon SH (2009) Detoxification of simulated textile wastewater using a membraneless electrochemical reactor with immobilized peroxidase. J Hazard Mat 162 (2–3):1014–1018

Enayatzamir K, Alikhani HA, Yakhchali B, Tabandeh F, Rodriguez-Couto S (2010) Decolouration of azo dyes by *Phanerochaete chrysosporium* immobilised into alginate beads. Environ Sci Pollut Res 17:145–153

Farias S, Mayer DA, de Oliveira D, Guelli U, de Souza SMA, Ulson de Souza AA (2017) Free and Ca-alginate beads immobilized horseradish peroxidase for the removal of reactive dyes: an experimental and modeling study. Appl Biochem Biotechnol 182(4):1290–1306

Fatima A, Husain Q (2007) A role of glycosyl moieties in the stabilization of bitter gourd (*Momordica charantia*) peroxidase. Int J Biol Macromol 41(1):56–63

Fatima A, Husain Q (2008) Purification, and characterization of a novel peroxidase from (*Momordica charantia)* bitter gourd. Protein Pept Lett 15(4):377–384

Fatima A, Husain Q, Khan RH (2007) Structural and functional studies of a novel peroxidase from bitter gourd (*Momordica charantia*): a comparison with horseradish peroxidase. J Mol Catal B Enzym 47(1–2):66–71

Ferrer I, Dezotti M, Durán N (1991) Decolorization of kraft effluent by free and immobilized lignin peroxidases and horseradish peroxidase. Biotechnol Lett 13:577–582

Fruhwirth GO, Paar A, Gudelj M, Cavaco-Paulo A, Robra KH, Gübitz GM (2002) An immobilised catalase peroxidase from the alkalothermophilic Bacillus SF for the treatment of textile-bleaching effluents. Appl Microbiol Biotechnol 60(3):313–319

Galárraga JCV, dos Santos AF, Bassan JC, Goulart AJ, Monti R (2013) Bromophenol blue discoloration using peroxidase immobilized on highly activated corncob powder. Revista de Ciências Farmacêuticas Básica e Aplicada 34(3):321–326

Gholami-Borujeni F, Mahvi AH, Naseri S, Faramarzi MA, Nabizade R, Alimohammad M (2011) Application of immobilized horseradish peroxidase for removal and detoxification of azo dye from aqueous solution. Res J Chem Environ 15(2):217

Gholami-Borujeni F, Nejatzadeh BF, Mahvi AH (2013) Detoxification of dye from industrial wastewater by immobilized horseradish peroxidase. J Color Sci Technol 7(2):133–142

Hamid M, Rehman KU (2009) Potential applications of peroxidases. Food Chem 115 (4):1177–1186

Hu Z, Xu L, Wen X (2013) Mesoporous silicas synthesis and application for lignin peroxidase immobilization by covalent binding method. J Environ Sci 25:181–187

Husain Q (2006) Potential applications of the oxidoreductive enzymes in the decolorization and detoxification of textile and other synthetic dyes from polluted water: a review. Crit Rev Biotechnol 26(4):201–221

Husain Q (2010) Peroxidase mediated decolorization and remediation of wastewater containing industrial dyes: a review. Rev Environ Sci Biotechnol 9:117–140

Husain Q (2019) Immobilized peroxidases as an efficient tool for the remediation of phenolic compounds from polluted water. In: Bhargava BN, Chowdhary P (eds) Emerging & eco-friendly approaches for waste management. Springer, Singapore, pp 329–358

Husain M, Husain Q (2008) Applications of redox mediators in the treatment of organic pollutants by using oxidoreductive enzymes: a review. Crit Rev Environ Sci Technol 38:1–42

Husain Q, Husain M (2012) Peroxidases as a potential tool for the decolorization and removal of synthetic dyes from polluted water. In: Malik A, Grohmann E (eds) Environmental protection strategies for sustainable development. Springer, Dordrecht, pp 453–498
Husain Q, Jan U (2000) Detoxification of phenols and aromatic amines from polluted wastewater by using phenol oxidases. J Sci Ind Res 59:286–293
Husain Q, Qayyum S (2013) Biological and enzymatic treatment of bisphenol A and other endocrine disrupting compounds: a review. Crit Rev Biotechnol 33:260–292
Husain Q, Ulber R (2011) Immobilized peroxidase as a valuable tool in the remediation of aromatic pollutants and xenobiotic compounds: a review. Crit Rev Environ Sci Technol 41:770–804
Husain Q, Husain M, Kulshrestha Y (2009) Remediation and treatment of organopollutants mediated by peroxidases: a review. Crit Rev Biotechnol 29:94–119
Iqbal HMN, Asgher M (2013a) Decolorization applicability of sol-gel matrix immobilized manganese peroxidase produced from an indigenous white rot fungal strain *Ganoderma lucidum*. BMC Biotechnol 13:56
Iqbal HMN, Asgher M (2013b) Characterization and decolorization applicability of xerogel matrix immobilized manganese peroxidase produced from *Trametes versicolor* IBL-04. Protein Pept Lett 20(5):591–600
Jamal F, Goel T (2014) Diethyl aminoethyl cellulose immobilized pointed gourd (*Trichosanthes dioica*) peroxidase in decolorization of synthetic dyes. J Bioprocess Biotech 4:187
Jamal F, Qidwai T, Singh D, Pandey PK (2012) Biocatalytic activity of immobilized pointed-gourd (*Trichosanthes dioica*) peroxidase-concanavalin a complex on calcium alginate pectin gel. J Mol Catal B Enzym 74(1–2):125–131
Janović BS, Mićić Vićovac ML, Vujčić ZM, Vujčić MT (2017) Tailor-made biocatalysts based on scarcely studied acidic horseradish peroxidase for biodegradation of reactive dyes. Environ Sci Pollut Res Int 24(4):3923–3933
Jiang Y, Tang W, Gao J, Zhou L, He Y (2014) Immobilization of horseradish peroxidase in phospholipid-templated titania and its applications in phenolic compounds and dye removal. Enzym Microb Technol 55:1–6
Jin X, Li S, Long N, Zhang R (2018) A robust and stable nano-biocatalyst by co-immobilization of chloroperoxidase and horseradish peroxidase for the decolorization of azo dyes. J Chem Technol Biotechnol 93(2):489–497
Kadnikova EN, Kostic NM (2003) Effects of the environment on microperoxidase-11 and on its catalytic activity in oxidation of organic sulfides to sulfoxides. J Org Chem 68:2600–2608
Kalsoom U, Bhatti HN, Asgher M (2015) Characterization of plant peroxidases and their potential for degradation of dyes: a review. Appl Biochem Biotechnol 176(6):1529–1550
Karim Z, Husain Q (2009) Redox-mediated oxidation and removal of aromatic amines from polluted water by partially purified bitter gourd (*Momordica charantia*) peroxidase. Int Biodeter Biodegrad 63(5):587–593
Karim Z, Adnan R, Husain Q (2012) A β-cyclodextrin–chitosan complex as the immobilization matrix for horseradish peroxidase and its application for the removal of azo dyes from textile effluent. Int Biodeter Biodegrad 72:10–17
Khan N, Husain Q (2019) Continuous degradation of direct red 23 by calcium-pectate bound *Ziziphus mauritiana* peroxidase: identification of metabolites and degradation routes. Environ Sci Pollut Res 26(4):3517–3529
Kim GY, Lee K-B, Cho S-H, Shim J, Moon S-H (2005) Electroenzymatic degradation of azo dye using an immobilized peroxidase enzyme. J Hazard Mater 126(1–3):183–188
Kumar V, Misra N, Goel NK, Thakar R, Gupta J, Varshney L (2016) A horseradish peroxidase immobilized radiation grafted polymer matrix: a biocatalytic system for dye waste water treatment. RSC Adv 6:2974–2298
Lade HS, Waghmode TR, Kadam AA, Govindwar SP (2012) Enhanced biodegradation and detoxification of disperse azo dye Rubine GFL and textile industry effluent by defined fungal-bacterial consortium. Int Biodeterior Biodegr 72:94–107
Liu JZ, Wang TL, Ji LN (2006) Enhanced dye decolorization efficiency by citraconic anhydride-modified horseradish peroxidase. J Mol Catal B Enzym 41:81–86

Lopez C, Mielgo I, Moreira MT, Feijoo G, Lema JM (2002) Enzymatic membrane reactors for biodegradation of recalcitrant compounds. Application to dye decolourisation. J Biotechnol 99 (3):249–257

Lopez C, Moreira MT, Feijoo G, Lema JM (2004) Dye decolorization by manganese peroxidase in an enzymatic membrane bioreactor. Biotechnol Prog 20:74–81

Maddhinni VL, Vurimindi HB (2006) Degradation of azo dye with horse radish peroxidase (HRP). J Ind Inst Sci 86:507–514

Maki F, Yugo U, Yasushi M, Isao I (2006) Preparation of peroxidase-immobilized polymers and their application to the removal of environment-contaminating compounds. Bull Fib Tex Res Found 15:15–19

Malani RS, Khanna S, Moholkar VS (2013) Sonoenzymatic decolourization of an azo dye employing immobilized horse radish peroxidase (HRP): a mechanistic study. J Hazard Mater 256-257:90–97

Matto M, Husain Q (2008) Redox-mediated decolorization of Direct Red 23 and Direct Blue 80 catalyzed by bioaffinity-based immobilized tomato (*Lycopersicon esculentum*) peroxidase. Biotechnol J 3(9–10):1224–1231

Matto M, Husain Q (2009a) Decolorization of textile effluent by bitter gourd peroxidase immobilized on concanavalin A layered calcium alginate-starch beads. J Hazard Mater 164 (2–3):1540–1546

Matto M, Husain Q (2009b) Decolorization of direct dyes by immobilized turnip peroxidase in batch and continuous processes. Ecotoxicol Environ Saf 72(3):965–971

Matto M, Satar R, Husain Q (2009) Application of calcium alginate-starch entrapped bitter gourd (*Momordica charantia*) peroxidase for the removal of colored compounds from a textile effluent in batch as well as in continuous reactor. Appl Biochem Biotechnol 158(3):512–523

Mohan SV, Prasad KK, Rao NC, Sarma PN (2005) Acid azo dye degradation by free and immobilized horseradish peroxidase (HRP) catalyzed process. Chemosphere 58(8):1097–1105

Najafpoor AA, Davoudi SER (2017) Decolorization of synthetic textile wastewater using electrochemical cell divided by cellulosic separator. J Environ Health Sci Eng 15:11

Nguyen TA, Juang R-S (2013) Treatment of waters and wastewaters containing sulfur dyes: a review. J Chem Eng 219:109–117

Oliveira SF, da Luz JMR, Kasuya MCM, Ladeira LO, Correa A Jr (2016) Enzymatic extract containing lignin peroxidase immobilized on carbon nanotubes: potential biocatalyst in dye decolourization. Saud J Biol Sci. In Press, Corrected Proof-Note to users

Pandey VP, Rani J, Jaiswal N, Singh S, Awasthi M, Shasany AK, Tiwari S, Dwivedi UN (2017) Chitosan immobilized novel peroxidase from *Azadirachta indica*: characterization and application. Int J Biol Macromol 104(Pt B):1713–1720

Peralta-Zamora P, Gomes de Moraes S, Esposito E, Antunes R, Reyes J, Durán N (1998) Decolorization of pulp mill effluents with immobilized lignin and manganese peroxidase from *Phanerochaete chrysosporium*. Environ Technol 19(5):521–528

Pisklak TJ, Macías M, Coutinho DH, Huang RS, Balkus KJ (2006) Hybrid materials for immobilization of MP-11 catalyst. Topics Cat 38:269–278

Preethi S, Anumary A, Ashokkumar M, Thanikaivelan P (2013) Probing horseradish peroxidase catalyzed degradation of azo dye from tannery wastewater. Springerplus 2:341

Satar R, Husain Q (2009) Applications of Celite-adsorbed white radish (*Raphanus sativus*) peroxidase in batch process and continuous reactor for the degradation of reactive dyes. Biochem Eng J 46(2):96–104

Satar R, Husain Q (2011) Catalyzed degradation of disperse dyes by calcium alginate-pectin entrapped bitter gourd (*Momordica charantia*) peroxidase. J Environ Sci (China) 23 (7):1135–1142

Satar R, Jerah ABA, Husain Q (2012) Role of redox mediators in enhancing dye decolorization by using oxidoreductive enzymes. In: El Nemr A (ed) Non-conventional textile waste water treatment. Nova Sciences, New York, pp 123–160

Šekuljica NŽ, Prlainović NŽ, Lukic N, Jakovljevic AM, Grbabcic SZ, Mijin DZ, Knežević-Jugović ZD (2015) Immobilization of peroxidase from fresh horseradish extract for anthraquinone dye decolorization. Zastita Mater 6(3):335–339

Šekuljica NŽ, Prlainović NŽ, Jovanović JR, Stefanović AB, Djokić VR, Mijin DŽ, Knežević-Jugović ZD (2016a) Immobilization of horseradish peroxidase onto kaolin. Bioprocess Biosyst Eng 39 (3):461–472

Šekuljica NZ, Prlainović NZ, Stefanović AB, Jovanović JR, Jakovetić SM, Knežević-Jugović ZD, Mijin DZ (2016b) The oxidation of anthraquinone dye using HRP immobilized as a cross-linked enzyme aggregates. Adv Technol 5(2):18–26

Šekuljica NZ, Prlainović NZ, Jakovetić SM, Grbavcic SZ, Ognjanovic ND, Knežević-Jugović ZD, Mijin DZ (2016c) Removal of anthraquinone dye by cross-linked enzyme aggregates from fresh horseradish extract. Clean Soil Air Water 7(2):891–900

Shaffiqu TS, Roy JJ, Nair RA, Abraham TE (2002) Degradation of textile dyes mediated by plant peroxidases. Appl Biochem Biotechnol 102–103(1–6):315–326

Shaheen R, Asgher M, Hussain F, Bhatti HN (2017) Immobilized lignin peroxidase from *Ganoderma lucidum* IBL-05 with improved dye decolorization and cytotoxicity reduction properties. Int J Biol Macromol 103:57–64

Shakeri M, Shoda M (2008) Decolorization of an anthraquinone dye by the recombinant dye-decolorizing peroxidase (rDyP) immobilized on mesoporous materials. J Mol Catal B Enzym 54(1–2):42–49

Shakeri M, Shoda M (2010) Efficient decolorization of an anthraquinone dye by recombinant dye-decolorizing peroxidase (rDyP) immobilized in silica-based mesocellular foam. J Mol Catal B Enzym 62:277–281

Shim J, Kim GY, Yeon KH, Cho SH, Woo JJ, Moon SH (2007) Degradation of azo dye by an electroenzymatic method using horseradish peroxidase immobilized on porous support. Korean J Chem Eng 24:72

Silva MC, Torres JA, Sa LRV, Chagas PMB, Ferreira-Leitão VS, Correa A (2013) The use of soybean peroxidase in the decolourization of Remazol Brilliant Blue R and toxicological evaluation of its degradation products. J Mol Catal B Enzym 89:122–129

Sofia P, Asgher M, Shahid M, Randhawa MA (2016) Chitosan beads immobilized *Schizophyllum commune* IBL-06 lignin peroxidise with novel thermostability, catalytic and dye removal properties. The J Anim Plant Sci 26(5):1451–1463

Sun H, Jin X, Long N, Zhang R (2017) Improved biodegradation of synthetic azo dye by horseradish peroxidase cross-linked on nano-composite support. Int J Biol Macromol 95:1049–1055

Sun H, Jin X, Jiang F, Zhang R (2018) Immobilization of horseradish peroxidase on ZnO nanowires/macroporous SiO_2 composites for the complete decolorization of anthraquinone dyes. Biotechnol Appl Biochem 65(2):220–229

Vásquez C, Anderson D, Oyarzún M, Carvajal A, Palma C (2014) Method for the stabilization and immobilization of enzymatic extracts and its application to the decolorization of textile dyes. Biotechnol Lett 36(10):1999–2010

Veitch NC (2004) Horseradish peroxidase: a modern view of a classic enzyme. Phytochemistry 65 (3):249–259

Wasak A, Drozd R, Struk Ł, Grygorcewicz B (2018) Entrapment of DyP-type peroxidase from *Pseudomonas fluorescens* Pf-5 into Ca-alginate magnetic beads. Biotechnol Appl Biochem 65 (2):238–245

Yincan Z, Yan L, Xueyong G, Qiao W, Xiaoping X (2017) Decolorization of color index acid orange 20 buffer solution using horseradish peroxidase immobilized on modified PAN-beads. RSC Adv 7:18976–18986

Zhu N, Yuan H, Lou Z, Wang L, Zhang X (2012) Degradation pathway of the naphthalene azo dye intermediate a-diazo-2-naphthol-4-sulphonic acid using Fenton's reagent. Water Res 46:3859–3867

Screening, Optimization and Assembly of Key Pathway Enzymes in Metabolic Engineering

Yanfeng Liu and Long Liu

1 Introduction

Metabolic engineering is a key technology for constructing synthetic pathway of target products and improving cellular traits for biosynthesis of industrially important chemicals, pharmaceuticals, and biofuels (Philp et al. 2013; Woolston et al. 2013; Becker and Wittmann 2015). In order to optimize synthetic pathways and enhance cellular properties, metabolic engineering is carried out via design-build-test-learn cycle (McNerney et al. 2015). Based on significant advances of systems biology and synthetic biology, the methods and tools for systems and synthetic biology were applied to metabolic engineering for identifying rate-limiting steps to precisely control the metabolic network (Lee et al. 2012; Nielsen and Keasling 2016). Currently two of the most important examples of metabolic engineering are biomanufacturing of 1, 4-butanediol as biofuel by engineered *Escherichia coli* and producing artemisinic acid, precursor of anti-malaria pharmaceutical artemisinin, by engineered *Saccharomyces cerevisiae* (Lee et al. 2012; Paddon et al. 2013). Therefore, metabolic engineering is of great importance for solving the issues of decreasing fossil resource and limited natural plant resource.

Enzymes are key catalysts of cellular metabolic reactions. Abundance and activities of pathway enzymes determine metabolic flux, which directly relates to production rate of targets products (Gerosa and Sauer 2011). Therefore, strategies for modulating abundance and activities of pathway enzymes were developed, such as controlling enzyme abundance by engineering of genetic regulatory elements for transcriptional or translation altering gene expression level, and introducing

Y. Liu · L. Liu (✉)
Key Laboratory of Carbohydrate Chemistry and Biotechnology, Ministry of Education, Jiangnan University, Wuxi, China

Key Laboratory of Industrial Biotechnology, Ministry of Education, Jiangnan University, Wuxi, China
e-mail: longliu@jiangnan.edu.cn

Q. Husain, M. F. Ullah (eds.), *Biocatalysis*,
https://doi.org/10.1007/978-3-030-25023-2_8

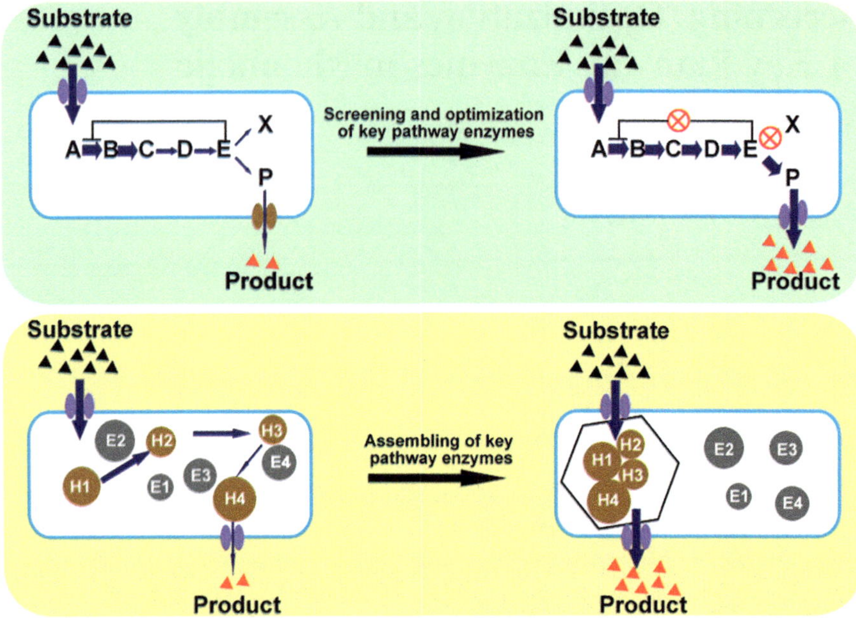

Fig. 1 Screening, optimization and assembling of key pathway enzymes in metabolic engineering. Feedback inhibition, low catalytic efficiency, undesired substrate specificity, and undesired activities of key pathway enzymes limit efficient target product synthesis. The limitations can be eliminated by screening and optimization of enzymes to obtain pathway enzymes with desired properties. Assembling of key pathway enzymes can further improve catalytic efficiency of biosynthetic pathways

mutagenesis into key pathway enzymes for enhanced catalytic efficiency (Alper et al. 2006; Leonard et al. 2010; Nowroozi et al. 2013). Enzyme screening and engineering is therefore an important research field in metabolic engineering for sustainable product efficiency.

In this chapter, we initially discuss how to screen key pathway enzymes from different organisms for enhanced production. Next, optimization of key pathway enzymes for alleviating feedback inhibition, improving catalytic efficiency, and altering substrate specificity is discussed. Finally, spatial organization engineering for balancing and strengthening synthetic pathways by assembling key pathway enzymes is taken into account. It is understood that combinatorial screening, optimization and assembly of key pathway enzymes will further facilitate efficiency of biochemical production and expand application of enzyme research in metabolic engineering (Fig. 1).

2 Screening Key Pathway Enzymes from Different Organisms for Enhanced Production

Efficient biosynthesis of target products relies on key pathway enzymes with high catalytic efficiency. Therefore, screening key pathway enzymes from different organisms is an important step for biosynthetic pathway construction. Such a screening mainly focuses on two aspects: (1) debottleneck of rate-limiting steps by testing enzymes from different organisms; (2) construction and fine-tuning of complete biosynthetic pathways.

Removal of rate-limiting steps in biosynthetic pathways is a key strategy for improving production of target products. Screening enzymes from different organisms that catalyze rate-limiting steps is widely used for identification of enzymes with improved catalytic efficiency. Screening enzymes can be carried out in combination with literature mining and experimental testing. For instance, in synthetic pathway of *N*-acetylglucosamine (GlcNAc), a functional sugar widely used as nutraceutical for maintaining joint health for elderly people; efficiency of glucosamine-6-P *N*-acetyltransferase (Gna1) is a limitation for GlcNAc production (Deng et al. 2005). In order to obtain high-level Gna1 expression with enhanced catalytic properties, the expression level of Gna1 from different organisms, including *Saccharomyces cerevisiae*, *Candida albicans*, and *Arabidopsis thaliana*, on GlcNAc production were evaluated (Deng et al. 2005). Gna1 from *S. cerevisiae* was demonstrated to be optimum for enhanced GlcNAc production. Moreover in the synthetic pathway of *N*-acetylneuraminic acid (NeuAc), which is widely used as a nutraceutical for facilitating infant brain development and maintaining brain health, with GlcNAc as one of important precursors, NeuAc synthase (NeuB) reportedly serves as an important pathway enzyme that catalyses NeuAc synthesis (Zhang et al. 2018). In one of the approaches to obtain efficient NeuAc synthesis, catalytic efficiency and substrate affinity of NeuB from various organisms were systematically compared based on data summarized from literatures. K_m values for ManNAc and PEP of NeuB from *Escherichia coli* K12 are lower than those of other organisms, which indicates its high substrate affinity (5.6 mM for ManNAc and 0.04 mM for PEP), respectively. Expressing NeuB from *E. coli* K12 in metabolically engineered *Bacillus subtilis* resulted in efficient NeuAc production (Zhang et al. 2018).

Construction and optimization of synthetic pathways requires balanced relative enzymatic activities of the whole pathway enzymes, especially for pathways with multiple steps, such as ester and natural plant products. High-level biological production of esters, which are widely used in flavors and fragrances industry, was facilitated by introducing enzymes from different organisms that catalyze ester biosynthesis (Rodriguez et al. 2014). For example, alcohol O-acyltransferase from *S. cerevisiae* was used for efficient isobutyl acetate production, reaching 17.2 g/L; and biosynthesis of a long-chain alkyl ester tetradecyl acetate was realized using the luminescence pathway from *Vibrio harveyi* (Rodriguez et al. 2014). Furthermore to achieve an efficient biological production of opioidsin (natural plant products used for pain management), from engineered *S. cerevisiae*, for the first time the complete opioid synthetic pathways were constructed by systematically screening the enzymes and their expression from various organisms, such as *S. cerevisiae* enzymes, mammalian cell enzymes (*Rattusnorvegicus*), bacterial enzymes (*Pseudomonasputida*), plant enzymes (*Eschscholziacalifornica, Coptis japonica, Papaver. bracteatum, Papaversomniferum*) (Galanie et al. 2015; Thodey et al. 2014). De novo synthesis of hydrocodone was demonstrated in engineered *S. cerevisiae*, which laid foundation of further optimization of opioid pathways for high-level production (0.3 μg/L).

3 Optimization of Key Pathway Enzymes for Improved Catalytic Properties

Though screening of natural enzyme is an important strategy for debottleneck of rate-limiting step and improvement of complete pathway efficiency, feedback inhibition and low catalytic efficiency of certain natural enzymes are associated with limited enzyme kinetic. Therefore, further strategies for optimizing key pathway enzymes for improved catalytic properties are considered to be significant. The undesired feedback inhibition, low catalytic efficiency and unsought substrate specificity are three key aspects of pathway enzymes that constrain efficient production of target products. To overcome such enzymatic property constraints of natural enzymes, optimization of key pathway enzymes is carried out by alleviating feedback inhibition, enhancing catalytic efficiency, and engineering substrate specificity.

Feedback inhibition is an important regulatory mechanism for metabolic flux control to avoid certain metabolite accumulation and maintain proper cellular resource allocation, and is often found in pathways that branch out of central carbon metabolism and amino acid biosynthetic pathways (Gerosa and Sauer 2011; Chen et al. 2012). Therefore, to enhance metabolic flux in the pathway branching out of central carbon metabolism and amino acid biosynthetic pathways, feedback inhibition needs to be removed or alleviated. For example, glucosamine (GlcN) is traditionally produced by extraction from crab or shrimp shell. However, the raw material may cause allergenic effects and limited supply may constrain GlcN production to meet the continuous demand for GlcN which is used as nutraceutical. Thus microbial fermentation for GlcN production was developed as a preferred method. In GlcN synthetic pathway, activity of GlcN-6-phosphate (GlcN-6-P) synthase, catalyzing GlcN-6-P formation from fructose-6-phosphate (Fru-6-P), is strong inhibited by GlcN-6-P, forming a rate-limiting step for enhancing GlcN production (Deng et al. 2005). Therefore, directed evolution was implemented for obtaining GlcN-6-P synthase with resistance of feedback inhibition. Mutant library of GlcN-6-P synthase was generated *via* error-prone PCR. GlcN-6-P synthase mutants, with resistance of feedback inhibition were screened based on plating mutant expressing *E. coli* with GlcN auxotrophic strain as an indicator. *E. coli* expressing mutant form of the enzyme with limited feedback inhibition should produce higher amount of GlcN which can be indicated by GlcN auxotrophic strain on plate. Based on multiple rounds of directed evolution, GlcN-6-P synthase mutants with improved resistance to feedback inhibition were obtained, leading to a 20-fold enhanced GlcN production.

In amino acid biosynthesis pathway, feedback inhibition plays important roles for controlling metabolic flux to produce proper amount of amino acid for cellular metabolism (Chen et al. 2014a, b). For instance, phosphoenolpyruvate carboxylase (PEPC) is the key enzyme in anaplerotic pathways controlling metabolic flux from glycolysis to aspartate and its derivative amino acid, such as lysine and threonine, which is inhibited by the binding of aspartate and malate. In order to overproduce aspartate, lysine, or threonine, deregulation of feedback inhibition mechanism of PEPC is a key step. Therefore, residues relating to the feedback inhibition were rationally selected for introducing mutation to reduce feedback inhibition. One of the point mutations (N917G) was introduced into the PEPC coding gene *ppc*. It was found that by expressing PEPC mutant in lysine-producing strain *Corynebacterium glutamicum*, significantly enhanced lysine production, which was 37% higher than that with the wild type PEPC was achieved (Chen et al. 2014a). In addition, a similar metabolic engineering strategy can also be used for introducing feedback inhibition in by-product synthetic pathways to dynamically down regulate the by-product formation via reducing key pathway enzymes activities by signal molecule binding. Thus to demonstrate the feasibility of introducing non-natural feedback inhibition for dynamic regulation of metabolic flux, *C. glutamicum* homoserine dehydrogenase (HSDH) that is naturally feedback inhibited by threonine and isoleucine was selected, for introducing feedback inhibition by non-natural inhibitor lysine. Firstly, binding sites of HSDH were predicted and verified by site-directed mutagenesis. The threonine binding pocket were then engineered to bind lysine by site-directed mutagenesis and screening (Chen et al. 2014b).

Low catalytic efficiency of synthetic pathway enzymes is another reason causing bottleneck in metabolic pathways. Optimization of enzyme catalytic efficiency is a useful strategy for enhancing kinetic properties of the enzyme. For example, levopimaradiene synthase and geranylgeranyl diphosphate synthase are rate-limiting enzymes in levopimaradiene synthesis pathway, which is diterpenoid precursor of the pharmaceutically important ginkgolides. Site-directed saturation mutagenesis was carried out to establish mutant library of levopimaradiene synthase and geranylgeranyl diphosphate synthase. Subsequently, mutants of levopimaradiene synthase and geranylgeranyl diphosphate synthase were systematically evaluated for kinetic properties and their effects on levopimaradiene production. Later by using mutants of levopimaradiene synthase and geranylgeranyl diphosphate synthase with enhanced catalytic efficiency in combination with improved precursor supply, levopimaradiene titer was reportedly increased 2600-fold, which was further improved to 700 mg/L via fed-batch fermentation in lab-scale bioreactor (Leonard et al. 2010).

Furthermore, engineering the substrate specificity is another important application for enzyme optimization in metabolic engineering, which determines the direction of metabolic reaction and yield of target product. Engineering of substrate specificity of pathway enzyme is widely used for bi-functional enzymes modification to strengthen reaction to the target product and repress reaction to the by-product. For instance, in benzylisoquinoline alkaloids (BIAs) synthetic pathway, tyrosinase has both tyrosine hydroxylase and L-3,4-dihydroxyphenylalanine (L-DOPA) oxidase activities. For production of BIAs, L-DOPA oxidase activity is undesirable, as it causes by-product melanin formation from L-DOPA and reduces BIAs yield from tyrosine. To obtain tyrosinase with only tyrosine hydroxylase activity, an enzyme-coupled biosensor-based high throughput screening method was developed, which couples intracellular L-DOPA concentration with betaxanthin-generating fluorescence by intracellularly expressing L-DOPA dioxygenase (DOD) (DeLoache et al. 2015). Using the constructed L-DOPA-responsive biosensor, an active tyrosine hydroxylase was identified from mutant library generated via error prone PCR, resulting in an enhanced L-DOPA yield of 2.8-fold, which laid the foundation for further biosynthesis of BIAs from important precursor L-DOPA. Another example is engineering of *N*-acetylglucosamine 2-epimerase (AGE), catalyzing the conversion between *N*-acetylglucosamine (GlcNAc) and *N*-acetylmannosamine (ManNAc) in *N*-acetylneuraminic acid (NeuAc) synthetic pathway, which is a nutraceutically useful compound for facilitating infant brain development and maintaining brain health for elder people. However, the reaction catalyzed by AGE from ManNAc to GlcNAc is undesired for NeuAc production. Therefore, structure guided rational engineering was carried out by simulating AGE structure for key amino acid residues selection and site-directed saturation mutagenesis (Chen et al. 2018). Expressing C372A mutant of AGE in combination with strengthening key enzyme expression led to significantly improved yield from GlcNAc (58.6%).

4 Assembling Key Pathway Enzymes for Balancing and Strengthening Synthetic Pathways

Protein-protein interactions (PPIs) of model microorganisms have been intensively investigated, which systematically identified genome scale PPIs and also revealed that PPIs are an important regulatory mechanism for cell metabolism (Arifuzzaman et al. 2006; Hu et al. 2009; Rajagopala et al. 2014). The Formation of protein complex is an important regulatory mechanism for controlling cellular process and metabolic flux in endogenous metabolism. It inspired metabolic engineers to design and construct spatial organization of key enzymes in a heterologous pathway for enhancing metabolic flux to target product. The co-clustering of multiple enzyme mimic functionality of substrate channeling and enhances catalytic efficiency as well as avoids intermediate diffusion (Zhang 2011; Castellana et al. 2014). Spatial organization assembling of metabolic enzymes mainly focus on the following three aspect: (1) fusion expression of two key enzymes to enhance catalytic efficiency, (2) construction of scaffold to co-localize key pathway enzymes, (3) targeting key pathway enzymes into bacterial micro compartment or yeast organelles.

Fusion expression of two adjacent enzymes by N-terminus of one enzyme to C-terminus of another enzyme with proper linker sequence into one protein complex is a direct approach for assembling key pathway enzymes. Theoretically, the distance of activity sites of assembled two enzymes may be shorten, which is beneficial for efficient passing of the intermediate from one enzyme to another. One of the most important factors that need to be considered is to avoid strong effects of fusion expression on the structure of individual enzymes. Therefore, to make sure that the functional expression of each enzyme is intact, the sequential order of enzymes and linker sequence needs to be carefully designed and tested. The effectiveness of assembling key pathway enzymes for mimicking substrate channeling was verified by fusing 4-coumarate-CoA ligase (4CL) and stilbene synthase (STS) for resveratrol biosynthesis in *S. cerevisiae* (Zhang et al. 2006). Compared with co-expressing 4CL and STS without fusion expression, fusing 4CL and STS improved resveratrol production by 3500-fold, reaching 5.25 μg/mL. Assembling key pathway enzymes by fusion expression is also applied into miltiradiene production, which is the precursor to tanshionones, a group of abietane-type norditerpenoids found in the Chinese medical plant *Salvia miltiorrhiza* with antibacterial, anti-inflammatory, and anticancer activities (Zhou et al. 2012). Labdadienyl/copalyl diphosphate synthase (SmCPS) and kaurene synthase-like (SmKSL) were fused for enhancing the efficiency of miltiradiene biosynthesis. In combination with modular pathway, engineering of miltiradiene metabolic pathway and fusion expression of SmCPS and SmKSL, miltiradiene titer was improved, reaching 365 mg/L.

However, fusion expression is normally used for assembly of two enzymes; more than two enzyme assembly usually cannot be used for enzyme co-localization by fusion expression with functional expression. Therefore, methods for assembly of multiple enzymes need to be developed. Synthetic protein scaffold, RNA scaffold and DNA scaffold were thus established for co-localizing multiple enzymes (Dueber et al. 2009; Conrado et al. 2012; Delebecque et al. 2012). The enzymes targeted for assembly were expressed with ligand fusion that can specifically bind to certain proteins, RNA sequences, or DNA sequences and can be co-localized onto scaffolds. By adjusting the ratio of co-localized enzymes on scaffolds the relative enzymatic activity in metabolic pathways can be regulated. Such an assembly of pathway enzymes for enhanced target product synthesis by synthetic protein scaffold was firstly demonstrated in *E. coli* for mevalonate production (Dueber et al. 2009). By co-localizing acetoacetyl-CoA thiolase, hydroxy-methylglutaryl-CoA synthase, and hydroxymethylglutaryl-CoA reductase with the ratio of 1:2:2, mevalonate titer increased 77-fold, reaching 5 mM. RNA- and DNA guided scaffold were subsequently developed to expand tools for assembly of pathway enzymes, which were successfully applied into production of resveratrol, threonine, GlcNAc, and glucaric acid, respectively (Conrado et al. 2012; Delebecque et al. 2012; Lee et al. 2013; Liu et al. 2014).

Although synthetic scaffold expanded the tools for enzyme assembly, co-localization may affect the flexibility of enzymes when catalyzing metabolic reactions. In order to resolve this issue, strategies for compartmentalization-engineering of metabolic pathways were developed and used for the production of branched-chain alcohols by targeting key enzymes into mitochondria in *S. cerevisiae* (Avalos et al. 2013). N-terminal mitochondrial localization signal sequence was used to target pathway enzymes of complete isobutanol biosynthesis pathways into mitochondria. Compared to the strategy of over-expressing enzymes of isobutanol biosynthesis in cytoplasm, targeting key enzymes of complete isobutanol biosynthesis pathways into mitochondria led to an approximately 2.5-fold increase of isobutanol titer. Although compartmentalization-engineering of metabolic pathways cannot be directly used in bacteria that have no organelle like *S. cerevisiae*, compartmentalization-engineering facilitated establishing synthetic bacterial micro compartments for co-localization of metabolic enzymes for target product synthesis (Lee et al. 2018). Expressing PduA∗, a shell protein of bacterial micro compartment with C-terminus modification, led to the designing of bacterial micro compartment in *E. coli*. Further, fusing de novo–designed complementary coiled-coil peptides that fuses with PduA∗ and pyruvate decarboxylase (Pdc) and alcohol dehydrogenase (Adh) resultedin targeting Pdc and Adh into synthetic micro compartment, which led to an enhanced ethanol production by 221% compared to those that do not harbor the micro compartment.

5 Concluding Remarks

Enzymes are key catalysts of biosynthetic pathways, which determine production efficiency of target product. To construct efficient metabolic pathway for bio-production of compounds of interest, key enzymes of the pathway need to be systematically screened, optimized, and assembled. Such a strategy eliminates undesired enzymatic kinetics such as feedback inhibition, low catalytic efficiency, undesired substrate specificity, and by product formation, and thus facilitates efficient metabolic pathway construction through metabolic engineering.

References

Alper H, Fischer C, Nevoigt E, Stephanopoulos G (2006) Tuning genetic control through promoter engineering (vol 102, pg 12678, 2005). Proc Natl Acad Sci U S A 103(8):3006–3006

Arifuzzaman M, Maeda M, Itoh A, Nishikata K, Takita C, Saito R, Ara T, Nakahigashi K, Huang HC, Hirai A (2006) Large-scale identification of protein–protein interaction of Escherichia coli K-12. Genome Res 16(5):686–691

Avalos JL, Fink GR, Stephanopoulos G (2013) Compartmentalization of metabolic pathways in yeast mitochondria improves the production of branched-chain alcohols. Nat Biotechnol 31 (4):335–341

Becker J, Wittmann C (2015) Advanced biotechnology: metabolically engineered cells for the bio-based production of chemicals and fuels, materials, and health-care products. Angew Chem Int Ed 54(11):3328–3350

Castellana M, Wilson MZ, Xu Y, Joshi P, Cristea IM, Rabinowitz JD, Gitai Z, Wingreen NS (2014) Enzyme clustering accelerates processing of intermediates through metabolic channeling. Nat Biotechnol 32(10):1011–1018

Chen L, Chen Z, Zheng P, Sun J, Zeng AP (2012) Study and reengineering of the binding sites and allosteric regulation of biosynthetic threonine deaminase by isoleucine and valine in Escherichia coli. Appl Microbiol Biotechnol 97(7):2939–2949

Chen Z, Bommareddy RR, Frank D, Rappert S, Zeng A-P (2014a) Deregulation of feedback inhibition of phosphoenolpyruvate carboxylase for improved lysine production in Corynebacterium glutamicum. Appl Environ Microbiol 80(4):1388–1393

Chen Z, Rappert S, Zeng A-P (2014b) Rational design of allosteric regulation of homoserine dehydrogenase by a non-natural inhibitor L-lysine. ACS Synth Biol 4(2):126–131

Chen X, Zhou J, Zhang L, Pu Z, Liu L, Shen W, Fan Y (2018) Development of an Escherichia coli-based biocatalytic system for the efficient synthesis of N-acetyl-D-neuraminic acid. Metab Eng 47:374–382

Conrado RJ, Wu GC, Boock JT, Xu H, Chen SY, Lebar T, Turnšek J, Tomšič N, Avbelj M, Koprivnjak T (2012) DNA-guided assembly of biosynthetic pathways promotes improved catalytic efficiency. Nucleic Acids Res 40(4):1879–1889

Delebecque CJ, Silver PA, Lindner AB (2012) Designing and using RNA scaffolds to assemble proteins in vivo. Nat Protoc 7(10):1797–1807

DeLoache WC, Russ ZN, Narcross L, Gonzales AM, Martin VJ, Dueber JE (2015) An enzyme-coupled biosensor enables (S)-reticuline production in yeast from glucose. Nat Chem Biol 11 (7):465–471

Deng MD, Severson DK, Grund AD, Wassink SL, Burlingame RP, Berry A, Running JA, Kunesh CA, Song L, Jerrell TA (2005) Metabolic engineering of Escherichia coli for industrial production of glucosamine and N-acetylglucosamine. Metab Eng 7(3):201–214

Dueber JE, Wu GC, Malmirchegini GR, Moon TS, Petzold CJ, Ullal AV, Prather KLJ, Keasling JD (2009) Synthetic protein scaffolds provide modular control over metabolic flux. Nat Biotechnol 27(8):753–759

Galanie S, Thodey K, Trenchard IJ, Filsinger Interrante M, Smolke CD (2015) Complete biosynthesis of opioids in yeast. Science 349(6252):1095–1100

Gerosa L, Sauer U (2011) Regulation and control of metabolic fluxes in microbes. Curr Opin Biotechnol 22(4):566–575

Hu P, Janga SC, Babu M, Díazmejía JJ, Butland G, Yang W, Pogoutse O, Guo X, Phanse S, Wong P (2009) Global functional atlas of Escherichia coli encompassing previously uncharacterized proteins. PLoS Biol 7(4):e96

Lee JW, Na D, Park JM, Lee J, Choi S, Lee SY (2012) Systems metabolic engineering of microorganisms for natural and non-natural chemicals. Nat Chem Biol 8(6):536–546

Lee JH, Jung S-C, Kang KH, Song J-J, Kim SC (2013) Improved production of l-threonine in Escherichia coli by use of a DNA scaffold system. Appl Environ Microbiol 79(3):774–782

Lee MJ, Mantell J, Hodgson L, Alibhai D, Fletcher JM, Brown IR, Frank S, Xue W-F, Verkade P, Woolfson DN, Warren MJ (2018) Engineered synthetic scaffolds for organizing proteins within the bacterial cytoplasm. Nat Chem Biol 14(2):142–147

Leonard E, Ajikumar PK, Thayer K, Xiao W-H, Mo JD, Tidor B, Stephanopoulos G, Prather KL (2010) Combining metabolic and protein engineering of a terpenoid biosynthetic pathway for overproduction and selectivity control. Proc Natl Acad Sci U S A 107(31):13654–13659

Liu Y, Zhu Y, Ma W, Shin H-D, Li J, Liu L, Du G, Chen J (2014) Spatial modulation of key pathway enzymes by DNA-guided scaffold system and respiration chain engineering for improved *N*-acetylglucosamine production by *Bacillus subtilis*. Metab Eng 24:61–69

McNerney MP, Watstein DM, Styczynski MP (2015) Precision metabolic engineering: the design of responsive, selective, and controllable metabolic systems. Metab Eng 31:123–131

Nielsen J, Keasling JD (2016) Engineering cellular metabolism. Cell 164(6):1185–1197
Nowroozi FF, Baidoo EE, Ermakov S, Redding-Johanson AM, Batth TS, Petzold CJ, Keasling JD (2013) Metabolic pathway optimization using ribosome binding site variants and combinatorial gene assembly. Appl Microbiol Biotechnol 98(4):1567–1581
Paddon C, Westfall P, Pitera D, Benjamin K, Fisher K, McPhee D, Leavell M, Tai A, Main A, Eng D (2013) High-level semi-synthetic production of the potent antimalarial artemisinin. Nature 496(7446):528–532
Philp JC, Ritchie RJ, Allan JE (2013) Biobased chemicals: the convergence of green chemistry with industrial biotechnology. Trends Biotechnol 31(4):219–222
Rajagopala SV, Sikorski P, Kumar A, Mosca R, Vlasblom J, Arnold R, Francakoh J, Pakala SB, Phanse S, Ceol A (2014) The binary protein-protein interaction landscape of Escherichia coli. Nat Biotechnol 32(3):285–290
Rodriguez GM, Tashiro Y, Atsumi S (2014) Expanding ester biosynthesis in Escherichia coli. Nat Chem Biol 10(4):259–265
Thodey K, Galanie S, Smolke CD (2014) A microbial biomanufacturing platform for natural and semisynthetic opioids. Nat Chem Biol 10(10):837–844
Woolston BM, Edgar S, Stephanopoulos G (2013) Metabolic engineering: past and future. Annu Rev Chem Biomol Eng 4:259–288
Zhang Y-HP (2011) Substrate channeling and enzyme complexes for biotechnological applications. Biotechnol Adv 29(6):715–725
Zhang Y, Li SZ, Li J, Pan X, Cahoon RE, Jaworski JG, Wang X, Jez JM, Chen F, Yu O (2006) Using unnatural protein fusions to engineer resveratrol biosynthesis in yeast and mammalian cells. J Am Chem Soc 128(40):13030–13031
Zhang X, Liu Y, Liu L, Wang M, Li J, Du G, Chen J (2018) Modular pathway engineering of key carbon-precursor supply-pathways for improved N-acetylneuraminic acid production in Bacillus subtilis. Biotechnol Bioeng 115(9):2217–2231
Zhou YJ, Gao W, Rong Q, Jin G, Chu H, Liu W, Yang W, Zhu Z, Li G, Zhu G, Huang L, Zhao ZK (2012) Modular pathway engineering of diterpenoid synthases and the mevalonic acid pathway for miltiradiene production. J Am Chem Soc 134(6):3234–3241

Designing of Artificial Metalloenzymes

Srishti Jha, Abdul Arif Khan, and Mohd. Tashfeen Ashraf

1 Introduction

Designing of artificial metalloenzymes (ArMs) has emerged as a frontline area of research in contemporary enzymology. ArMs serve as a bridge between organo-catalysis, transition metal catalysis and biocatalysis, and have immense potential to synergize synthetic biology with preparative chemistry (Rosati and Roelfes 2010). It has long been known that about half of the enzymes found in nature are metallo-enzymes (Lu et al. 2009; Finkelstein 2009). Metalloenzymes have been explored extensively in catalyzing thermodynamically exigent reactions such as nitrogenase mediated reduction of nitrogen to ammonia, oxidation of water during photo-synthesis and conversion to hydrocarbons with concomitant release of molecular oxygen. ArMs, with features as summarized in Fig. 1, find a broad range of industrial applications including the production of biofuels, electricity, drugs and other value added products (Lu et al. 2009; Happe and Hemschemeier 2014).

Invention of ArMs took place in late 1970s by Wilson and Whitesides who incorporated biotinylated rhodium cofactor into biotin binding protein avidin. This conferred enantioselectivity for hydrogenation reaction and simultaneously it was also found that racemic product was produced in the absence of second coordination sphere/protein scaffold (Wilson and Whitesides 1978). However, the progress in this area decelerated due to limited availability of tools for protein designing/engineering and organometallic syntheses. In last few decades owing to significant advancement in the areas of structural, biophysical and computational biology, a resurgence of interest in ArMs has taken place (Schwizer et al. 2018). During this period, the field

S. Jha · M. T. Ashraf (✉)
School of Biotechnology, Gautam Buddha University, Greater Noida, India
e-mail: Tashfeen@gbu.ac.in

A. A. Khan
Department of Pharmaceutics, College of Pharmacy, King Saud University, Riyadh, Kingdom of Saudi Arabia

Q. Husain, M. F. Ullah (eds.), *Biocatalysis*,
https://doi.org/10.1007/978-3-030-25023-2_9

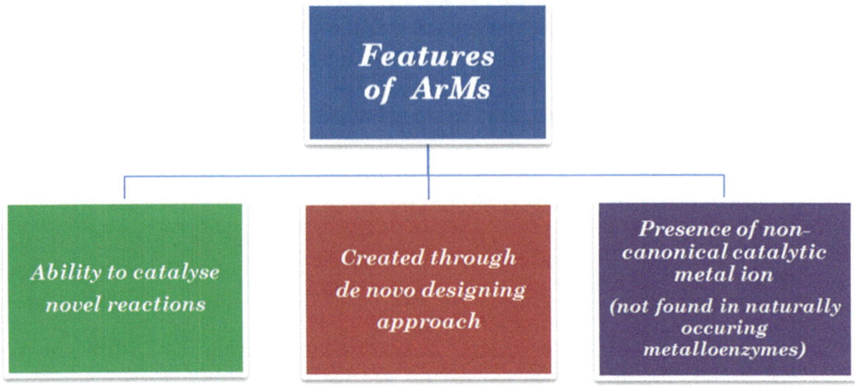

Fig. 1 Artificial metalloenzyme possess of one or more of these features

of designing of ArMs has slowly been transformed from solely structural to a more of functional designing (Yu et al. 2014) and from modifications of sphere to more of designing of second coordination sphere and beyond.

While the first coordination sphere refers to interactions between central metal cofactor and surrounding molecules, the second coordination sphere refers to molecules and ions attched in various ways to the first coordination sphere (Bos and Roelfes 2014).

Following parameters are important for selection of bioscaffolds that provide second coordination sphere:

- Bioscaffold must be stable over a wide range of pH and temperature.
- Bioscaffold must endow characteristic potential to forbear different organic solvents because reactants often require specific quantity of organic solvent for catalysis.
- Metal ion binding site of the biomolecular scaffold must be of substantial size to accommodate transition metal cofactor and still be able to provide space for reactants to participate in reaction. Few examples of successfully created, functionally active bioscaffolds include streptavidin (Sav), apo-myoglobin, BSA etc.
- Building blocks of biomolecular scaffold, whether amino acids or nucleotides, must be chosen carefully. It has been observed that complications arise frequently while using ArMs that consists of DNA scaffold. The reason for such complications is that the DNA scaffold is susceptible to undergo oxidative strand scission.
- Scaffold selection should be made carefully, depending on whether it consists of preexisting metal cofactor binding site or whether it require modifications/ reengineering or whether the newly constructed active site need to be introduced in bioscaffold (Rosati and Roelfes 2010; Bos and Roelfes 2014).

1.1 Significance of Second Coordination Sphere in Catalysis

1. Reaction Rate The rate of reaction is a consequence of activity of both metal ion (cofactor) and bioscaffold. Bioscaffolds have been studied to evaluate how they influence the rate of reaction, positively as well as negatively. Positive increase in rate of reaction (from two folds to two orders of magnitude) by ArMs make them more preferred over transition metal catalysts. This results in providing an efficient system, with reduced cost and time. However in some cases, negative regulation may also occur. It has been observed that in case of ArMs constructed using supramolecular anchoring strategies, at diminutive concentrations, the site of equilibrium between unbound and bound catalyst may become inimical, as a consequence of which, due to the presence of enormous quantity of unbound catalyst, racemic product gets generated, which subsequently leads to lower enantiomeric excess (ee). DNA bioscaffolds based asymmetric catalytic systems are an exception to this phenomenon and the reason for this is because the complex of metallocofactor bound with DNA scaffold is substantially faster as compared to unbound catalyst (Rosati and Roelfes 2010)

2. Enantioselectivity Chirality provided by bioscaffold, allows the entry of substrate and further determines enantiomeric characteristic of products being formed.

3. Chemoselectivity The bioscaffold has been shown to regulate the polarity of the surrounding microenvironment. This confirms that the bioscaffold has a role in determining the chemoselectivity of reaction (Rosati and Roelfes 2010).

4. Shape and Size Selectivity A variety of ArMs designed to date have been shown to provide catalytic system having a wide scope of substrates. This unique characteristic could be considered as a constituent of chemocatalysis. However, in some of the cases, it has been observed that the second coordination sphere imposes demarcations on substrates or on combination of reactant and substrate for the reaction catalysis to proceed. It has also been observed that the lack of accurate compatibility between the active site and substrate leads to a complete loss of functionality along with enantiomeric selectivity (Rosati and Roelfes 2010).

2 Designing of ArMs

Figure 2 gives the outline of an ArM (Rosati and Roelfes 2010). The construction of ArMs invariably involves the incorporation of one or more transition metal ions into apoprotein which usually consists of more than 50 amino acid residues (Jeschek et al. 2018). Following are three basic prerequisites to be considered for designing of ArMs (Bos and Roelfes 2014):

1. Selection of transition metal cofactor capable of catalyzing the desired reaction.
2. Selection of anchoring strategy for binding of transition metal ion to the bioscaffold.
3. Selection of bioscaffold (second coordination sphere): protein, peptide or nucleotide scaffold.

For selection of metal cofactor it is essential to consider that the reactivity of the metal cofactor requires being orthogonal to the second coordination sphere i.e. it needs to be inert towards chemical reactivity provided by the biomolecular scaffold. Moreover, the nature of metal ion catalyst needs to be water tolerant in order to ensure activity of ArMs in aqueous environment.

One major advantage offered by ArMs is that they have the ability to catalyse asymmetric reactions within aqueous solution (Pàmies et al. 2015) and inorganic solvents. Furthermore, there are other advantages associated with employment of ArMs in catalytic reactions; some of them include: high efficiency, high enantioselectivity, high stability, high turnover number and confer novel functionality to enzyme (Jeschek et al. 2018). For effectuating novel functionality to artificial metalloenzyme, either of the following designing approaches can be taken into consideration:

1. Synthetic cofactor approaches
2. Designing/Redesigning of enzyme
3. Repurposing of natural enzymes based on promiscuity and directed evolution.

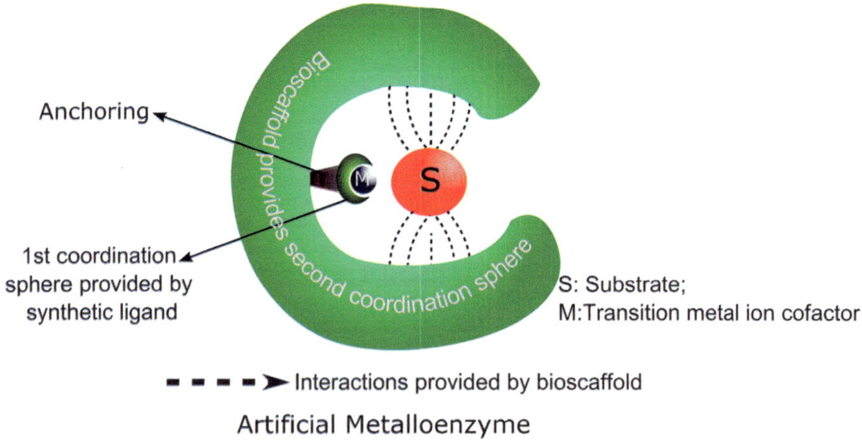

Fig. 2 Pictorial representation of artificial metalloenzyme

3 Anchoring Strategies

In order to assure accurate localization of metallocofactor into bioscaffold, certain anchoring strategies have been developed (Fig. 3) (Rosati and Roelfes 2010), which include:

1. Covalent Anchoring Evocative of well-known bioconjugation methodologies (Hermanson and Preceded by: Hermanson 1997; Stephanopoulos and Francis 2011), covalent anchoring involves a highly productive irreversible reaction which takes place between functional group of cofactor and constituents of bioscaffold which may be amino acids in case of proteins/peptides or nucleotides in case of DNA scaffold (Schwizer et al. 2018). This process requires bioscaffold to undergo post biosynthetic modifications and additionally it also requires reactive group of cofactor to be orthogonal to bioscaffold in order to achieve highly selective modifications during designing of ArMs (Bos and Roelfes 2014). The major disadvantage of this strategy is the involvement of an essential purification step in the middle of the process (Rosati and Roelfes 2010). The development of ArMs by utilization of covalent anchoring strategy involves some major reactions which include:

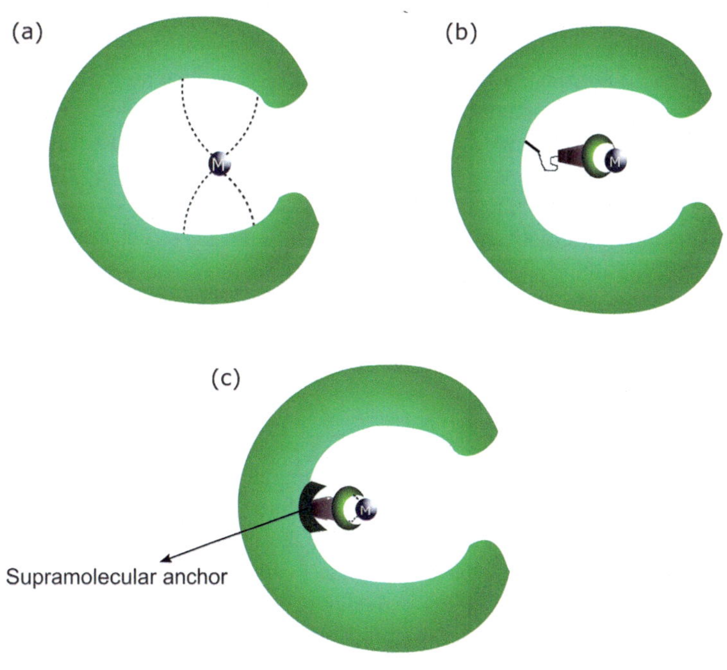

Fig. 3 Three anchoring procedures employed for developing artificial metalloenzymes (**a**) dative anchoring, (**b**) covalent anchoring, and (**c**) supramolecular anchoring

(a) Formation of disulphide bond between metallocofactor that has been substituted with sulfuric electrophile and cysteine residue of protein/peptide bioscaffold (Schwizer et al. 2018).
(b) Nucleophilic substitution reaction carried out by cysteine or any other activated residue present within second coordination sphere (provided by bioscaffold) which targets an electrophilic group (for example, alpha halocarbonyl, maleimide, etc.) of metallocofactor (Schwizer et al. 2018).
(c) Huisgen [3 + 2] cycloaddition reaction which takes place between a not-naturally occuring amino acid consisting of terminal azide or alkyne group or metallocofactor substituted with alkyne or azide group (Yang et al. 2014; Schwizer et al. 2018).

2. Supramolecular Anchoring This anchoring approach leverages the feature of 'self-assembly' of artificial metalloenzymes, which facilitate high speed and effective optimization during designing process (Rosati and Roelfes 2010). It exploits the attribute of certain proteins of displaying high affinity towards a defined set of ligands (Schwizer et al. 2018). For example, ArMs designed through biotin/streptavidin technology is based on supramolecular anchoring strategy. Biotin has a very strong affinity for protein called avidin and a molecule bound to biotin will move inside and would tend to get localized at a particular site.Most of the ArMs designed to date by replacement of natural metallocofactor with artificial/synthetic cofactor in native active site of protein scaffold (e.g. myoglobin) are all based on supramolecular anchoring strategy. DNA bioscaffold based ArMs that have been created for catalyzing asymmetric reactions with subsequently high efficiency, are another example for this strategy. One of the major challenges faced during utilization of this anchoring strategy is that the development of a heterogeneous system, due to which each and every catalytic site being present in a different micro-environment, displays a distinctively unique activity and also selectivity. Another important point is that since the binding of catalyst depends on selectivity and strength of supramolecular anchoring, it is very difficult to predict where the binding of catalyst will occur. This consequentially to leads active site structure to be comparatively less defined. Therefore, it is important to ensure strong binding of metallocofactor with the biomolecular scaffold, in order to avoid occurrence of catalysis outside the second coordination sphere/bioscaffold (Rosati and Roelfes 2010; Bos and Roelfes 2014).

3. Dative Anchoring This anchoring strategy is based on coordinate bond formation between a nucleophile residue of protein/peptide bioscaffold (e.g. Cys, His, Asp, Glu, Ser, etc.) and catalytically active, coordinately unsaturated metallocofactor (Schwizer et al. 2018). One of the major advantages associated with this approach is that the coordinate bonding enables accurate positioning of metal cofactor inside biomolecular scaffold (Rosati and Roelfes 2010). Although this anchoring strategy had been found to operate in most of the naturally occurring metalloenzymes, still it is immensely difficult to construct an ArM by replacement of native cofactor by synthetic cofactor by utilization of this anchoring strategy because a distinctive metallocofactor requires a uniquely distinct coordination microenvironment conditions (Bos and Roelfes 2014).

4 De novo Designing of Metalloenzymes: Bottom-Up Approach

De novo designing of metalloenzymes, also referred to as bottom-up approach (Fig. 4), is mainly concerned with construction of a unique polypeptide chain of sequence which is not similar to any of the protein sequences found in nature and that also possess the ability to fold accurately into a well-defined three dimensional structure which subsumes the ability to bind precisely to the metal ion (cofactor). In recent years, this designing approach has been shown to be very useful in unraveling those structural characteristics of metalloenzymes that could not gather the attention of researchers till date. This designing strategy provides complete control over structural as well as functional attributes of metalloenzyme. This strategy is currently being exploited extensively for both reasons of creation of novel artificial metalloenzymes and also for the creation of ArMs that possess the ability to mimic natural metalloenzymes in terms of either structural or functional characteristics (Lu et al. 2009).

To date, the major area of research concerned with de novo designing of ArMs has focused on incorporation of metal cofactor binding sites into artificially constructed alpha helical polypeptide structures (which was one of the initially designed protein structures through utilization of de novo designing approach). Alpha helical bundles are most frequently occurring scaffolds found in various naturally occurring haem group containing proteins (DeGrado et al. 1999; Reedy and Gibney 2004; Lu et al. 2009; Rosati and Roelfes 2010). Besides, haem cofactor, various other metal cofactors had also been designed and incorporated into alpha helical scaffolds in order to produce efficient biomimetic catalytic systems. Although, in comparison to alpha helical scaffolds metalloenzymes containing beta sheet scaffolds had been known to support much more rigorous and pre organized metal ion binding sites, but because of the fact that beta sheet structures

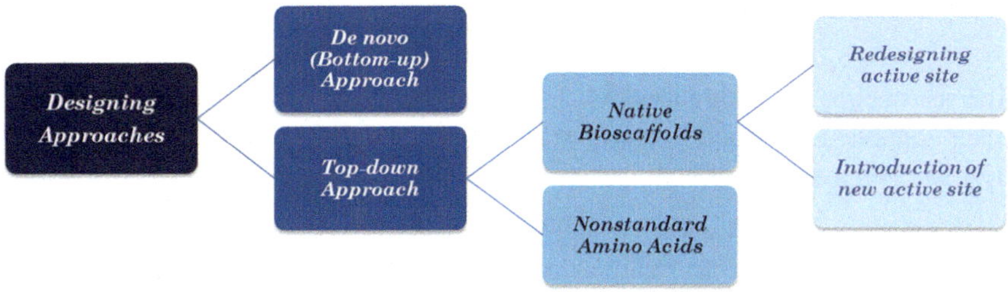

Fig. 4 Overview of De novo (Bottom up) and top-down approaches for designing artificial metalloenzymes

are less frequently found in nature as compared to alpha helical scaffolds, de novo designing of ArMs having beta sheet structures has not been widely studied (Williams 1995; Lu et al. 2009). One of the examples of ArMs designed using de novo approach employing SCADS computational tool (Kono and Saven 2001) and that comprises of beta sheet-rich scaffold is redox-active rubredoxin biomimetic system RM1 which has the ability to bind Fe ions and conduct reversible cycle between Fe^{2+} and Fe^{3+} oxidation states and thus mimics the naturally found rubredoxin (Vikas Nanda et al. 2005).

There are various online tools that can be utilised for de novo designing of metalloenzymes. Although Rosetta software suite is one preferred suite of programs for enzyme design, there exist other alternatives as well. For instance, the matching of the ideal *theozyme* (theoretical enzyme constructed by computing the optimal geometry for transition-state stabilization by functional groups) into an existing protein scaffold can be also carried out with OptGraft, Scaffold-Selection and ProDA. A different strategy is employed by the SABER program. Instead of placing the ideal active site into selected proteins, this program searches for proteins already presenting the catalytic residues in the adequate relative positions, thus requiring the introduction of a lower number of mutations. One of the major advantages of de novo approach is that various artificially designed metalloenzymes have been reported to display an intriguing characteristic potential of stabilizing anomalous metal coordination states. In some cases, it has been observed that the binding of metal cofactor can regulate protein folding and assembly of polypeptide structure from a misfolded or random coil or comparatively less folded condition to a functionally active native state. This approach holds enormous potential for creation of multifarious ArMs that are functionally active and structurally stable in nature.

The only major limitation of this approach is the lack of knowledge of protein folding mechanism. As a consequence of which, for in silico designing of metalloenzyme, advance computational calculations need to be performed (Rosati and Roelfes 2010).

5 Top-Down Approach

5.1 Designing of ArMs Using Native Bioscaffolds

The reason why designing of ArMs using native bioscaffold has been the most frequently applied approach is that it provides a myriad of advantages over de novo designing approach. Following are some of these advantages:

1. In nature, similar or rather exactly same bioscaffolds can be found in various metalloenzymes having significantly different metal cofactor incorporated within and performing diversified functions. Further, the presence of similar/same bioscaffolds in different metalloenzymes also imply that these natural scaffolds inherently bestow intriguing attributes of robustness and flexibility and thus, both

structure and functions of these bioscaffolds can be easily altered as per requirement, i.e. the native bioscaffolds are susceptible to mutations and even after undergoing multiple rounds of mutations, these scaffolds maintain their thermodynamic stability and folding patterns. This also indicates that the usage of this strategy offers researchers to escape concerns and complications associated with the maintenance of enzyme stability while introducing new metal cofactor binding site into bioscaffold (Lu et al. 2009; Rosati and Roelfes 2010).

2. As the computational tools/databases present to date, for example PDB (Protein Data Bank), contains diminutive information for de novo designed proteins as compared to information available for native proteins, researchers often prefer to design ArMs using native scaffolds (Berman et al. 2000).
3. The use of naturally occurring bioscaffolds provide researchers with significant advantage to choose from a profusion of bioscaffold choices available in nature.
4. In contrast to de novo designed protein bioscaffolds, natural protein scaffolds are facile to crystallization, which makes structural (three dimensional) characterization of protein scaffolds possible (Lu et al. 2009).
5. Site directed mutagenesis can be applied for functional characterization of protein scaffolds. For instance, the knockout/loss-of-function mutations introduced usually in conserved regions lead to identification of essential biomolecular markers responsible for a specific function (Lu et al. 2009).

5.1.1 Redesigning Active Site/Metal Cofactor Binding Site in Bioscaffold

The purpose of redesigning the metal binding site is either to impart novel functionality or to impart specificity to metal cofactor for binding to protein scaffold. Different mutagenesis techniques can be used for redesigning of metalloenzymes. In order to impart the desired novel functionality to metalloenzyme, its metal cofactor binding site requires to be altered in such a way that the geometry of metal binding site along with substrate binding site of metalloenzyme gets changed (Lu et al. 2009; Lewis and Ellis-Guardiola 2018). Metalloenzymes can be successfully redesigned by introduction of amino acid residues that have the characteristic tendency to bind metal ions, into the preexisting active site of enzyme that originally may or may not be metal binding proteins. Novel metal binding sites that are incorporated into protein scaffold may be mononuclear, homonuclear or heteronuclear in nature. Heteronuclear sites naturally had been found in multiple proteins that carry out biological functions like degradation of lignin fibers, reduction of nitric oxide, etc. It has been observed that the designing of heteronuclear sites is far more challenging as compared to designing of homonuclear metal binding sites because of the requirement of accurate localization and binding of multiple metal cofactors together. The designing of heteronuclear metal binding site and its incorporation into protein bioscaffold is usually aided by computational designing approach. Complications often arise during designing of metalloenzyme by introduction of metal clusters or dinuclear metal binding sites into the protein scaffold. This process of introduction of dinuclear metal binding site is mostly achieved through application

Table 1 Benefits of redesigning

S. No.	Redesigning strategy: benefits
1.	To decipher structural characteristics of bioscaffold that may have a role in gain or any change in function.
2.	Applications in metal sensing technology.
3.	Better understanding of protein-protein interaction mechanisms.

of loop directed mutagenesis technique (Franklin and Welch 2005; Greisen and Khare 2014; Jeschek et al. 2018). Usually, the redesigning of metal binding site is based upon structural disparity that exists between target protein and template protein. This strategy is usually directed by empirical designing approach which is centralized upon prior experience and knowledge, rational designing approach which makes use of computational tools and bioinformatics, or combinatorial approach for selection (Lu et al. 2009).

Redesigning Strategy: Benefits

1. The major advantage associated with this strategy is that it can be employed to elucidate the structural characteristics of protein scaffold present in a particular metalloenzyme that may have role in gain or any change in function (Table 1).
2. Another advantage associated with this strategy is that the selectivity for binding to particular metal ion imparted by redesigning of active site can be used in metal sensing applications such as construction of metal sensors (Lu et al. 2009; Hu et al. 2014).
3. Owing to the complexity associated with non-covalent interactions, a limited understanding of protein-protein interactions has not enabled control over these interactions. Designing of metalloenzymes active site works in providing insight into protein–protein interaction mechanism and also allows for regulation of these interactions (Lu et al. 2009; Yu et al. 2014).

5.1.2 Introduction of New Active Site

Introduction of new active site into bioscaffold which lacks any metal cofactor binding pocket is an alternative strategy for designing of ArMs using native bioscaffolds. Although being complicated, this approach can be regarded as significantly important strategy as it can overcome the limitation of extrapolating common structural characteristics found in both target protein and template protein (where metal binding site has to be introduced), which has not been possible in case of redesigning of metal ion binding site strategy.

A common method for introduction of metal binding site is to design the active site based on homology of template and target protein structures. In case of absence of structural homology, computational designing tools such as SWISS-MODEL, MODELLER, PHYRE and METAL-SEARCH are frequently used.

The major advantage of using this approach is that it extensively expands the number of options available for bioscaffolds that can be used. However, it cannot be predicted that how the introduction of newly designed metal binding site may impact stability and folded configuration of protein bioscaffold, because this introduction may even lead to dislocation/disruption of intermolecular and/or intramolecular interactions (Lu et al. 2009; Rosati and Roelfes 2010; Bos and Roelfes 2014).

5.2 *Designing of Novel Metalloenzymes via Insertion of New Amino Acids*

Designing of native bioscaffold is one of the effective tools that has enabled the synthesis of novel ArMs. However, as one of the major objectives in developing ArMs has been creating enzymes with new functionalities, this led to this approach of introducing new amino acid residues not found in protein scaffolds of naturally occurring metalloenzymes (Lu 2005). Following are some of the main methods (Fig. 5) employed for designing of ArMs with newer functions:

1. *Solid Phase Synthesis*: Designing of bioscaffolds by using this methodology permits complete synthesis of protein scaffold. This technique allows introduction of any artificial amino acid residues. Major limitations of this strategy include size and cost constraints which limits the use of this technique in designing proteins that are no larger than 60–100 amino acids in length (Merrifield 1997; Lu et al. 2009).
2. *Native Chemical Ligation*: This strategy is based on formation of covalent interactions between artificial peptides. This strategy can be adopted for the production of proteins that are larger in size but cost effectiveness still remains an issue in case of bulky proteins (Dawson et al. 1994)
3. *Expressed Protein Ligation*: This technique involves production of recombinant proteins from microbial cells such as bacteria, and it further involves conjugation

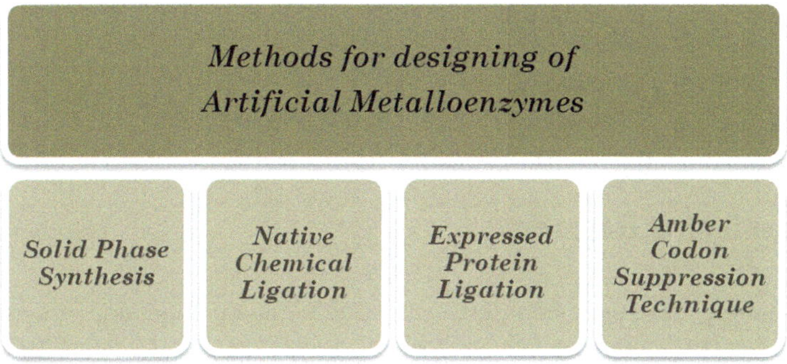

Fig. 5 Snapshot of methods commonly used in designing of artificial metalloenzymes

of expressed polypeptide chain with that of artificial protein/peptide chain constituting incorporation of not-naturally-occurring amino acids, that has been produced using some other strategy (e.g. chemical synthesis). This strategy has significantly reduced the cost involved in production of very large proteins, but the drawback associated with this technique is that it frequently requires the presence of cysteine residue at the site of ligation. In order to increase the efficiency of ligation, the amino acids other than cysteine have also been studied but they often provide less ligation efficiency (Muir 2003).

4. *Amber Codon Suppression Technique*: This method involves recognition of 'UAG' amber stop codon sequence based upon orthogonal tRNA and/or aminoacyl transfer RNA synthase, for in vivo and in vitro translation and incorporation of new, not-naturally occurring amino acid residues into protein scaffold (Noren et al. 1989; Wang et al. 2001; Lu et al. 2009; Jeschek et al. 2018).
5. *Other Techniques*: Auxotroph development (Ikeda et al. 2003), chemical modification of amino acid residues (Dongfeng Qi et al. 2001), cavity complementation (Barrick 1995) are some of the methods that are not restricted by length of protein. However, these methods are comparatively less versatile in terms of amino acids that can be incorporated efficiently (Lu et al. 2009).

Designing of ArMs Using Not-Naturally-Occurring Amino Acids: Advantages

1. This strategy can be applied to study the role of individual amino acid residue incorporated within protein scaffold and its effects on catalytic activity and reactivity of artificially designed metalloenzymes, and how this affects the reaction catalyzed by the enzyme. For example, the substitution of levo-enantiomer for dextro-enantiomer has been found to allow for fine regulation between structures of designed ArMs.
2. Incorporation of not-naturally-occurring amino acid residues mediate recognition of hydrophobic residues that have role in alteration of reduction potential associated with protein which leads to formation of novel metalloenzymes having reduction potentials exceeding the reduction potential of naturally occurring proteins.
3. Bioscaffolds constructed by incorporation of not-naturally occurring amino acids, provides significant advantage over native scaffolds, as these scaffolds are amenable to introduction of metal cofactors that are not naturally found in metalloenzymes in nature (Lu et al. 2009).

6 Concluding Remarks

An accurately designed ArM must necessarily incorporate properties of steric compatibility of primary coordination sphere along with precise geometric configuration. Although significant advancements have been made in field of protein design, still to date it is quite challenging to design a structurally stable and functionally active artificial metalloenzyme. One of the major challenges is creation of metal ion clusters and sites having unique geometric configuration. Another challenge is to design metal binding site at the interface of proteins. It has also been observed that the incorporation of changes in primary coordination sphere has never proven to be sufficient for developing a functionally active metalloenzyme. The reason why designing of metalloenzymes have proven to be more challenging as compared to designing of protein-only enzymes is that in addition to designing the polypeptide the designing of artificial metalloenzyme also require the selection of metal ions from a myriad of options with high variability in their geometric configurations.

Some of the issues that pose a challenge in designing ArMs include (Lu et al. 2009):

1. Creation of new metallocofactor binding site or modification of already existing metal binding site located at the interfacing region of proteins. It has been observed that it is quite challenging to couple redox reactions associated with metal ion binding pocket to conformational changes, charge separation or proton transfer.
2. Creation of metal ion binding sites by utilization of membrane associated proteins.
3. Forego the use of metallochaperons for designing of ArMs.

Since the designing of metal ion binding site often requires metallochaperons, it has been observed that for developing functionally active ArMs, the interaction of metal cofactor with apoprotein is not always sufficient and in such cases functionality is conferred by additional interactions provided by metallochaperons to the metal cofactor along with that of apoprotein.

The advantage of designing of ArMs over non-metalloenzymes is that the cofactor binding pockets exhibit characteristic magnetic properties and are chromophoric in nature due to which the spectroscopic characterization of ArMs become feasible; this therefore enhances the efficiency of the designing procedure (Lu et al. 2009).

Therefore, in spite of challenges, the designing of ArMs hold immense promise for developing novel enzymes for catalyzing chemical transformations unheard of so far. Though efficiency enhancement of known metalloenzymes is another promising area for application of ArMs, the lack of any mature, proven technology calls for more research in this area so as to develop better tools, both In silico and wet lab-based, for designing ArMs with desired functionality.

References

Barrick D (1995) Depletion and replacement of protein metal ligands. Curr Opin Biotechnol 6: 411–418. https://doi.org/10.1016/0958-1669(95)80070-0

Berman HM, Westbrook J, Feng Z et al (2000) The protein data bank. Nucleic Acids Res 28: 235–242

Bos J, Roelfes G (2014) Artificial metalloenzymes for enantioselective catalysis. Curr Opin Chem Biol 19:135–143
Dawson PE, Muir TW, Clark-Lewis I, Kent SB (1994) Synthesis of proteins by native chemical ligation. Science 266:776–779. https://doi.org/10.1126/SCIENCE.7973629
DeGrado WF, Summa CM, Pavone V et al (1999) De novo design and structural characterization of proteins and metalloproteins. Annu Rev Biochem 68:779–819. https://doi.org/10.1146/annurev.biochem.68.1.779
Finkelstein J (2009) Metalloproteins. Nature 460:813–813. https://doi.org/10.1038/460813a
Franklin SJ, Welch JT (2005) The helix-turn-helix as a scaffold for chirmeric nuclease design. Comment Inorg Chem 26:127–164. https://doi.org/10.1080/02603590500201188
Greisen P, Khare SD (2014) Computational redesign of metalloenzymes for catalyzing new reactions. Humana, New York, NY, pp 265–273
Happe T, Hemschemeier A (2014) Metalloprotein mimics – old tools in a new light. Trends Biotechnol 32:170–176. https://doi.org/10.1016/j.tibtech.2014.02.004
Hermanson GT, Preceded by: Hermanson GT (1997) Bioconjugate techniques. Gen Pharmacol Vasc Syst 28:167. https://doi.org/10.1016/S0306-3623(97)89864-3
Hu C, Chan SI, Sawyer EB et al (2014) Metalloprotein design using genetic code expansion. Chem Soc Rev 43:6498–6510. https://doi.org/10.1039/C4CS00018H
Ikeda Y, Kawahara S-I, Taki M et al (2003) Synthesis of a novel histidine analogue and its efficient incorporation into a protein in vivo. Protein Eng Des Sel 16:699–706. https://doi.org/10.1093/protein/gzg084
Jeschek M, Panke S, Ward TR (2018) Artificial metalloenzymes on the verge of new-to-nature metabolism. Trends Biotechnol 36:60–72. https://doi.org/10.1016/j.tibtech.2017.10.003
Kono H, Saven JG (2001) Statistical theory for protein combinatorial libraries. Packing interactions, backbone flexibility, and the sequence variability of a main-chain structure. J Mol Biol 306:607–628. https://doi.org/10.1006/jmbi.2000.4422
Lewis JC, Ellis-Guardiola K (2018) Preparation of artificial metalloenzymes. In: Artificial metalloenzymes and metalloDNAzymes in catalysis. Wiley-VCH Verlag GmbH & Co. KGaA, Weinheim, pp 1–40
Lu Y (2005) Design and engineering of metalloproteins containing not-naturally occuring amino acids or non-native metal-containing cofactors. Curr Opin Chem Biol 9:118–126. https://doi.org/10.1016/J.CBPA.2005.02.017
Lu Y, Yeung N, Sieracki N, Marshall NM (2009) Design of functional metalloproteins. Nature 460:855–862. https://doi.org/10.1038/nature08304
Merrifield B (1997) Concept and early development of solid-phase peptide synthesis. Methods Enzymol 289:3–13. https://doi.org/10.1016/S0076-6879(97)89040-4
Muir TW (2003) Semisynthesis of proteins by expressed protein ligation. Annu Rev Biochem 72:249–289. https://doi.org/10.1146/annurev.biochem.72.121801.161900
Nanda V, Rosenblatt MM, Osyczka A et al (2005) De novo design of a redox-active minimal rubredoxin mimic. J Am Chem Soc 127(16):5804–5805. https://doi.org/10.1021/JA050553F
Noren CJ, Anthony-Cahill SJ, Griffith MC, Schultz PGA (1989) General method for site-specific incorporation of not-naturally occuring amino acids into proteins. Science (80–) 244:182–188
Pàmies O, Diéguez M, Bäckvall J-E (2015) Artificial metalloenzymes in asymmetric catalysis: key developments and future directions. Adv Synth Catal 357:1567–1586. https://doi.org/10.1002/adsc.201500290
Qi D, Tann C-M, Haring D, Distefano MD (2001) Generation of new enzymes via covalent modification of existing proteins. Chem Rev 101(10):3081–3111. https://doi.org/10.1021/CR000059O
Reedy CJ, Gibney BR (2004) Heme protein assemblies. Chem Rev 104:617–649
Rosati F, Roelfes G (2010) Artificial metalloenzymes. ChemCatChem 2:916–927
Schwizer F, Okamoto Y, Heinisch T et al (2018) Artificial metalloenzymes: reaction scope and optimization strategies. Chem Rev 118:142–231. https://doi.org/10.1021/acs.chemrev.7b00014

Stephanopoulos N, Francis MB (2011) Choosing an effective protein bioconjugation strategy. Nat Chem Biol 7:876–884. https://doi.org/10.1038/nchembio.720
Wang L, Brock A, Herberich B, Schultz PG (2001) Expanding the genetic code of Escherichia coli. Science (80–) 292:498–500. https://doi.org/10.1126/science.1060077
Williams RJP (1995) Energised (entatic) states of groups and of secondary structures in proteins and metalloproteins. Eur J Biochem 234:363–381. https://doi.org/10.1111/j.1432-1033.1995.363_b.x
Wilson ME, Whitesides GM (1978) Conversion of a protein to a homogeneous asymmetric hydrogenation catalyst by site-specific modification with a diphosphinerhodium(I) moiety. J Am Chem Soc 100:306–307. https://doi.org/10.1021/ja00469a064
Yang H, Srivastava P, Zhang C, Lewis JC (2014) A general method for artificial metalloenzyme formation through strain-promoted azide-alkyne cycloaddition. Chembiochem 15:223–227. https://doi.org/10.1002/cbic.201300661
Yu F, Cangelosi VM, Zastrow ML et al (2014) Protein design: toward functional metalloenzymes. Chem Rev 114:3495–3578. https://doi.org/10.1021/cr400458x

Anti-inflammatory and Antidote Drug Discovery with Secreted Phospholipase A_2

Ramakrishnan Chandrasekaran, Atanu Bhattacharjee, and Velmurugan Devadasan

1 Introduction

Secreted phospholipase A_2 ($sPLA_2$) initiates the arachidonic acid (AA) pathway (Fig. 1) by catalyzing hydrolysis at sn-2 position of membrane phospholipids (Gelb et al. 1995). AA pathway leads to production of biologically important lysophospholipids (Rivera and Chun 2008) and various pro-inflammatory compounds known as eicosanoids (Funk 2001).

PLA_2 enzyme shows several fold increased activity when its substrate (phospholipids) is in aggregated form (Jain and Cordes 1973; Verger and De Haas 1973; Verger et al. 1973; Verheij et al. 1981) and it is a common phenomenon known as interfacial activation of $sPLA_2$ at the lipid-water interface. This is facilitated by the interfacial recognition site (IRS), the common structural feature of all family members (Jain and Berg 1989). The role of $sPLA_2$ enzymes become vital as the phospholipids and their metabolites are responsible for many cellular functions, particularly, inflammation and signal transduction and thereby involved in a different physiological functions and diseases such as asthma, arthiritis, sepsis, atherosclerosis, etc. In this view, a compound that alters the function of a particular type of $sPLA_2$ of human has many pharmacological importances. This will be an effective control mechanism inflammatory pathway while there are many cyclooxygenase 2 (cox2) inhibitors playing a significant role.

R. Chandrasekaran
Department of Biotechnology, Bhupat and Jyoti Mehta School of Biosciences, Indian Institute of Technology Madras, Chennai, Tamil Nadu, India

A. Bhattacharjee
Department of Biotechnology and Bioinformatics, North-Eastern Hill University, Shillong, Meghalaya, India

V. Devadasan (✉)
Centre of Advanced Study in Crystallography and Biophysics, University of Madras, Chennai, Tamil Nadu, India

Q. Husain, M. F. Ullah (eds.), *Biocatalysis*,
https://doi.org/10.1007/978-3-030-25023-2_10

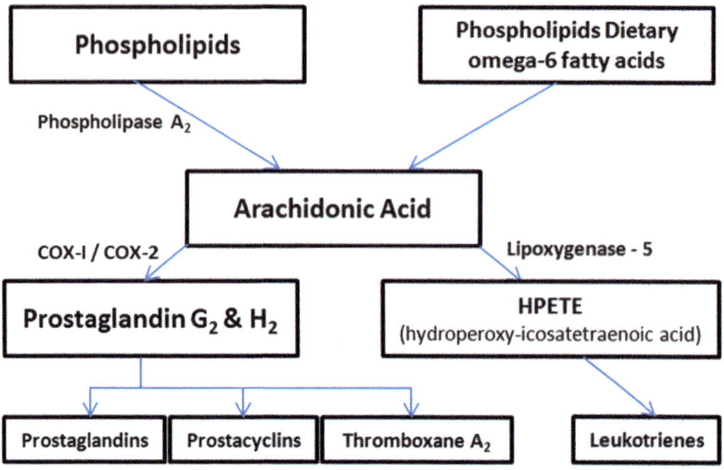

Fig. 1 Inflammatory pathway leading to production of the eicosanoids

1.1 Classification of sPLA$_2$

Continuous advent of sPLA$_2$s of different sources resulted into nine classes including few that deserve high pharmacological importance. In general, sPLA2s are classified into I, II, III, V, X, and XII groups with the similar structural characteristics including conserved active site residues (His and Asp) and a calcium binding loop. Sequence homology is also employed as an important determinant for classification. A traditional scheme of classification can be seen in the literature (Balsinde et al. 1999; Davidson and Dennis 1990; Dennis 1994, 1997). Further expansion in the classification and details related to non-redundant function of each group and sub-group of sPLA2s is also available (Six and Dennis 2000). In this chapter, the structural characteristics, function of human and snake venom sPLA2s and importance of their inhibition are highlighted.

Number of crystal structures of the free and complex forms of sPLA2 is increased, tremendously. Together, number of group specific inhibitors of different classes of sPLA2s is also increased. Based on the organism source and sequence alignment, the sPLA2 family proteins are classified into nine groups (I–VII, X and XII) and they are well characterized by biochemical and structural studies. In mammals, sPLA2s expressed by different organs are classified into I, IIA, IIC, V, X and XII and they govern important cell functions including inflammatory responses, signalling, etc. Over-expression of these sPLA2s triggers inflammation, neurological disorders (like Alzheimer's disease, ischemia, etc.) and neuropsychiatric disorders. Likewise, snake and bee venoms contain sPLA2s (belonging to groups IA, IIA, IIB and III) as the major functional protein constituent that are responsible for acute inflammation related disorders and few of them show neurotoxic or myotoxin effects.

Table 1 Group or type specific details of selected $sPLA_2$ enzymes

Type	Source	Nature	Substrate	Disorder(s)
IA	*Naja naja* Krait	Acidic Basic	PC PC	Neurological, myocardial, etc.
IB	Mammalian organs	NA	PS, PG	Inflammation
IIA	Mammalian (SF, Hnps) Viper and rattle snake venom	NA	PG, PS	Signaling, Atherosclerosis
IIB	Gabon viper venom	Anionic	PG, PS	Inflammation
II	Saw scaled/*Echis carinatus*	Acidic	PG, PS	Neurological, myocardial, etc.
V, X, XII	Mammalian organs	ZI	PC	Atherosclerosis

PC phosphatidylcholine, *PS* phosphatidylserine, *PG* phosphatidylglycerol, *SF* synovial fluid, *Hsnp* human secretory non-pancreatic, *NA* not available, *ZI* Zwitter ionic

Briefly, the $sPLA_2$s classified under the group I are found as a major composition of the old world snake venoms (Stephens and Myers 1898) and human pancreas and they potentially involve in progression of many inflammatory diseases. The group II $sPLA_2$s are also known to involve in the inflammatory process in higher organisms. Particularly, about ten distinct subtypes of PLA2s are found to be expressed in mammalian cells with lipolytic activity (Lambeau and Gelb 2008), classified into the groups IB, II (A,C,D,E,F), III, V, X and XIIA and interestingly have unique characteristic in substrate/inhibitor recognition despite they share a common structural fold. Similarity, there is a non-redundant role of IB, IIA, V and X PLA2s found in mammalian tissues. The group IA and group IIB PLA_2s are found in snake venom (krait and naja naja, and viper, respectively). Please refer Table 1 for more details.

Mammalian $sPLA_2$s (group I, II, V, X and XII) have evolutionarily conserved three dimensional structure and share a common domain (CDD: cd00125) (Nevalainen et al. 2012; Valentin and Lambeau 2000). Similarly, the venom sPLA2s also share the similar structural characteristics (Kini 2003). Though the structural features are comparable between these two major sources, only venom $sPLA_2$s show diverse toxic effects and deserve pharmacological importance during onset of neuronal apoptosis, Alzheimer's ischemia, sepsis, arithritis, and also cancers. Toxicity of venom sPLA2s is due to their ability to penetrate any membrane and digest the phospholipids.

1.2 Structure and Function of $sPLA_2$s

The members of $sPLA_2$ superfamily share a common fold (particularly I, II, V, X) which consists of three helices (H1, H2, and H3), two short helices (SH4 and SH5) and a β-wing with two strands aligned in an anti-parallel fashion (Dennis 1994; Dijkstra et al. 1981; Fremont et al. 1993). Additionally, His48 and Asp49 (catalytic diad), calcium binding loop, surface loop, Tyr52 and Asp99 are catalytically important and conserved among the subtypes. The overall fold is stabilized through seven disulfide bridges which are also found conserved. The H1 helix (located at the N-terminal) is rich of hydrophobic amino acids and interacts with the membrane for interfacial activation (Dijkstra et al. 1981, 1984). Figure 2 depicts the conserved amino acid residues among the selected groups through multiple sequence alignment.

In spite of a conserved fold, the sequence similarity between any pair of members shows a maximum of 55.46% identity while it is as least as 23.91%. Considerable differences observed at the sequence level (Ward et al. 1998; Jeyaseelan et al. 2000) makes the complexity in designing subtype specific inhibitors. Invariably, all the sPLA2 types are activated at the interface of water and condensed lipid which was confirmed through biochemical and simulation studies. A membrane induced conformational changes in sPLA2 also observed at the interacting surface of the enzyme (i-face or interfacial recognition site or IRS) (Ward et al. 1998; Wee et al. 2008). In all the subtypes with a significant pharmacological importance (IA, IB, IIA, V and X), the i-face is constituted by mostly aromatic amino acids and hence an increased affinity towards the zwitter-ionic phase. The conformational changes in the enzyme as well as in the solvent-lipid interface both equally contribute to the interfacial activation (Arni and Ward 1996; Ramirez and Jain 1991; Winget et al. 2006). The major constituent of the i-face, N-terminal helix undergo a twist-like motion and increases the rate of reaction of the enzyme towards the aggregated substrates, micelles and membrane phospholipids (Demaret and Brunie 1992). The unique structural characteristics of each subtype are explained under Sect. 4.1.

2 Pharmacological Importance

In this section, the sPLA2s are grouped into two based on mammalian and snake venom sources. Among them, group IIA sPLA2 is produced by both the sources and it is involved in the reaction leading to the production of prostaglandins (PGs, particularly PGE2).

2.1 Pharmacological Importance of Mammalian $sPLA_2s$

There are 11 isoforms are found in mammalian, of which, the group I, II, V and X are well characterized sPLA2s for their involvement in signal transduction (inflammation and cancer) through digesting the membrane phospholipids, membrane remodelling through non-cellular lipid components (tissue injury and atherosclerosis) (Karabina et al. 2006; Kimura-Matsumoto et al. 2008; Menschikowski et al. 2006; Murakami and Kudo 2003; Rosengren et al. 2006; Webb 2005), defence against microbes and food metabolism (obesity) (Murakami et al. 2011). These group members are structurally related through a conserved calcium binding loop and a catalytic diad with histidine and aspartic acid. However,

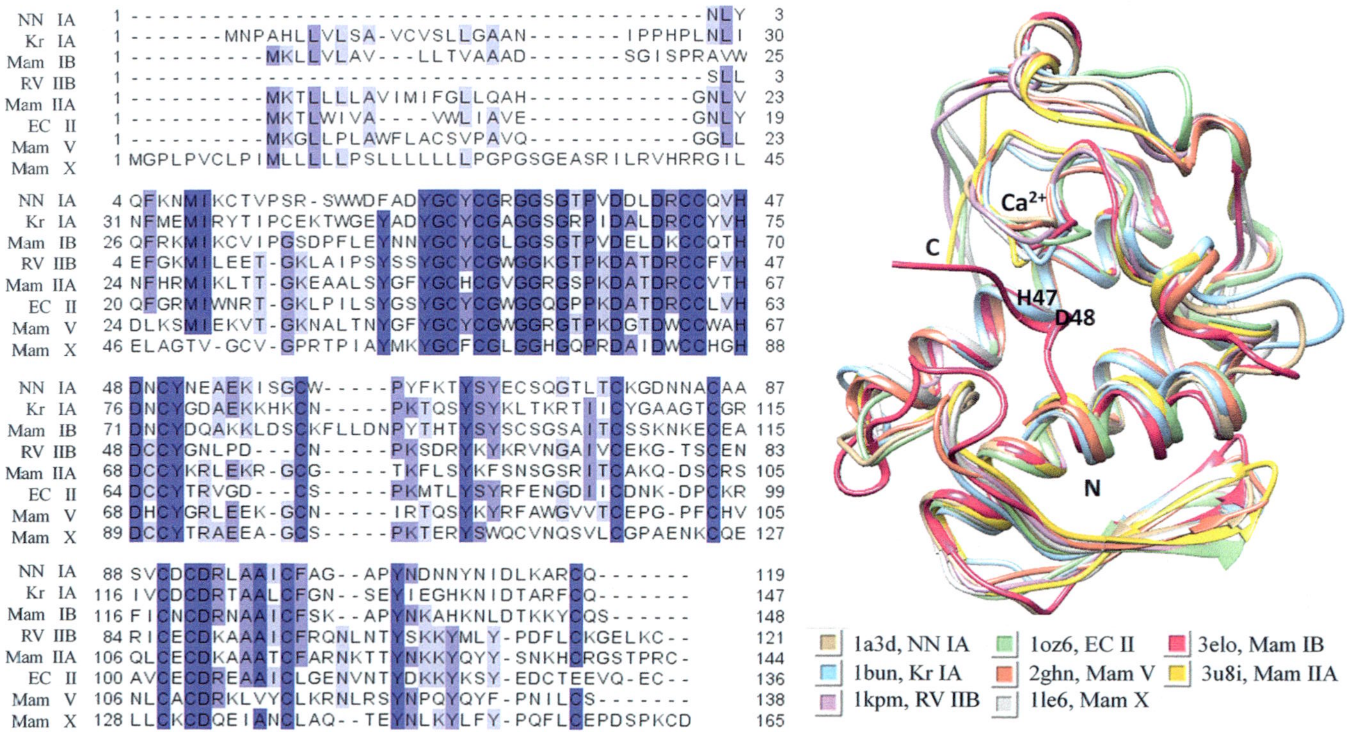

Fig. 2 Multiple sequence alignment (MSA; using clustalW) of the eight sPLA2s from venom (NN = Naja naja, Kr = Krait, RV = Russell's viper, EC = Echis carinatus) and human (Mam) sources. The cartoon diagram shows the structural similarity corresponding to the MSA. Important structural elements are labelled (N = N-terminal helix, Ca^{2+} = Calcium binding loop, H47 and D48 = Catalytic diad with histidine and aspartic acid, C = C terminal). Superposition was made using the UCSF Chimera software. The bar diagram shows a number of structures available in the PDB (as of September, 2018)

variations in amino acid composition of the N-terminal helix (Fig. 2, MSA) and interfacial binding residues make difference in their interfacial recognition. Evidently, the group V and X members are capable of hydrolyzing phosphatidylcholine (PC) as well as anionic phospholipids, while the group I is intermediate. Moreover, this makes a huge difference in the efficiency of producing the eicosanoids. The group X has more potency compared to the other isoforms (Bezzine et al. 2000) while the groups IIA and V have less potency depending on the function of cytosolic PLA2s (Murakami et al. 1998, 1999).

In a nutshell, the group IIA, V and X increase the risk of atherosclerosis (Rosenson and Gelb 2009). Group V $sPLA_2$ hydrolyzes the lung surfactants and thereby it leads to the acute respiratory distress syndrome (ARDS) (Ohtsuki et al. 2006). This was affirmed by treating the group V transgenic mice with broad spectrum sPLA2 inhibitors. There is an evidence for accumulation of its product involved in the type II diabetes (Wootton et al. 2007) and many metabolic disorders. The group X sPLA2s are not in concurrent with V. However, both the group V and X are involved in asthma and pneumonia as their expression were seen in the airway tissues (Hallstrand et al. 2007; Masuda et al. 2005; Munoz et al. 2007). The group X is specifically involved in negative regulation of corticosteroid production in adrenal gland (Li et al. 2010). Conversely, the group IIA is known to involve in arthritis as its over-expression was seen in the inflammatory states (Seilhamer et al. 1989) and induces the autoimmune response (Boilard et al. 2010). Moreover, an elevated group IIA level in plasma is found as a unique factor of heart diseases (Mallat et al. 2005).

Above data from the literature confirms that the pharmacological significance of each group of sPLA2 is tissue specific and isoform specific. Hence, a two-dimensional challenge must be faced for structure-based drug design if any of these subtypes is considered as the target. The isoform specificity must be achieved in concurrent with a successful targeted delivery towards a particular tissue type/ organ.

2.2 Pharmacological Importance of the Snake Venom sPLA2s

$sPLA_2$ is a major protein constituent of snake venom while the other enzymes such as serine proteases, disintegrins, metalloproteases, L-amino acid oxidase, hyaluronidase, etc., are also present (Warrell 2010). In general, PLA_2s of many snake venom exhibit neurotoxic or cytotoxic activity. The *viper* $sPLA_2$ causes haemorrhage in liver, kidney and lung as well as neurotoxic symptoms. Multiple functional sites in snake venom sPLA2s is the key factor for toxicity (Kini 2003).

Neurotoxicity is promoted by blocking neuromuscular transmission in the skeletal muscles. It leads to respiratory depression due to paralysis and neuromuscular weakness, and death (Ranawaka et al. 2013). Particularly, sPLA2s of the kraits, elapids, and coral snakes (belong to Elapidae family), and vipers and rattlesnakes

(Viperidae family) show neurotoxicity. *Acute necrosis* of skeletal muscle leads to a cardiac arrest (Lomonte and Rangel 2012). A permanent tissue loss was observed during the myonecrosis induced by the venom of sea snakes (Otero et al. 2002) and Viperidae (Gutierrez and Ownby 2003). Additionally, increase in creatine kinase level leads to a renal failure and myoglobinuria (Gutierrez and Ownby 2003). Involvement of venom PLA2 in necrosis is confirmed by site directed mutagenesis of Tyr117, Arg118, Tyr119, Lys122, and Phe125 (Chioato et al. 2007). *Anticoagulant* effect elicited by the venom $sPLA_2$ is primarily due to hydrolysis of phospholipids (Kini 2006) while there are some isoforms which inhibit coagulation in the absence of phospholipids (Saikia et al. 2011). Hence, the uncertainty exists in anticoagulant effect. Alternatively, there are evidences for venom sPLA2 mediated deactivation mechanism of conversion of FX to FXa and prothrombin to thrombin (Stefansson et al. 1990), which are necessary for blood coagulation.

Though there are many debates on the consideration for sPLA2s as druggable target, there are evidences for effective use of PLIs in snake venom envenomation treatment (Perumal Samy et al. 2012).

3 Inhibitors of $sPLA_2$

As the number of snake bite reports continues, more efficacious antivenom compounds (antidotes) must be discovered or designed to cure envenoming and reduce the risk. Presently, the sPLA2 inhibitors (PLIs) are abundant in nature and there are many synthetic organic compounds reported. Many of the non-steroidal anti-inflammatory drugs (NSAIDs) are also proved as PLIs. From the co-crystal structures available in the PDB, the inhibitors are categorized into peptides, NSAIDs, natural compounds, substrate mimics, indoles, etc. Detailed information on the PLIs is discussed broadly in the following subsections.

3.1 Natural/Plant Sources with Anti-venom Activities

Traditionally, the native plants were used to treat envenomation, effectively (Mukherjee 2012). Some of the important secondary metabolites such as aristolochic acid, benzoic acid derivatives, rosmarinic acid, diterpenes from *Aristolochia* sp., *Cordia vernenecea, Baccharis trimera*, respectively are known for their venom neutralization capabilities (Singh et al. 2017). Flavones (Mors et al. 2000; Perumal Samy et al. 2012; Soares et al. 2005), acids (Aung et al. 2010; Mors et al. 2000), steroids (Mors et al. 2000; Strauch et al. 2013), alkaloids (Mukherjee et al. 2008), terpenoids (Venkatesan et al. 2014), and coumestans (Melo et al. 2010) are important classes of antidotes. Animals showing a natural resistance to snake venom were also studied in details to understand the biochemical responses (Thwin and Gopalakrishnakone 1998). The sera of such animals have been studied for their

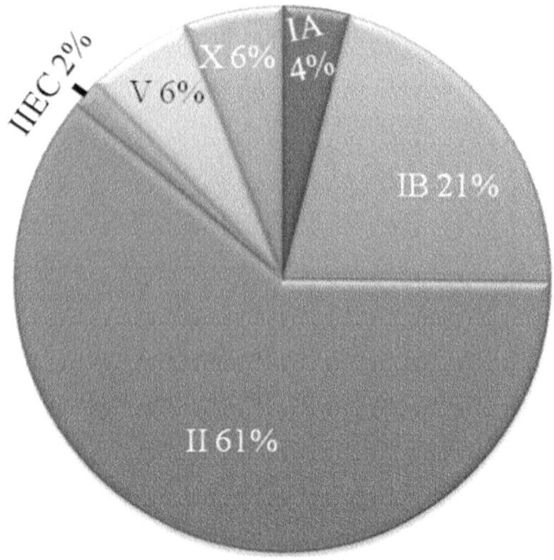

Fig. 3 A total of 450 diverse set of PLIs with experimental activity data (IC50/Ki/ED50 etc.). It is prepared by eliminating highly similar compounds from the total PLIs available in the Chembl database. There is no clear information of isoform specificity

effective anti-toxic effects of snake venom. In this, the snake itself acts as a good source as there are predators snakes for others. Most of the systemic endogenous inhibitors of sPLA2s are proteins from blood of other species/subspecies which potentially neutralize the toxic sPLAs entering the blood stream (Campos et al. 2016). As the scope of this chapter is confined towards the discussion on small molecule inhibitors to provide a ground for structure-based drug design, it is not appropriate to discuss more on macromolecular inhibitors. Many anti-inflammatory compounds have also been reported from marine natural sources.

3.2 *Small Organic/Chemical/Synthetic Molecule Inhibitors*

There are widely used chemical databases such as Pubchem (Kaiser 2005), Zinc (Sterling and Irwin 2015), Chembl (Gaulton et al. 2012), Bindingdb (Liu et al. 2007), Drugbank (Wishart et al. 2006), etc. to retrieve target specific inhibitors. As of mid of 2018, Chembl listed approximately 2800 inhibitors with experimental activity for the sPLA2 family. Figure 3 depicts the percentage of inhibitors group wise for 450 chemically diverse set prepared. Due to cross links between the databases, considerable redundancy exists overall.

4 Structure-Based Drug Design

Structure-based drug design (SBDD) approach is a robust method to search or design novel inhibitors of the target protein based on the three dimensional structure. To date, there are many methods such as molecular docking, high-throughput virtual screening (HTVS, pharmacophore-based and structure-based), quantitative structure-activity relationship (QSAR), etc. to study the binding mode of single to millions of compounds. For sPLA2 enzyme, there are many crystal structures available that make SBDD more feasible.

4.1 Crystal Structures of sPLA2s

Both to understand the mechanism of action of sPLA2s in atomic detail and for SBDD, availability of crystal structures provides a strong knowledge-base. The current search result for sPLA2s in PDB is shown in Fig. 2 as a bar diagram. It includes the crystal structures from human/mammalian and snake venom sources. Among these, sPLA2s of mammalian (group IIB, V and X) and Russel's viper are in majority. Besides, there are free wild type and mutant structures (Rajakannan et al. 2002; Sekar et al. 2003) determined to explain the mechanism of interfacial recognition. Among the venom $sPLA_2s$, *viper* $sPLA_2$ has some unique structural features characterized by biochemical studies and X-ray crystal structures are available in the PDB. Mainly, it is a asymmetric dimer with an rmsd of 0.71 Å between the two subunits (Chandra et al. 2002) and hence a difference in orientation of Trp31 which is located at the gateway of the active site exists. The dimer interface is stabilized through many intermolecular interactions involving and Trp31(A), Leu2(A), Leu3 (A), Leu119(A), Ala101(B), Gln108(B) and Arg43(B) of the subuint A and Arg43 (B), Phe46(B), and Val47(B) of subunit B (Chandra et al. 2002; Yuan et al. 1990). These interactions dictate the orientation of Trp31 in subunit A with an open conformation whereas it is closed in subunit B. The closed (buried) conformation is influenced by the surrounding water molecules. Hence, the subunit A is accessible for most of the PLIs and they will not enter into the active site of the subunit B. This crystallographic inference was validated using MD simulation studies (Ramakrishnan et al. 2010). Despite the conserved fold, cobra sPLA2 forms a trimeric structure (Fremont et al. 1993). Recent studies resulted in deposition of the structure of Russel's viper (group IIA) sPLA2 in complex with p-coumaric acid, resveratrol, spermidine, corticosterone and gramine derivative (Shukla et al. 2015). Among the mammalian sources, there are seven entries belonging to the group X sPLA2 while there is no experimentally determined structure for group V. Similarly, for group IIA of human/mammalian there are 17 entries. Number of entries mentioned here includes all, wild type, mutant, free, inhibitor-bound, etc.

4.2 Molecular Modeling

Molecular modeling studies play important role in the prediction of three dimensional structure while the experimentally determined structure is absent, for example, the homology model of group V mammalian sPLA2 (2GHN). Besides, the availability of biochemical and biophysical data on the function of this enzyme, lack of the structural data hamper the computational methods to find the novel inhibitors as well as to assess the off-target binding while working with the other isoforms. Due to evolutionarily conserved fold, the prediction becomes less complex for any subtype of secreted PLA2s of any species. Here, homology modeling (Chothia and Lesk 1986; Marti-Renom et al. 2000) will play an inevitable role in the case of prediction of three dimensional structure of sPLA2s genome wide. However, additional procedures must be followed to optimize the side chains and loops in order to alleviate the close contacts and structural constrains imposed by the method from the template(s).

4.2.1 Molecular Docking, Pharmacophore Modeling and Virtual Screening

These in silico methods are primarily used to reduce the time and cost investments in the routine laboratory procedures. Molecular docking methods are being used widely in support of experimental activity data to explain the mode of interactions at atomic detail. There are many methods (scoring functions) with their unique performance and some of the widely used methods are Autodock (Morris et al. 2009; Trott and Olson 2010), Dock (Moustakas et al. 2006), Glide (Friesner et al. 2004), Gold (Jones et al. 1997), etc. Further, it is extended to screen large chemical libraries or a limited target focussed libraries to shortlist few compounds for further experimental validation. Most of these programs have many features to perform high throughput virtual screening and active site flexible docking (induced fit docking). On the other hand, availability of diverse set of small molecule inhibitors enables the researcher to build the pharmacophore model by aligning the three dimensional structure of the inhibitors in 3D space and finding the common structural-features (pharmacophore features) such as hydrogen bond acceptors/donors, hydrophobic, aromatic, positively/negatively charged, etc. A typical application of pharmacophore modeling in snake venom sPLA2 can be seen in the literature (Ramakrishnan et al. 2014).

4.2.2 Molecular Dynamics Simulations

Molecular dynamics (MD) simulations are employed for conformational sampling, binding energy calculations, assessing the stability of the enzyme-inhibitor/substrate complex, etc. It is a semi-empirical force field based method to compute the potential associated with the simulation system. Some of the widely used simulation software are AMBER, CHARMM, Gromacs (Van Der Spoel et al. 2005), NAMD (Phillips et al. 2005), TINKER (Ponder and Richards 1987; Ponder et al. 2010) and Desmond (Bowers et al. 2006; Lippert et al. 2007). For SBDD, MD is widely used to generate a series of conformations called ensembles with different active site scaffold, which is very tedious to capture using experimental techniques. Moreover, some statistical methods such as MM/GBSA, MM/PBSA, etc. are widely used to calculate the binding free energies based on the simulation trajectories. The structure-function relationships of many sPLA2s were studied using MD simulations. Particularly, MD simulations studies to understand the association of different sPLA2s with different phospholipid membranes (Bucher et al. 2013; Mouchlis et al. 2015; Qin et al. 2013; Wee et al. 2008) are informative in the absence of experimental results.

5 Computational sPLA$_2$ Enzymology

In the growing phase of pharmacological importance of sPLA2s, it is essential to study their catalytic mechanism at atomic detail to explain the subtype specificity for different phospholipids. As of now there is no rigorous studies made for the sPLA2 family members. Besides, there are some clear kinetics data available to some extent (Berg et al. 2001) as there is no coverage over all the subtypes. This recruits the computational enzymology to fill the gaps.

5.1 Catalytic Reaction of sPLA2

sPLA2s hydrolyze the membrane phospholipids at the sn2 acyl ester bond which is initiated by a proton transfer from the catalytic water molecule to the His48. This makes the water a strong nucleophile to cleave the ester bond and produces lysophospholipid and a fatty acid (Fig. 4).

5.2 Specificity

As there are ten distinct groups of sPLA2s with a conserved functional domain, the specificity of each subtype is understudy to ascertain the nature of substrate and the structure-function relationship associated with the selective binding of a particular group of sPLA2 with its cognate substrate (membrane phospholipids). Recently, there are biochemical and biophysical studies focused on the substrate specificity of some of the isoforms (Murakami et al. 2015; Yamamoto et al. 2017). However, the main limitation in the sPLA2 structural biology is lack of structures for each type in complex with their cognate substrates. Also, a concrete evidence to ascertain the cognate enzyme-substrate pair is in the infant stage. Complexity exists to confirm such a pair due to multi-specificity i.e., sPLA2s tend to bind more than one lipid

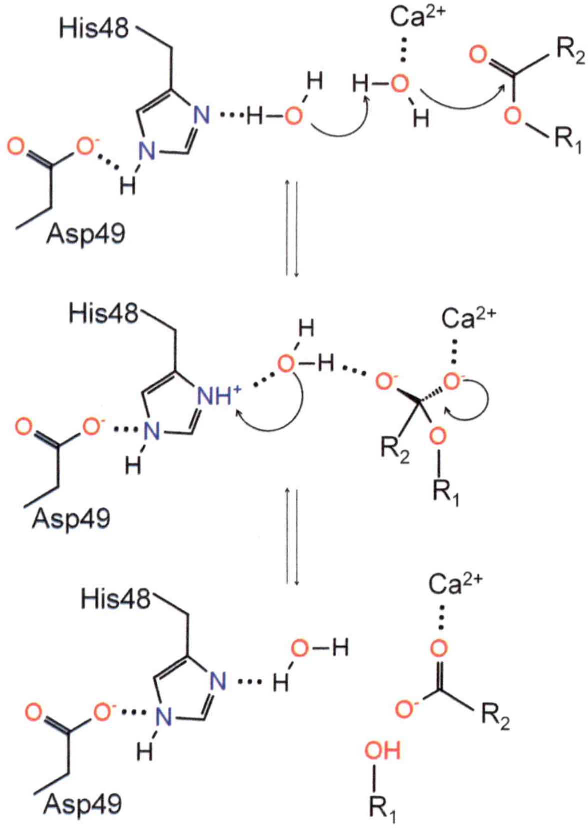

Fig. 4 Stepwise reaction catalyzed by the secreted phospholipase A2. Image abstrated from the Wikepedia (https://en.wikipedia.org/wiki/Phospholipase_A2)

type. Moreover, the optimum activity could be seen only at the critical concentration of aggregated substrates (*aka* critical micelle concentration, CMC).

5.3 Computational Advancements

In the present scenario, it is worth to revive some of the important computational methods to study the enzymology and explain the molecular mechanism at atomic details. This will provide robust predictive models for identification of inhibitors with more specificity. Also, one can measure the binding affinity quantitatively for a series of substrates towards a particular isoform and rank them. The following subsections explain the basics of some of the important computational chemistry methods and their applications in enzyme-catalyzed reactions.

5.3.1 Quantum Mechanics/Molecular Mechanics (QM/MM) Methods

QM/MM (couple potential) methods were developed primarily to achieve the accuracy and speed of the calculations (Brunk and Rothlisberger 2015; Warshel and Levitt 1976). Initially, the enzymes were treated with low levels of QM theory like density functional theory (DFT) to overcome the computational intensive calculations. Initially, the accuracy was compromised up to 10 kcal/mol in comparison with the experimental data due to the involvement of the semi-emperical potential from the molecular mechanics counterpart and due to application in a system with large number of atoms. In the further developments (Manby et al. 2006), the accuracy is comparable with the experiment for enzyme-catalysed reactions (Claeyssens et al. 2006) and provided a scope for calculating the transition state and estimate the activation energy association with different states of the reaction path.

5.3.2 Empirical Valance Bond

Empirical valance bond (EVB) theory is a method for calculating the binding free energies along the enzymatic reaction and this approach was developed in early 1980 (Warshel and Weiss 1980, 1981). This method provided accurate results for histo-aspartic protease from the *plasmodium falciparum* (Bjelic and Aqvist 2004; Kamerlin and Warshel 2010). Alternate to EVB, an approximate valence bond (AVB) method was employed to study the proton transfer from the catalytic water molecule to His48, a critical/first step of the multistep catalysis by sPLA2 (Bala et al. 2000).

6 Summary

According to the report from the WHO in the early of 2018, the snake bite becomes one of the neglected public health issues for tropical and sub-tropical countries. In countries like India, Africa and many parts of Asia, there are millions of snake bite case reports every year which includes 10–15% of deaths and permanent disabilities. In this scenario, flood of data from the biochemical and biophysical studies shed light on many aspects of the secreted PLA2 biology such as the evolution, function, structure, etc., which helped the community to control the diseases posed by imbalances in the PLA2 metabolites as well as envenomation. Still, there is a long way to go to understand the sPLA2 biology in genome wide. As there are many isoforms existing and new isoforms are being discovered on approximately yearly basis, there is an increase in the gap between the number of known sPLA2 subtypes and the number of sPLA2 characterized completely. For example, among the number of cocrystal structures available for the both snake venom and mammalian sources, there no complex structures with any of the substrate types such as phosphatidylcholine (PC), phosphatidylserine (PS) and phosphatidylglycerol, even to date. Eventually, all these challenges open the avenue to conduct computational studies on the sPLA2s with their cognate and non-cognate substrates to unveil the specificity and to calculate the kinetic parameters associated with. Moreover, there is no computational study conducted so far to calculate the binding parameters at the water-lipid interface.

References

Arni RK, Ward RJ (1996) Phospholipase A2 – a structural review. Toxicon 34:827–841. https://doi.org/10.1016/0041-0101(96)00036-0

Aung HT, Nikai T, Niwa M, Takaya Y (2010) Rosmarinic acid in Argusia argentea inhibits snake venom-induced hemorrhage. J Nat Med 64:482–486. https://doi.org/10.1007/s11418-010-0428-3

Bala P, Grochowski P, Nowinski K, Lesyng B, McCammon JA (2000) Quantum-dynamical picture of a multistep enzymatic process: reaction catalyzed by phospholipase A(2). Biophys J 79:1253–1262. https://doi.org/10.1016/S0006-3495(00)76379-X

Balsinde J, Balboa MA, Insel PA, Dennis EA (1999) Regulation and inhibition of phospholipase A2. Annu Rev Pharmacol Toxicol 39:175–189. https://doi.org/10.1146/annurev.pharmtox.39.1.175

Berg OG, Gelb MH, Tsai MD, Jain MK (2001) Interfacial enzymology: the secreted phospholipase A(2)-paradigm. Chem Rev 101:2613–2654

Bezzine S et al (2000) Exogenously added human group X secreted phospholipase A(2) but not the group IB, IIA, and V enzymes efficiently release arachidonic acid from adherent mammalian cells. J Biol Chem 275:3179–3191

Bjelic S, Aqvist J (2004) Computational prediction of structure, substrate binding mode, mechanism, and rate for a malaria protease with a novel type of active site. Biochemistry 43:14521–14528. https://doi.org/10.1021/bi048252q

Boilard E et al (2010) A novel anti-inflammatory role for secretory phospholipase A2 in immune complex-mediated arthritis. EMBO Mol Med 2:172–187. https://doi.org/10.1002/emmm.201000072

Bowers KJ, Dror RO, Shaw DE (2006) The midpoint method for parallelization of particle simulations. J Chem Phys 124:184109. https://doi.org/10.1063/1.2191489

Brunk E, Rothlisberger U (2015) Mixed quantum mechanical/molecular mechanical molecular dynamics simulations of biological systems in ground and electronically excited states. Chem Rev 115:6217–6263. https://doi.org/10.1021/cr500628b

Bucher D, Hsu YH, Mouchlis VD, Dennis EA, McCammon JA (2013) Insertion of the Ca(2)(+)-independent phospholipase A(2) into a phospholipid bilayer via coarse-grained and atomistic molecular dynamics simulations. PLoS Comput Biol 9:e1003156. https://doi.org/10.1371/journal.pcbi.1003156

Campos PC, de Melo LA, Dias GLF, Fortes-Dias CL (2016) Endogenous phospholipase A2 inhibitors in snakes: a brief overview. J Venom Anim Toxins Incl Trop Dis 22:37. https://doi.org/10.1186/s40409-016-0092-5

Chandra V, Jasti J, Kaur P, Betzel C, Srinivasan A, Singh TP (2002) First structural evidence of a specific inhibition of phospholipase A2 by α-tocopherol (vitamin E) and its implications in inflammation: crystal structure of the complex formed between phospholipase A2 and α-tocopherol at 1.8 Å resolution. J Mol Biol 320:215–222. https://doi.org/10.1016/S0022-2836(02)00473-4

Chioato L, Aragao EA, Lopes Ferreira T, Medeiros AI, Faccioli LH, Ward RJ (2007) Mapping of the structural determinants of artificial and biological membrane damaging activities of a Lys49 phospholipase A2 by scanning alanine mutagenesis. Biochim Biophys Acta 1768:1247–1257. https://doi.org/10.1016/j.bbamem.2007.01.023
Chothia C, Lesk AM (1986) The relation between the divergence of sequence and structure in proteins. EMBO J 5:823–826
Claeyssens F et al (2006) High-accuracy computation of reaction barriers in enzymes. Ange Chem 45:6856–6859. https://doi.org/10.1002/anie.200602711
Davidson FF, Dennis EA (1990) Evolutionary relationships and implications for the regulation of phospholipase A2 from snake venom to human secreted forms. J Mol Evol 31:228–238
Demaret J-P, Brunie S (1992) A unique structural feature of a phospholipase A2 is probed by molecular dynamics. J Mol Graph 10:257–260. https://doi.org/10.1016/0263-7855(92)80079-S
Dennis EA (1994) Diversity of group types, regulation, and function of phospholipase A2. J Biol Chem 269:13057–13060
Dennis EA (1997) The growing phospholipase A2 superfamily of signal transduction enzymes. Trends Biochem Sci 22:1–2
Dijkstra BW, Kalk KH, Hol WG, Drenth J (1981) Structure of bovine pancreatic phospholipase A2 at 1.7A resolution. J Mol Biol 147:97–123
Dijkstra BW, Kalk KH, Drenth J, de Haas GH, Egmond MR, Slotboom AJ (1984) Role of the N-terminus in the interaction of pancreatic phospholipase A2 with aggregated substrates. Properties and crystal structure of transaminated phospholipase A2. Biochemistry 23:2759–2766
Fremont DH, Anderson DH, Wilson IA, Dennis EA, Xuong NH (1993) Crystal structure of phospholipase A2 from Indian cobra reveals a trimeric association. Proc Natl Acad Sci 90:342–346
Friesner RA et al (2004) Glide: a new approach for rapid, accurate docking and scoring. 1. Method and assessment of docking accuracy. J Med Chem 47:1739–1749. https://doi.org/10.1021/jm0306430
Funk CD (2001) Prostaglandins and leukotrienes: advances in eicosanoid biology. Science 294:1871–1875. https://doi.org/10.1126/science.294.5548.1871
Gaulton A et al (2012) ChEMBL: a large-scale bioactivity database for drug discovery. Nucl Acids Res 40:D1100–D1107. https://doi.org/10.1093/nar/gkr777
Gelb MH, Jain MK, Hanel AM, Berg OG (1995) Interfacial enzymology of glycerolipid hydrolases: lessons from secreted phospholipases A2. Annu Rev Biochem 64:653–688. https://doi.org/10.1146/annurev.bi.64.070195.003253
Gutierrez JM, Ownby CL (2003) Skeletal muscle degeneration induced by venom phospholipases A2: insights into the mechanisms of local and systemic myotoxicity. Toxicon 42:915–931. https://doi.org/10.1016/j.toxicon.2003.11.005
Hallstrand TS, Chi EY, Singer AG, Gelb MH, Henderson WR Jr (2007) Secreted phospholipase A2 group X overexpression in asthma and bronchial hyperresponsiveness. Am J Respir Crit Care Med 176:1072–1078. https://doi.org/10.1164/rccm.200707-1088OC
Jain MK, Berg OG (1989) The kinetics of interfacial catalysis by phospholipase A2 and regulation of interfacial activation: hopping versus scooting. Biochim Biophys Acta 1002:127–156
Jain MK, Cordes EH (1973) Phospholipases. I. Effect of n-alkanols on the rate of enzymatic hydrolysis of egg phosphatidylcholine. J Membr Biol 14:101–118
Jeyaseelan K, Armugam A, Donghui M, Tan N-H (2000) Structure and phylogeny of the venom group I phospholipase A2 gene. Mol Biol Evol 17:1010–1021
Jones G, Willett P, Glen RC, Leach AR, Taylor R (1997) Development and validation of a genetic algorithm for flexible docking. J Mol Biol 267:727–748. https://doi.org/10.1006/jmbi.1996.0897
Kaiser J (2005) Science resources. Chemists want NIH to curtail database. Science 308:774. https://doi.org/10.1126/science.308.5723.774a

Kamerlin SC, Warshel A (2010) At the dawn of the 21st century: Is dynamics the missing link for understanding enzyme catalysis? Proteins 78:1339–1375. https://doi.org/10.1002/prot.22654

Karabina SA et al (2006) Atherogenic properties of LDL particles modified by human group X secreted phospholipase A2 on human endothelial cell function. FASEB J 20:2547–2549. https://doi.org/10.1096/fj.06-6018fje

Kimura-Matsumoto M et al (2008) Expression of secretory phospholipase A2s in human atherosclerosis development. Atherosclerosis 196:81–91. https://doi.org/10.1016/j.atherosclerosis.2006.08.062

Kini RM (2003) Excitement ahead: structure, function and mechanism of snake venom phospholipase A2 enzymes. Toxicon 42:827–840. https://doi.org/10.1016/j.toxicon.2003.11.002

Kini RM (2006) Anticoagulant proteins from snake venoms: structure, function and mechanism. Biochem J 397:377–387. https://doi.org/10.1042/BJ20060302

Lambeau G, Gelb MH (2008) Biochemistry and physiology of mammalian secreted phospholipases A2. Annu Rev Biochem 77:495–520. https://doi.org/10.1146/annurev.biochem.76.062405.154007

Li X, Shridas P, Forrest K, Bailey W, Webb NR (2010) Group X secretory phospholipase A2 negatively regulates adipogenesis in murine models. FASEB J 24:4313–4324. https://doi.org/10.1096/fj.10-154716

Lippert RA et al (2007) A common, avoidable source of error in molecular dynamics integrators. J Chem Phys 126:046101. https://doi.org/10.1063/1.2431176

Liu T, Lin Y, Wen X, Jorissen RN, Gilson MK (2007) BindingDB: a web-accessible database of experimentally determined protein-ligand binding affinities. Nucl Acids Res 35:D198–D201. https://doi.org/10.1093/nar/gkl999

Lomonte B, Rangel J (2012) Snake venom Lys49 myotoxins: From phospholipases A(2) to non-enzymatic membrane disruptors. Toxicon 60:520–530. https://doi.org/10.1016/j.toxicon.2012.02.007

Mallat Z et al (2005) Circulating secretory phospholipase A2 activity predicts recurrent events in patients with severe acute coronary syndromes. J Am Coll Cardiol 46:1249–1257. https://doi.org/10.1016/j.jacc.2005.06.056

Manby FR, Werner HJ, Adler TB, May AJ (2006) Explicitly correlated local second-order perturbation theory with a frozen geminal correlation factor. J Chem Phys 124:94103. https://doi.org/10.1063/1.2173247

Marti-Renom MA, Stuart AC, Fiser A, Sanchez R, Melo F, Sali A (2000) Comparative protein structure modeling of genes and genomes. Annu Rev Biophys Biomol Struct 29:291–325. https://doi.org/10.1146/annurev.biophys.29.1.291

Masuda S, Murakami M, Mitsuishi M, Komiyama K, Ishikawa Y, Ishii T, Kudo I (2005) Expression of secretory phospholipase A2 enzymes in lungs of humans with pneumonia and their potential prostaglandin-synthetic function in human lung-derived cells. Biochem J 387:27–38. https://doi.org/10.1042/BJ20041307

Melo PA et al (2010) Ability of a synthetic coumestan to antagonize Bothrops snake venom activities. Toxicon 55:488–496. https://doi.org/10.1016/j.toxicon.2009.09.021

Menschikowski M, Hagelgans A, Siegert G (2006) Secretory phospholipase A2 of group IIA: is it an offensive or a defensive player during atherosclerosis and other inflammatory diseases? Prostaglandins Other Lipid Mediat 79:1–33. https://doi.org/10.1016/j.prostaglandins.2005.10.005

Morris GM, Huey R, Lindstrom W, Sanner MF, Belew RK, Goodsell DS, Olson AJ (2009) AutoDock4 and AutoDockTools4: Automated docking with selective receptor flexibility. J Comput Chem 30:2785–2791. https://doi.org/10.1002/jcc.21256

Mors WB, Nascimento MC, Pereira BM, Pereira NA (2000) Plant natural products active against snake bite – the molecular approach. Phytochemistry 55:627–642

Mouchlis VD, Bucher D, McCammon JA, Dennis EA (2015) Membranes serve as allosteric activators of phospholipase A2, enabling it to extract, bind, and hydrolyze phospholipid substrates. Proc Natl Acad Sci U S A 112:E516–E525. https://doi.org/10.1073/pnas.1424651112

Moustakas DT, Lang PT, Pegg S, Pettersen E, Kuntz ID, Brooijmans N, Rizzo RC (2006) Development and validation of a modular, extensible docking program: DOCK 5. J Comput Aided Mol Des 20:601–619. https://doi.org/10.1007/s10822-006-9060-4

Mukherjee AK (2012) Green medicine as a harmonizing tool to antivenom therapy for the clinical management of snakebite: the road ahead. Indian J Med Res 136:10–12

Mukherjee AK, Doley R, Saikia D (2008) Isolation of a snake venom phospholipase A2 (PLA2) inhibitor (AIPLAI) from leaves of Azadirachta indica (Neem): mechanism of PLA2 inhibition by AIPLAI in vitro condition. Toxicon 51:1548–1553. https://doi.org/10.1016/j.toxicon.2008.03.021

Munoz NM, Meliton AY, Arm JP, Bonventre JV, Cho W, Leff AR (2007) Deletion of secretory group V phospholipase A2 attenuates cell migration and airway hyperresponsiveness in immunosensitized mice. J Immunol 179:4800–4807

Murakami M, Kudo I (2003) New phospholipase A(2) isozymes with a potential role in atherosclerosis. Curr Opin Lipidol 14:431–436. https://doi.org/10.1097/01.mol.0000092629.86399.d6

Murakami M, Shimbara S, Kambe T, Kuwata H, Winstead MV, Tischfield JA, Kudo I (1998) The functions of five distinct mammalian phospholipase A2S in regulating arachidonic acid release. Type IIa and type V secretory phospholipase A2S are functionally redundant and act in concert with cytosolic phospholipase A2. J Biol Chem 273:14411–14423

Murakami M et al (1999) Different functional aspects of the group II subfamily (Types IIA and V) and type X secretory phospholipase A(2)s in regulating arachidonic acid release and prostaglandin generation. Implications of cyclooxygenase-2 induction and phospholipid scramblase-mediated cellular membrane perturbation. J Biol Chem 274:31435–31444

Murakami M, Taketomi Y, Sato H, Yamamoto K (2011) Secreted phospholipase A2 revisited. J Biochem 150:233–255. https://doi.org/10.1093/jb/mvr088

Murakami M, Sato H, Miki Y, Yamamoto K, Taketomi Y (2015) A new era of secreted phospholipase A(2). J Lipid Res 56:1248–1261. https://doi.org/10.1194/jlr.R058123

Nevalainen TJ, Cardoso JC, Riikonen PT (2012) Conserved domains and evolution of secreted phospholipases A(2). FEBS J 279:636–649. https://doi.org/10.1111/j.1742-4658.2011.08453.x

Ohtsuki M et al (2006) Transgenic expression of group V, but not group X, secreted phospholipase A2 in mice leads to neonatal lethality because of lung dysfunction. J Biol Chem 281:36420–36433. https://doi.org/10.1074/jbc.M607975200

Otero R et al (2002) Complications of Bothrops, Porthidium, and Bothriechis snakebites in Colombia. A clinical and epidemiological study of 39 cases attended in a university hospital. Toxicon 40:1107–1114

Perumal Samy R, Gopalakrishnakone P, Chow VT (2012) Therapeutic application of natural inhibitors against snake venom phospholipase A(2). Bioinformation 8:48–57

Phillips JC et al (2005) Scalable molecular dynamics with NAMD. J Comput Chem 26:1781–1802. https://doi.org/10.1002/jcc.20289

Ponder JW, Richards FM (1987) An efficient newton-like method for molecular mechanics energy minimization of large molecules. J Comput Chem 8:1016–1024. https://doi.org/10.1002/jcc.540080710

Ponder JW et al (2010) Current status of the AMOEBA polarizable force field. J Phys Chem B 114:2549–2564. https://doi.org/10.1021/jp910674d

Qin SS, Yu YX, Li QK, Yu ZW (2013) Interaction of human synovial phospholipase A2 with mixed lipid bilayers: a coarse-grain and all-atom molecular dynamics simulation study. Biochemistry 52:1477–1489. https://doi.org/10.1021/bi3012687

Rajakannan V et al (2002) Observation of additional calcium ion in the crystal structure of the triple mutant K56,120,121M of bovine pancreatic phospholipase A2. J Mol Biol 324:755–762

Ramakrishnan C, Subramanian V, Velmurugan D (2010) Molecular dynamics study of secretory phospholipase A2 of Russell's viper and bovine pancreatic sources. J Phys Chem B 114:13463–13472. https://doi.org/10.1021/jp102073f

Ramakrishnan C, Joshi V, Joseph JM, Vishwanath BS, Velmurugan D (2014) Identification of novel inhibitors of Daboia russelli phospholipase A2 using the combined pharmacophore modeling approach. Chem Biol Drug Design 84:379–392. https://doi.org/10.1111/cbdd.12332

Ramirez F, Jain MK (1991) Phospholipase A2 at the bilayer interface. Proteins: Structure, Function, and Bioinformatics 9:229–239

Ranawaka UK, Lalloo DG, de Silva HJ (2013) Neurotoxicity in snakebite—the limits of our knowledge. PLoS Negl Trop Dis 7:e2302. https://doi.org/10.1371/journal.pntd.0002302

Rivera R, Chun J (2008) Biological effects of lysophospholipids. Rev Physiol Biochem Pharmacol 160:25–46. https://doi.org/10.1007/112_0507

Rosengren B, Jonsson-Rylander AC, Peilot H, Camejo G, Hurt-Camejo E (2006) Distinctiveness of secretory phospholipase A2 group IIA and V suggesting unique roles in atherosclerosis. Biochim Biophys Acta 1761:1301–1308. https://doi.org/10.1016/j.bbalip.2006.06.008

Rosenson RS, Gelb MH (2009) Secretory phospholipase A2: a multifaceted family of proatherogenic enzymes. Curr Cardiol Rep 11:445–451

Saikia D, Thakur R, Mukherjee AK (2011) An acidic phospholipase A(2) (RVVA-PLA(2)-I) purified from Daboia russelli venom exerts its anticoagulant activity by enzymatic hydrolysis of plasma phospholipids and by non-enzymatic inhibition of factor Xa in a phospholipids/Ca(2 +) independent manner. Toxicon 57:841–850. https://doi.org/10.1016/j.toxicon.2011.02.018

Seilhamer JJ, Pruzanski W, Vadas P, Plant S, Miller JA, Kloss J, Johnson LK (1989) Cloning and recombinant expression of phospholipase A2 present in rheumatoid arthritic synovial fluid. J Biol Chem 264:5335–5338

Sekar K et al (2003) Crystal structures of the free and anisic acid bound triple mutant of phospholipase A2. J Mol Biol 333:367–376

Shukla PK, Gautam L, Sinha M, Kaur P, Sharma S, Singh TP (2015) Structures and binding studies of the complexes of phospholipase A2 with five inhibitors. Biochim Biophys Acta 1854:269–277. https://doi.org/10.1016/j.bbapap.2014.12.017

Singh P, Yasir M, Hazarika R, Sugunan S, Shrivastava R (2017) A review on venom enzymes neutralizing ability of secondary metabolites from medicinal plants. J Pharmacopunct 20:173–178. https://doi.org/10.3831/KPI.2017.20.020

Six DA, Dennis EA (2000) The expanding superfamily of phospholipase A(2) enzymes: classification and characterization. Biochim Biophys Acta 1488:1–19

Soares AM et al (2005) Medicinal plants with inhibitory properties against snake venoms. Curr Med Chem 12:2625–2641

Stefansson S, Kini RM, Evans HJ (1990) The basic phospholipase A2 from Naja nigricollis venom inhibits the prothrombinase complex by a novel nonenzymatic mechanism. Biochemistry 29:7742–7746

Stephens JWW, Myers W (1898) The action of cobra poison on the blood: a contribution to the study of passive immunity. J Pathol Bacteriol 5:279–301. https://doi.org/10.1002/path.1700050307

Sterling T, Irwin JJ (2015) ZINC 15 – ligand discovery for everyone. J Chem Inform Model 55:2324–2337. https://doi.org/10.1021/acs.jcim.5b00559

Strauch MA et al (2013) Antiophidic activity of the extract of the Amazon plant Humirianthera ampla and constituents. J Ethnopharmacol 145:50–58. https://doi.org/10.1016/j.jep.2012.10.033

Thwin MM, Gopalakrishnakone P (1998) Snake envenomation and protective natural endogenous proteins: a mini review of the recent developments (1991-1997). Toxicon 36:1471–1482

Trott O, Olson AJ (2010) AutoDock Vina: improving the speed and accuracy of docking with a new scoring function, efficient optimization, and multithreading. J Comput Chem 31:455–461. https://doi.org/10.1002/jcc.21334

Valentin E, Lambeau G (2000) Increasing molecular diversity of secreted phospholipases A(2) and their receptors and binding proteins. Biochim Biophys Acta 1488:59–70

Van Der Spoel D, Lindahl E, Hess B, Groenhof G, Mark AE, Berendsen HJ (2005) GROMACS: fast, flexible, and free. J Comput Chem 26:1701–1718. https://doi.org/10.1002/jcc.20291

Venkatesan C et al (2014) Antivenom activity of triterpenoid (C34H68O2) from Leucas aspera Linn. against Naja naja naja venom induced toxicity: antioxidant and histological study in mice. Hum Exp Toxicol 33:336–359. https://doi.org/10.1177/0960327113494901

Verger R, De Haas GH (1973) Enzyme reactions in a membrane model. 1. A new technique to study enzyme reactions in monolayers. Chem Phys Lipids 10:127–136

Verger R, Mieras MC, de Haas GH (1973) Action of phospholipase A at interfaces. J Biol Chem 248:4023–4034

Verheij HM, Slotboom AJ, de Haas GH (1981) Structure and function of phospholipase A2. Rev Physiol Biochem Pharmacol 91:91–203

Ward RJ, Alves AR, Ruggiero Neto J, Arni RK, Casari G (1998) A SequenceSpace analysis of Lys49 phopholipases A2: clues towards identification of residues involved in a novel mechanism of membrane damage and in myotoxicity. Protein Eng 11:285–294

Warrell DA (2010) Snake bite. Lancet 375:77–88. https://doi.org/10.1016/S0140-6736(09)61754-2

Warshel A, Levitt M (1976) Theoretical studies of enzymic reactions: dielectric, electrostatic and steric stabilization of the carbonium ion in the reaction of lysozyme. J Mol Biol 103:227–249

Warshel A, Weiss RM (1980) An empirical valence bond approach for comparing reactions in solutions and in enzymes. J Am Chem Soc 102:6218–6226. https://doi.org/10.1021/ja00540a008

Warshel A, Weiss RM (1981) Empirical valence bond calculations of enzyme catalysis. Ann N Y Acad Sci 367:370–382

Webb NR (2005) Secretory phospholipase A2 enzymes in atherogenesis. Curr Opin Lipidol 16:341–344

Wee CL, Balali-Mood K, Gavaghan D, Sansom MS (2008) The interaction of phospholipase A2 with a phospholipid bilayer: coarse-grained molecular dynamics simulations. Biophys J 95:1649–1657. https://doi.org/10.1529/biophysj.107.123190

Winget JM, Pan YH, Bahnson BJ (2006) The interfacial binding surface of phospholipase A2s. Biochim Biophys Acta 1761:1260–1269. https://doi.org/10.1016/j.bbalip.2006.08.002

Wishart DS et al (2006) DrugBank: a comprehensive resource for in silico drug discovery and exploration. Nucl Acids Res 34:D668–D672. https://doi.org/10.1093/nar/gkj067

Wootton PT et al (2007) Tagging SNP haplotype analysis of the secretory PLA2-V gene, PLA2G5, shows strong association with LDL and oxLDL levels, suggesting functional distinction from sPLA2-IIA: results from the UDACS study. Hum Mol Genet 16:1437–1444. https://doi.org/10.1093/hmg/ddm094

Yamamoto K, Miki Y, Sato H, Murase R, Taketomi Y, Murakami M (2017) Secreted phospholipase A2 specificity on natural membrane phospholipids. Methods Enzymol 583:101–117. https://doi.org/10.1016/bs.mie.2016.09.007

Yuan W, Quinn DM, Sigler PB, Gelb MH (1990) Kinetic and inhibition studies of phospholipase A2 with short-chain substrates and inhibitors. Biochemistry 29:6082–6094. https://doi.org/10.1021/bi00477a028

Clinical Significance of Enzymes in Disease and Diagnosis

Mohammad Fahad Ullah, Showket H. Bhat, Muhammad Tariq, and Faisel M. Abuduhier

1 Altered Enzymes, Variants and Cleaved Products in Cancer

Alterations in the expression and activities of certain enzymes in serum and tissues provide an indication of malignant environment, serving as a footprint of cancer disease (Ullah and Aatif 2009). Diagnostic detection and measurement of these enzymes, their variants and products are thus essential elements for successful cancer disease management in patients with malignant burden. The level of an active enzyme in physiological function is strictly regulated by gene expression (rate of enzyme synthesis), turnover (rate of enzyme degradation) and covalent modifications (reversible activation and inactivation of proenzymes). In cancer cells which acquire characteristics unique to their needs like high metabolic rate, such regulatory pathways fail to recognize the natural restrictions leading to an abnormal pattern of expression and activities which is different from the normal cells. These alterations have potential to trigger specific biochemical processes and generate a supportive milieu to aid neoplastic cells over normal cells, contributing to the progression and aggressiveness of the disease (Stefanini 1985). Moreover, low molecular weight molecules also appear as biomarker from enzyme-generated fragments of proteolytic activity (Liang and Chan 2007). It is considered that such markers should be sensitive and specific to the tumor, precisely reflecting tumor burden and thus can be correlated with the desired therapeutic outcomes.

M. Fahad Ullah (✉) · S. H. Bhat · M. Tariq · F. M. Abuduhier
Faculty of Applied Medical Science, Prince Fahd Research Chair, Department of Medical Laboratory Technology, University of Tabuk, Tabuk, Saudi Arabia
e-mail: m.ullah@ut.edu.sa

Q. Husain, M. F. Ullah (eds.), *Biocatalysis*,
https://doi.org/10.1007/978-3-030-25023-2_11

1.1 Serine Proteases

Serine proteases, human kallikreins (hKs) have been found to be substantially dysregulated in several cancers including ovarian, pancreatic, neuroendocrine and prostate, where an active role in cancer metastasis and invasion has been suggested (Borgono et al. 2004; Paliouras et al. 2007). Over -expression of kallikreins have been observed in ovarian carcinoma tissues/serum and high KLK4 and KLK5 mRNAs expression serve as indicators of poor prognostic outcome in grade 1 and grade 2 tumors (Kim et al. 2001). KLK11 also show an over expressed pattern which is unique in neuroendocrine carcinomas (Bhattacharjee et al. 2001), whereas KLK10 is one of the most highly and specifically over expressed genes in pancreatic cancer compared with normal and benign pancreas tissues (Iacobuzio-Donahue et al. 2003). Prostate specific antigen (PSA) or hK3 is known to have high diagnostic significance in prostate cancer; elevation of the blood level of PSA 4–10 ng/ml is strongly associated with increased cancer risk (Figler et al. 2007). Further, hK11and hK14 also compliment the prostate cancer prognosis along with PSA as these have been shown to be over expressed in prostate malignancy and present an enhanced expression pattern as the disease advances from earlier to late stages (Yousef et al. 2003; Jemal et al. 2004). Serine proteases such urokinase plasminogen activator (uPA) are involved in the degradation of basement membrane and extracellular matrix, leading to cancer cell invasion and metastasis. High endogenous levels of uPA and its receptor (uPAR) are associated with advanced metastatic cancers such as those of esophageal (Shiomi et al. 2000), gastric (Ji et al. 2005) and endometrial origin (Memarzadeh et al. 2002).

1.2 Cysteine Cathepsins

Cysteine cathepsins (CCs) are lysosomal proteases of papain family that have been found to be upregulated in various human cancers (Turk 2006; Mohamed and Sloane 2006) and have been shown to be prognostic indicators of many types of tumors (Palermo and Johanna 2008). In normal cells, cysteine cathepsins are usually located in lysosomes, whereas during cancer progression these proteolytic enzymes move to the cell surface, from where they are secreted into the extracellular milieu (Turk and Guncar 2003) to promote tumor invasion (Gocheva and Johanna 2007). It was reported that an enhanced expression of cathepsin B correlates with good prognostic value in various cancer types including lung, breast, ovarian, brain, head and neck cancer, melanoma and in premalignant lesions situated within colon, thyroid, liver, and prostate (Jedeszko and Sloane 2004; Kos et al. 2000). Similarly, cathepsin L activity has been observed in multiple tumor types and seems to act as a prognostic indicator of shorter survival rates in patients with breast, colorectal and head and neck cancers (Berdowska 2004). Recently a high resolution magnetic resonance imaging (MRI) technique has been developed as a promising probe to map cathepsin expression in vivo, in a tumor model using poly-L-glutamate (PLG) (Haris et al. 2014). This high-resolution method marks the differences in the chemical exchange saturation transfer (CEST) signals of PLG in the native form and cathepsin mediated cleaved form. The method is based on the chemical exchange saturation transfer between protons attached to nitrogen in amine moieties with bulk water protons. In principle, on account of very slow exchange rates of backbone amide protons, PLG in its native form does not exhibit perceptible CEST under normal conditions. However, when cleaved by the lysosomal enzymes such as cathepsins, resulting in individual glutamate (Glu) moieties and smaller Glu peptides; these readily exhibit CEST effect from their amine protons, thereby quantifying tumor protease expression (Fig. 1).

1.3 Thioredoxin Reductase

Thioredoxin reductase (TrxR) is a member of the selenium-containing pyridine nucleotide-disulfide oxidoreductase family with antioxidant functions and it is known to play an important role in tumor growth, progression and metastasis (Selenius et al. 2010; Dong et al. 2016). Studies have reported that in patients with thyroid cancer and non-small cell lung cancer (NSCLC), high expression of TrxR in cancer tissue and elevated serum levels of the enzyme exists, respectively (Lincoln et al. 2010; Kang et al. 2017). These studies have shown a linear correlation between TrxR1 activity and pathological stage, as a function of malignancy and cancer progression. Further an assessment of serum TrxR in addition to PSA has been reported to be of substantial benefit for a more accurate differential diagnosis of benign prostate hyperplasia (BPH) and prostate cancer (PCa), with a study showing significantly increased levels of TrxR in serum of patients diagnosed with PCa compared with BPH and healthy subjects (Gad et al. 2015). In patients with EGFR wild type and ALK (anaplastic lymphoma kinase) negative advanced NSCLC, the lower serum TrxR1 activity has been found to be associated with long progression-free survival and overall survival compared to patients with higher serum activity (Chen et al. 2017). The study suggested the potential application of serum TrxR1 activity as an independent prognostic factor for EGFR wild type and ALK negative advanced NSCLC patients. Similarly, in patients with hepatocellular carcinoma, the median levels of serum thioredoxin were reported to be significantly higher than that of healthy subjects and patients with cirrhosis and chronic liver diseases. Moreover, the study also demonstrated a significant correlation between thioredoxin concentrations and tumor size; and the level of thioredoxin was shown to increase as liver disease progressed from tumor stage I–IV (Li et al. 2015).

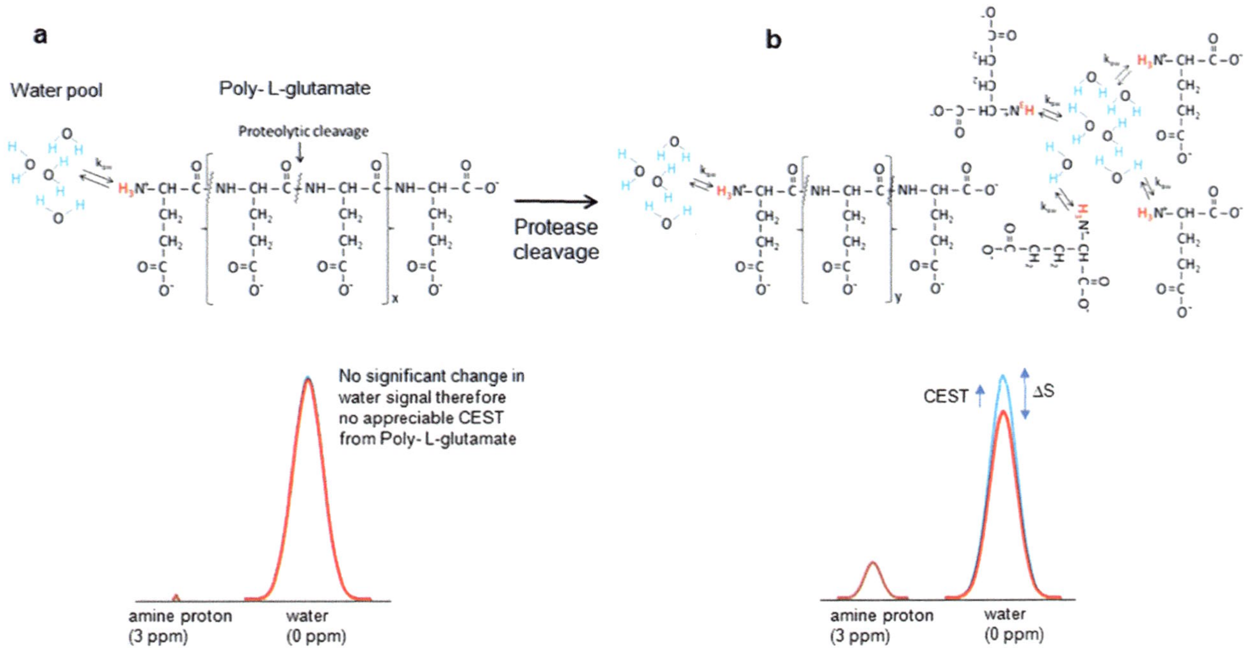

Fig. 1 Poly L-glutamate (PLG) as a protease activated CEST contrast agent. (**a**) Molecular structure of PLG shows three amine protons in exchange with the bulk water. Multiple amide protons are also present in the backbone of PLG. Due to the presence of significantly lower number of exchangeable amine protons, the saturation transfer of these protons does not reduce bulk water signal noticeably and therefore no appreciable CEST contrast is observed from PLG. No CEST contrast is observed from amide protons due to its slow exchange rate and saturation power and duration used in the current study. (**b**) Proteolytic cleavage of PLG into monomers (i.e. glutamate) and/or smaller peptide fragments exposes fair number of exchangeable amine protons. The exchange of saturated magnetization of these amine protons with bulk water significantly reduces the bulk water signal and results in appreciable CEST contrast. Reproduced from the original source (Haris et al. 2014) under the terms of Creative Common Attribution License

1.4 Alkaline Phosphatase

Alkaline phosphatase (ALP) is a ubiquitous enzyme present in all tissues with its isoforms clinically useful in bone and liver metastasis (Fizazi et al. 2015). An abnormal ALP has been reported in a high proportion of breast cancer patients with bone and/or liver metastases, and it was effective in distinguishing patients with relapse (Crivellari et al. 1995). It was also suggested that CA15–3 along with ALP, together provide a reasonable cue as an early predictor of recurrence of breast cancer (Keshaviah et al. 2007). Serum ALP levels have been reported to be elevated in patients with metastatic colorectal cancer (CRC). A study evaluating this relation with the degree of disease severity found serum ALP levels significantly elevated in 74% of patients with liver metastasis and in 33% without liver metastasis (Saif et al. 2005). Further when the levels of carcinoembryonic antigen (CEA) were compared in tandem with ALP in a random sample of 18 patients, a positive correlation was demonstrated between the increasing levels of CEA and enhanced ALP levels. Tomlins et al. (2007) have reported much higher expression of ALP in metastatic disease compared with primary prostate cancer samples. It is believed that tumor cells express bone-specific markers to survive in the bone microenvironment during metastatic colonization (Rucci and Teti 2010). An enhanced ALPL expression in PCa cells from metastatic sites have been reported to have a regulatory function as it has been found that a reduction in tumor-derived ALPL expression or ALP activity resulted in an increased rate of cell death, mesenchymal-to-epithelial transition and reduced migration (Rao et al. 2017). The study demonstrated a strong association between *ALPL* expression in prostate cancer cells and metastasis or disease-free survival, suggesting the role of alkaline phosphatase in disease progression and its significance as a prognostic marker.

1.5 Cyclooxygenases

The cyclooxygenases (COXs) are a family of cytoplasmic enzymes, which catalyze the rate-limiting step of prostaglandin biosynthesis from arachidonic acid; which include ubiquitously expressed COX-1 (homeostasis), inducible COX-2 isoform (inflammation and cancer) and COX-3 (expressed in brain and spinal cord) (Zarghi and Arfaei 2011). COX-2 over expression is associated with an increased production of PGE2, which is known to modulate cell proliferation, cell death, and tumor invasion in several cancers including colon, breast, and lung (Sobolewski et al. 2010). Studies have found COX-2 expression to be a consistent feature of all stages of breast cancer, with presence in premalignant lesions (dysplasia and atypia) that attains greater prominence as the disease progresses to invasive and metastatic stages; however COX-2 is usually not detectable in normal (non-inflamed) mammary tissues (Parrett et al. 1997; Harris et al. 2014). In colorectal adenocarcinomas, COX-2 is over expressed in approximately 80% of the cases and it is believed to be involved in invasiveness, apoptotic resistance, and increased tumor angiogenesis; which are associated with decreased cancer patient survival (Wang and Dubois 2010). COX-2 expression has also been shown to be a significant prognostic factor in patients with advanced NSCLC receiving celecoxib with chemotherapy (Edelman et al. 2008) and similarly COX-2 expression was also reported to be a powerful predictor of disease free survival among patients treated with adjuvant platinum-based chemotherapy (Shimizu et al. 2015). Several studies have shown that COX-2 has an important role in the growth and metastasis of head and neck cancer (HNC) through important signaling pathways (Hsu et al. 2015) and further indicated that COX-2 expression could act as a prognostic factor for patients with HNC (Yang et al. 2016). A meta-analysis of 2465 patients from 25 studies demonstrated that over expression of COX-2 might be associated with poor overall survival in esophagus cancer as such an enhanced expression was found to be related to invasiveness, metastasis, and advanced stages; favoring COX-2 as a potential prognostic biomarker for esophagus cancer (Hu et al. 2017).

1.6 Thymidine Kinase

Thymidine Kinase (TK1) is associated with the DNA synthesis (S phase) and thus its expression is essential for proliferating cells. The level of TK1 is very low in resting cells, which enhances substantially at late G1 to late S-phase during the cell cycle in proliferating cells and tumor cells (Aufderklamm et al. 2012). It has been reported that the levels of TK1 in the peripheral blood of malignant carcinoma patients are significantly enhanced, indicating their potential roles as serological marker for cancer diagnosis (Chen et al. 2011). A health screening study engaging 8135 individuals reported 1.1% of individuals with elevated TK1 levels, and a further evaluation found that nearly 90% of those with elevated TK1 had pre-cancerous diseases (Huang et al. 2011). A recent study investigating the importance of TK1, CEA and AFP (alpha-fetoprotein)as diagnostic and prognostic marker in cancer patients (56,286 cases- including liver, cervix and lung cancer), demonstrated that serum TK1 correlated well with tumor growth rate and was more sensitive than CEA and AFP for diagnosis of malignancies, whereas a combination of these markers increased the sensitivity by about 30% (Wang et al. 2016). A comparative study examining the expression of TK1 in 46 colorectal cancer patients and 46 healthy controls, found median serum levels of TK1 to be 3.33 (0.78–5.78) pmol/L and 0.99 (0.34–4.46) pmol/L in colorectal cancer patients and healthy controls respectively, thereby demonstrating its clinical value as diagnostic biomarker for colorectal cancer (Zhu et al. 2017). In breast cancer cases it was demonstrated that TK1 levels were significantly enhanced in atypical ductal hyperplasia, ductal carcinoma in situ, and invasive ductal carcinoma compared to normal breast tissue (Guan et al. 2009). Serum samples from breast cancer patients (124) and healthy individuals (53) when analyzed for serum TK1 activity and TK1 protein levels, significant correlations were reported between the serum TK1 activity and TK1 protein levels in healthy and breast cancer patients (Kumar et al. 2016). In healthy group, TK1 activity values ranged from 0.6 to 3.1 pmol/min/ml whereas in breast cancer patients, the STK1 activity levels were significantly higher in the range of 0.9–48 pmol/min/ml. Moreover majority of the serum samples from healthy subjects had TK1 protein levels near or below the detection limit (0.17 ng/ml) and 40% had detectable TK1 protein levels in the range of 0.17–0.33 ng/ml. However, significantly higher serum TK1 protein levels were found in sera from breast cancer patients compared to healthy control group and the protein concentration ranged from 0.17 to 9.9 ng/ml. Alegre et al. (2014) have also shown significantly elevated TK1 in serum from patients with stage I and stage II lung cancer and it was further reported that TK1 concentration was a more sensitive and accurate indicator of lung cancer than TK1 activity. Similar finding was reported by Bi et al. (2016) where the study revealed that TK1 expression was positively correlated with NSCLC TNM staging and the extent of NSCLC cell differentiation (grade III > II > I).

2 Enzymes as Significant Regulator of Myocardial Diseases

Myocardial diseases are primary disorders of the cardiac muscle with cardiomyopathies reported to be among the most frequent causes of sudden death (Esteban 2006). It is believed that a significant proportion of cases of sudden death can be prevented via early diagnosis and identification of high-risk individuals. In this context, the understanding of certain enzymatic factors have been potentially sought to improve the diagnosis and clinical outcome.

2.1 Enzymes in Mitochondrial Energetics Causing Myocardial Disease

The normal function of the human myocardium requires the proper generation and utilization of energy. Regulation of cardiac energetics in the fetal as well as adult heart is depended on the enzymatic activities of different molecules. Large amounts of energy in the form of ATP are consumed by heart that is continuously provided by oxidative phosphorylation in mitochondria and, to a lesser extent, by glycolysis (Chinnery 2014). For this continuous generation and supply of ATPs to cardiac myocytes, a very complex network of enzymes and signaling pathways are involved to maintain this stream by a controlled metabolic activities in the mitochondria, (collectively termed as mitochondrial energetics) (Munnich et al. 2012; El-Hattab and Scaglia 2015). When these metabolic processes underperform, heart muscle defects can occur with or without abnormalities of other organ systems, collectively termed as mitochondrial diseases (Chinnery 2014; Munnich et al. 2012; Gilbert-Barness 2004).

Mitochondrial diseases are clinically and genetically heterogeneous resulting from dysfunction of the mitochondrial respiratory chain. In failing hearts, various errors are investigated due to improper utilization of substrates and faulty relation of intermediate metabolism resulting in oxidative stress and energy deficiency. These defects in energy metabolism are quite common in heart muscle diseases (cardiomyopathies). Cardiomyopathies occur in approximately one-third of pediatric population with mitochondrial errors (El-Hattab and Scaglia 2016). These can be associated with defects in electron transport chain (ETC) complexes subunits and their assembly factors, mitochondrial tRNAs, rRNAs, ribosomal proteins, translation factors, mtDNA maintenance, and CoQ10 synthesis. Other mitochondrial errors with cardiomyopathies include Barth syndrome and its associated 3-methylglutaconic aciduria disorders, and Friedreich ataxia (El-Hattab and Scaglia 2015).

Barth syndrome is an X-linked defect with cardiomyopathy, skeletal myopathy, growth retardation, neutropenia, and increased urinary levels of 3-methylglutaconic acid. It is caused by mutations in the TAZ gene encoding tafazzin, a phospholipid transacylase. This enzyme is located in the inner mitochondrial membrane and plays an important role in the remodeling of cardiolipin. Left ventricular non-compaction and dilated cardiomyopathies are common in Barth syndrome, whereas hypertrophic cardiomyopathy appears to be less common. Additional cardiac manifestations of Barth syndrome are arrhythmia (including supraventricular and ventricular tachycardia) and sudden death (Clarke et al. 2013; Jefferies 2013). Other disorders in Barth syndromes that might be associated with cardiomyopathy are caused by mutations in *DNAJC19*, *TMEM70*, and *AGK* (Wortmann et al. 2013). Mutations in *TMEM70* (mitochondrial complex V deficiency), encodes a protein helping in the insertion of ATP synthase (complex V) into the mitochondrial membrane, results in mitochondrial disease with hypertrophic cardiomyopathy (Cízková et al. 2008). Sengers syndrome, caused by mutations in AGK, is accompanied by 3-methylglutaconic aciduria and is characterized by hypertrophic cardiomyopathy, cataracts, myopathy, exercise intolerance, and lactic acidosis. The AGK encodes acylglycerol kinase involved in the assembly of ANT1, a mitochondrial adenine nucleotide transporter (Haghighi et al. 2014). Recently Kennedy et al. (2016) identified biallelic missense mutations in mitochondrial inorganic pyrophosphatase (*PPA2*) as cause of mitochondrial cardiomyopathy and sudden cardiac death. This finding highlights a critical role of PPA2 in mitochondrial function. Smyd1 is a muscle-specific histone methyltransferase, with roles in growth and differentiation regulation of skeletal and cardiac muscles (Tracy et al. 2018). Its knockdown in models results in global down regulation of mitochondrial proteins involved in oxidative phosphorylation, with consequences of aberrant mitochondrial respiration capacity. Collectively, Smyd1 is key regulator of mitochondrial energetic and cardiac energetic (Warren et al. 2018).

For normal heart function, there should be a balance between reactive oxygen species and cellular antioxidant defense mechanisms. There are experimental evidences that any imbalance between these can deteriorate the myocardial function at cellular level. Thus intracellular enzymatic pathways are involved in the elimination of reactive oxygen species in the left ventricular myocardium (Kaul et al. 1995; Singh et al. 1995; Singal and Kirshenbaum 1990; Romero-Alvira et al. 1996). Defects in enzymes involved in oxidative phosphorylation produces high levels of mitochondrial free radicals and this causes a form of familial dilated cardiomyopathy (Pitkanen and Robinson 1996). These enzyme systems are composed of left ventricular enzyme activities, mRNA and protein levels of the hydrogen peroxide scavenging enzymes catalase (CAT) and glutathione peroxidase (GPX), and the superoxide anion scavenging enzymes mitochondrial (Mn-SOD) and cytosolic (Cu/Zn-SOD) superoxide dismutases. CAT activity is a post-transcriptional mechanism; however decrease in its levels may lead dilated cardiomyopathy (reviewed in Bäumer et al. 2000).

2.2 *Extracellular Matrix and Proteases in Tissue Repair*

Mammalian heart is covered with myocardium and this cover is composed of cardiomyocytes and populations of non-cardiomyocytes enmeshed within extracellular matrix (ECM) proteins. ECM is a major player in this myocardial injury, repair and remodeling involving different proteases and growth factors, and transduce signaling cascades. In this context cardiomyocyte injury and resulting necrosis is associated with induction and activation of proteases in the infarcted region. These proteases being activated by ECM machinery cause fragmentation of the native ECM, resulting in release of matrikines, bioactive peptides that activate an inflammatory macrophage phenotype and may also modulate responses of fibroblasts and vascular endothelial cells (Frangogiannis 2017).

2.3 *Enzymes in Genetics of Acquired Forms of Heart Failure*

G protein-coupled receptor kinases (GRKs) are classically known in regulating the activity of the largest known class of membrane receptors maintaining diverse biological processes in the human body at cellular levels. These activities include phosphorylation of non-receptor targets and kinase-independent molecular interactions. GRK2, GRK3, and GRK5 are the predominant members expressed in the heart. Their canonical and non-canonical actions within cardiac and other tissues have significant implications for cardiovascular function in healthy animals and for the development and progression of disease. The β-1 receptor (ADRB1) is the dominant beta-adrenergic receptor in the heart. ß-1 receptors mediate the chronotropic, inotropic and lusitropic effects of catecholamines and its genetic variants have acted as modifiers of heart failure risk and prognosis. Some variants have shown that an elevated adenylyl cyclase signaling enhanced inotropy and also increased the susceptibility to beta-blockers (Dorn 2nd 2010; Lopes and Elliott 2013). Down-regulation and desensitization of myocardial β-adrenergic receptor is regulated by GRKs. It is hypothesized that gain-of-function genetic alleles could be responsible for desensitization and protection against the hyperactive adrenergic system in heart failure and loss-of-function alleles increases the risk for progressive heart failure. A known allele, single nucleotide polymorphism (SNP) Gln41Leu in GRK5 (rs17098707) increases desensitization mimicking the beta-blockade effect (Liggett et al. 2008). G protein-coupled receptor kinase-2 (GRK2) is a regulator of GPCRs, particularly β-adrenergic receptors (ARs), and has a pivotal role in HF progression and development (Cannavo et al. 2018).

2.4 Diagnostic Enzymes in Heart Diseases

Cardiac biomarkers has become the frontline diagnostic tools for myocardial infarction (MI) and other heart conditions, and has greatly expedited the process of diagnosis for clinician and ultimately prompt treatment planning. These biomarkers have reduced the mortality rate (Mythili and Malathi 2015).

World Health Organization (WHO) had defined the term myocardial infarction by the presence of following at least characteristics (World Health Organization 1972): (1) Symptoms of acute ischemia (or chest pain), (2) development of Q waves in electrocardiogram (ECG) and (3) elevated levels of certain enzymes in the serum [combination of total creatine kinase (CK), CK-myocardial band (MB), aspartate aminotransferase (AST) and lactate dehydrogenase (LDH)].

2.4.1 Creatine Kinase (CK)

Creatine kinase, also known as creatine phosphokinase (CPK) is a muscle enzyme. The enzyme exists in three isozyme forms. The CK-MB type is specific to myocardial cells, whereas CK-MM and CK-BB are specific to skeletal muscle and brain tissue, respectively (Bloomberg et al. 1975). CK was first indicated as a cardiac biomarker in 1979. The CK level increases approximately 3–4 h after MI and remains elevated for 3–4 days. The total CK contains 95% of the CK-MM fraction, clinicians follow the use of the relative index score (RI) as follows (Gerhardt et al. 1991; Wu et al. 1992).

$$\text{RI of CK} - \text{MB} = [\text{CK} - \text{MB (ng/ml)}/\text{Total CK (U/l)}] \times 100$$

2.4.2 Myeloperoxidase (MPO)

MPO is a metalloproteinase produced by the polymorphonuclear leukocytes and macrophages. MPO initiates the reaction producing reactive oxygen species such as hypochlorous acid (HOCl) from hydrogen peroxide (H_2O_2) and chloride anion.

These oxygen species help the development of atheroma and plaque rupture (Khan et al. 2007). An increased level of MPO is a marker of plaque instability (Wang et al. 2007). Also MPO serves as a predictive marker for unknown adverse cardiovascular stages (Cavusoglu et al. 2007).

2.4.3 Pregnancy-Associated Plasma Protein A (PAPPA)

PAPP-A is a zinc-binding metalloproteinase with active role during the rupture of an atherosclerotic plaque (Oxvig et al. 1994). The syncytiotrophoblasts of the placenta, as well as the fibroblasts, vascular endothelial cells and vascular smooth muscle cells usually produce PAPP-A. It is potentially proatherosclerotic protein and acts as a specific activator of insulin-like growth factor I (IGF-I) (Lawrence et al. 1999). In atherosclerosis, it has been associated with plaque progression and instability, thus used as a marker of acute coronary syndromes. PAPP-A is found in unstable plaques as well as circulating levels are elevated in acute coronary syndromes reflecting the instability of atherosclerotic plaques (Bayes-Genis et al. 2001).

3 Diagnostic Enzymes in Gastrointestinal and Associated Disorders

Diagnostic enzymes indicated for clinical investigation of disorders associated with digestive system (GI and associated organs), commonly originate from stomach, pancreas and most notably, liver. There are several enzymes that are typically used as specific markers in the clinical laboratory to diagnose such disorders, since these enzymes are relatively easy to assay using automated techniques in standard blood examination.

3.1 Gastric Disorders

In several studies of gastrointestinal reflux disease (GERD), salivary pepsin has been evaluated as a diagnostic marker. In a recent study salivary samples were collected from 250 patients with symptoms suggestive of GERD and 35 asymptomatic healthy volunteers at different time points including postprandial. The results were demonstrative that pepsin detection had a sensitivity of 73% and a specificity of 88.3% for diagnosing of GERD making such a test promising for the clinical diagnosis (Du et al. 2017). Pepsin detection in saliva has also been suggested as a diagnostic marker for laryngo pharyngeal reflux (LPR), although there have been no clinical trials to date with adequate numbers of controls and patients, that can establish it as a clinical marker for diagnosis (Sifrim 2015). However, saliva being a non-invasive sample, the detection of pepsin in saliva would be of immense clinical use for the diagnosis of reflux diseases.

3.2 Pancreatic Inflammation

Pancreas secretes a bundle of enzymes that aid in digestive functions, and among these; few have been implicated in the diagnosis of pancreatitis. Lower than normal levels of pancreatic amylase and lipase are routinely detected in patients with chronic pancreatitis. In the course of worsening chronic inflammation as the secretory function of pancreas decreases, the levels of pancreatic serum lipase and amylase fall, and therefore this functional reduction is exploited for clinical diagnosis of chronic pancreatitis. Detection in the fall of either of the two enzymes below the normal range, but not combined detection, has high specificity for the diagnosis of chronic pancreatitis (Oh et al. 2017). Fecal detection of elastase-1 has also been shown to aid the diagnosis of pancreatic insufficiency. Being low in specificity, it rather helps in the screening of chronic pancreatitis (Löhr 2016). In a similar context, the sudden rise in acute serum amylase and/or lipase is a standard clinical diagnostic marker for acute pancreatitis. However in this case, lipase probably offers a larger diagnostic window than amylase since it remains in circulating blood for longer period of time (Ismail and Bhayana 2017).

3.3 Liver Diseases

More recently Clinical Services and Standards of British Society of Gastroenterology under the auspices of the liver section of the BSG strongly recommended the measurement of alanine transaminase (ALT), alkaline phosphatase (ALP), gamma-glutamyltransferase (GGT), and aspartate transaminase (AST) in serum for screening and diagnosis of liver dysfunction (Newsome et al. 2018). It has been observed that clinicians are often in a situation where the reports do not correspond with the clinical condition of the patient which appears as an obstacle to interpret the liver function test (LFT). In this regard an attempt has been made to study and understand the LFT with a simplified algorithm based interpretation (Thapa and Walia 2007). Altered serum levels of liver enzymes reflect liver damage or disturbances in bile flow. Most often the clinicians seek to examine for serum ALP, AST, ALT and less commonly GGT as part of panel to assess the functionality of liver. Table 1 shows the abnormal levels of these four enzymes and the clinical conditions to which these are commonly associated (Giannini et al. 2005).

It is known that GGT is a plasma membrane protein which plays a central role in glutathione homeostasis and therefore helps to protect the cells against oxidative insults. It is primarily present in kidney, liver, and pancreatic cells. However, small amounts are also known to be present in other tissues. The renal tissue has the

Table 1 Serum hepatic enzymes are often altered in most liver diseases/injuries

Enzyme (s)	Normal range (IU/L)	Altered serum level	Commonly associated condition/ disease
		>75 × upper limit[a]	Ischemic hepatitis
ALT	0–45 IU/L	>10 × upper limit[a]	Acute viral hepatitis
AST	0–35 IU/L	>6–7 × upper limit[a]	Alcoholic liver disease[b]
		Mild[a]	Non-alcoholic liver disease
ALP	20–120	>6–7 × upper limit	Alcoholic liver disease
			Cholestasis
GTT	0–48	Mild	Non-alcoholic liver disease
			Drug induced liver injury

The table shows the most commonly associated clinical condition with the respective altered ranges of specific enzymes or their combinations

[a]The altered levels are for both given enzymes (ALT, AST) in combination

[b]AST/ALT ratio >1–2

highest level of GGT. However, the primary source for serum GGT is hepatobiliary system, and therefore, elevated serum GGT activity appear in most liver diseases and hepatobillary complications (Whitfield 2001). The alteration of GGT in serum has probably been undervalued for hepatic dysfunction and therefore may not always find a place in the standard panel for liver function (Everhart and Wright 2013). Highest elevations of GTT are seen in intra- or post-hepatic biliary obstruction. GGT is sensitive towards liver injury particularly by insult caused due to alcohol (Pratt and Kaplan 2000) and certain drugs (Rosalki et al. 1971). Non-steroidal anti-inflammatory drugs as well as penicillin derivatives are the most common cause for drug related elevation of serum GGT and have been reviewed elsewhere (Aragon and Younossi 2010). GGT levels are also elevated in serum of patients suffering from metabolic syndrome and such elevated levels have been correlated to an increased risk for cardiovascular diseases (Ege et al. 2013; Bozbaş et al. 2011).

Liver synthesizes two aminotransferases namely aspartate aminotransferase a (AST) and alanine aminotransferase (ALT) which play central role in amino acid metabolism in hepatocytes. ALT is highly specific to liver as compared to AST which is also diffusely present in the heart, skeletal muscle, kidneys, brain and red blood cells. Therefore, an injury to liver—acute or chronic, eventually leads to elevated levels of two aminotransferases synthesized in liver viz., ALT and AST. It is understood that very high levels of ALT and AST are most commonly associated with ischemic or toxic liver injury (Cuperus et al. 2017). In such cases AST will usually peak before ALT, and the two are rapidly cleared once the peaks are achieved. For differential diagnosis, it is very important to have an insight into the pattern of change in serum levels of the two enzymes along with lactate dehydrogenase (marker of ischemic damage) as well as bilirubin levels (Dufour et al. 2000). A moderate increase in the serum ALT and AST may be indicative of alcoholic liver disease. In such patients the AST levels are more generally elevated lesser than 6–7 times, AST/ALT ration tends to be >1–2 (Cohen and Kaplan 1979).

Furthermore, rise in GGT addition to the said ratio is highly suggestive of alcohol abuse. Howsoever a moderate to higher elevation of liver enzymes, in particular AST, may not be always because of hepatic injury but rather because of obstruction in the extra hepatic biliary ducts. Such elevated levels of AST in serum quickly tend to normalize once the obstruction is cleared (Dufour et al. 2000). Probably the most common cause of mildly elevated serum ALP and AST presented in the clinics is nonalcoholic fatty liver disease and in such cases GGT levels are commonly elevated as well to up to nearly three times the upper reference limit. Likelihood of non-alcoholic fatty liver is further increased if the case is presented with associated metabolic syndrome and/or insulin resistance (Jansen 2004). Other conditions that generally present with mild increases in aminotransferases includes autoimmune hepatitis, Wilson's disease and alpha-1 antitrypsin deficiency which have been reviewed elsewhere (Harrison et al. 2002).

The enzyme alkaline phosphatase is present in high concentration in the epithelia of hepatic bile ducts. Any obstruction in bile flow causes an increase in the synthesis and secretion of the enzyme leading to elevated serum ALP. However, enzyme is not specific to liver. Other sources are tissues like bone, kidneys and intestines. More recently few cohort studies have also demonstrated serum cholinesterase as a valuable marker for staging of cirrhosis in chronic liver diseases; although long-term, large cohorts have been warranted (Ramachandran et al. 2015; Abbas and Abbas 2017) before the enzyme can be incorporated in the clinics.

Acknowledgments We thankfully acknowledge the authors and the publishers of the articles from where the Fig. 1 has been reproduced, and note that since a brief overview on the subject was attempted, some worthwhile studies and relevant observations may have been inadvertently missed.

References

Abbas M, Abbas Z (2017) Serum cholinesterase: a predictive biomarker of hepatic reserves in chronic hepatitis D. World J Hepatol 9(22):967

Alegre MM, Weyant MJ, Bennett DT et al (2014) Serum detection of thymidine kinase 1 as a means of early detection of lung cancer. Anticancer Res 34:2145–2151

Aragon G, Younossi Z (2010) Evaluation of asymptomatic elevation of liver enzymes. Cleve Clin J Med 77(3):195–204

Aufderklamm S, Todenhöfer T, Gakis G, Kruck S, Hennenlotter J, Stenzl A, Schwentner C (2012) Thymidine kinase and cancer monitoring. Cancer Lett 316(1):6–10

Bäumer AT, Flesch M, Wang X, Shen Q, Feuerstein GZ, Böhm M (2000) Antioxidative enzymes in human hearts with idiopathic dilated cardiomyopathy. J Mol Cell Cardiol 32(1):121–130

Bayes-Genis A, Conover CA, Overgaard MT, Bailey KR, Christiansen M, Holmes DR Jr, Virmani R, Oxvig C, Schwartz RS (2001) Pregnancy-associated plasma protein A as a marker of acute coronary syndromes. N Engl J Med 345(14):1022–1029

Berdowska I (2004) Cysteine proteases as disease markers. Clin Chim Acta 342:41–69

Bhattacharjee A, Richards WG, Staunton J et al (2001) Classification of human lungcarcinomas by mRNA expression profiling reveals distinct adenocarcinomasubclasses. Proc Natl Acad Sci U S A 98:13790–13795

Bi MH, Han W, Liu JJ, Wang HY, Gao ZY, Tian WL (2016) Clinical significance of serum thymidine kinase 1 (TK1) expression in patients with non-small cell lung cancer. Int J Clin Exp Med 9:8536–8542
Bloomberg DJ, Kimber WD, Burke MD (1975) Creatine kinase isoenzymes. Predictive value in the early diagnosis of acute myocardial infarction. Am J Med 59:464–469
Borgono CA, Michael IP, Diamandis EP (2004) Human tissue kallikreins: physiologicroles and applications in cancer. Mol Cancer Res 2:257–280
Bozbaş H, Yıldırır A, Karaçağlar E, Demir O, Ulus T, Eroğlu S, Aydınalp A, Özin B, Müderrisoğlu H (2011) Increased serum gamma-glutamyltransferase activity in patients with metabolic syndrome. Turk Kardiyol Dern Ars 39(2):122–128
Cannavo A, Komici K, Bencivenga L, D'amico ML, Gambino G, Liccardo D, Ferrara N, Rengo G (2018) GRK2 as a therapeutic target for heart failure. Expert Opin Ther Targets 22(1):75–83
Cavusoglu E, Ruwende C, Eng C, Chopra V, Yanamadala S, Clark LT, Pinsky DJ, Marmur JD (2007) Usefulness of baseline plasma myeloperoxidase levels as an independent predictor of myocardial infarction at two years in patients presenting with acute coronary syndrome. Am J Cardiol 99:1364–1368
Chen ZH, Huang SQ, Wang Y et al (2011) Serological thymidine kinase 1 is a biomarker for early detection of tumours – a health screening study on 35,365 people, using a sensitive chemiluminescent dot blot assay. Sensors (Basel) 11:11064–11080
Chen G, Chen Q, Zeng F et al (2017) The serum activity of thioredoxinreductases 1 (TrxR1) is correlated with the poor prognosis in EGFR wild-type and ALK negative non-small cell lung cancer. Oncotarget 8:115270–115279
Chinnery PF (2014) Mitochondrial disorders overview. In: Pagon RA, Adam MP, Ardinger HH, Wallace SE, Amemiya A, Bean LJ et al (eds) GeneReviews®. University of Washington Seattle, Seattle, WA
Cízková A, Stránecký V, Mayr JA, Tesarová M, Havlícková V, Paul J et al (2008) TMEM70 mutations cause isolated ATP synthase deficiency and neonatal mitochondrial encephalocardiomyopathy. Nat Genet 40:1288–1290
Clarke SLN, Bowron A, Gonzalez IL, Groves SJ, Newbury-Ecob R, Clayton N et al (2013) Barth syndrome. Orphanet J Rare Dis 8:23
Cohen JA, Kaplan MM (1979) The SGOT/SGPT ratio—an indicator of alcoholic liver disease. Dig Dis Sci 24(11):835–838
Crivellari D, Price KN, Hagen M et al (1995) Routine tests during follow-up of patients after primary treatment for operable breast cancer. International (Ludwig) Breast Cancer Study Group (IBCSG). Ann Oncol 6:769–776
Cuperus FJC, Drenth JPH, Tjwa ET (2017) Mistakes in liver function test abnormalities and how to avoid them. UEG Educ 17:1–5
Dong C, Zhang L, Sun R et al (2016) Role of thioredoxinreductase 1 in dysplastic transformation of human breast epithelial cells triggered by chronic oxidative stress. Sci Rep 6:36860
Dorn GW 2nd (2010) Adrenergic signaling polymorphisms and their impact on cardiovascular disease. Physiol Rev 90:1013–1062
Du X, Wang F, Hu Z, Wu J, Wang Z, Yan C, Zhang C, Tang J (2017) The diagnostic value of pepsin detection in saliva for gastro-esophageal reflux disease: a preliminary study from China. BMC Gastroenterol 17(1):107
Dufour DR, Lott JA, Nolte FS, Gretch DR, Koff RS, Seeff LB (2000) Diagnosis and monitoring of hepatic injury. I. Performance characteristics of laboratory tests. Clin Chem 46(12):2027–2049
Edelman MJ, Watson D, Wang X et al (2008) Eicosanoid modulation in advanced lung cancer: cyclooxygenase-2 expression is a positive predictive factor for celecoxib + chemotherapy – Cancer and Leukemia Group B Trial 30203. J Clin Oncol 26:848–855
Ege MR, Guray U, Guray Y, Demirkan B, Kisacik H (2013) Serum γ-glutamyltransferase levels correlate with epicardial adipose tissue thickness in patients with coronary artery disease. Angiology 64(1):21–25

El-Hattab AW, Scaglia F (2015) Mitochondrial disorders. In: Lee B, Scaglia F (eds) Inborn errors of metabolism: from neonatal screening to metabolic pathways. Oxford University Press, New York, NY, pp 180–202

El-Hattab AW, Scaglia F (2016) Mitochondrial Cardiomyopathies. Front Cardiovasc Med 3:25

Esteban MTT (2006) Myocardial disease as a cause of sudden cardiac death syndrome in children and young adults: recognition and preventive strategies needed. Rev Esp Cardiol 59:197–199

Everhart JE, Wright EC (2013) Association of γ-glutamyltransferase (GGT) activity with treatment and clinical outcomes in chronic hepatitis C (HCV). Hepatology 57(5):1725–1733

Figler BD, Reuther AM, Dhar N et al (2007) Preoperative PSA is still predictive of cancer volume and grade in late PSA era. Urology 70:711–716

Fizazi K, Massard C, Smith M et al (2015) Bone-related parameters are the main prognostic factors for overall survival in men with bone metastases from castration-resistant prostate cancer. Eur Urol 68(1):42–50

Frangogiannis NG (2017) Fibroblasts and the extracellular matrix in right ventricular disease. Cardiovasc Res 113(12):1453–1464

Gad HM, Elsisi AM, Ali AA (2015) Thioredoxinreductase as biomarker for differential diagnosis of prostate cancer. Az J Pharm Sci 52:140–149

Gerhardt W, Katus H, Ravkilde J et al (1991) S-troponin T in suspected ischemic myocardial injury compared with mass and catalytic concentrations of S-creatine kinase isoenzyme MB. Clin Chem 37:1405–1411

Giannini EG, Testa R, Savarino V (2005) Liver enzyme alteration: a guide for clinicians. CMAJ 172(3):367–379

Gilbert-Barness E (2004) Review: metabolic cardiomyopathy and conduction system defects in children. Ann Clin Lab Sci 34:15–34

Gocheva V, Johanna AJ (2007) Cysteine cathepsins and the cutting edge of cancerinvasion. Cell Cycle 6:60–64

Guan H, Sun Y, Zan Q, Xu M, Li Y, Zhou J, He E, Eriksson S, Wen W, Skog S (2009) Thymidine kinase 1 expression in atypical ductal hyperplasia significantly differs from usual ductal hyperplasia and ductal carcinoma in situ: a useful tool in tumor therapy management. Mol Med Rep 2:923–929

Haghighi A, Haack TB, Atiq M, Mottaghi H, Haghighi-Kakhki H, Bashir RA et al (2014) Sengers syndrome: six novel AGK mutations in seven new families and review of the phenotypic and mutational spectrum of 29 patients. Orphanet J Rare Dis 9:119

Haris M, Singh A, Mohammed I, Ittyerah R, Nath K, Nanga RPR et al (2014) In vivo magnetic resonance imaging of tumor protease activity. Sci Rep 4:6081

Harris RE, Casto BC, Harris ZM (2014) Cyclooxygenase-2 and the inflammogenesis of breast cancer. World J Clin Oncol 5:677–692

Harrison SA, Kadakia S, Lang KA, Schenker S (2002) Nonalcoholic steatohepatitis: what we know in the new millennium. Am J Gastroenterol 97(11):2714

Hsu JY, Chang KY, Chen SH, Lee CT, Chang ST, Cheng HC, Chang WC, Chen BK (2015) Epidermal growth factor-induced cyclooxygenase-2 enhances head and neck squamous cell carcinoma metastasis through fibronectin up-regulation. Oncotarget 6:1723–1739

Hu Z, Yang Y, Zhao Y, Huang Y (2017) The prognostic value of cyclooxygenase-2 expression in patients with esophageal cancer: evidence from a meta-analysis. Onco Targets Ther 10:2893–2901

Huang S, Lin J, Guo N, Zhang M, Yun X, Liu S, Zhou J, He E, Skog S (2011) Elevated serum thymidine kinase 1 predicts risk of pre/early cancerous progression. Asian Pac J Cancer Prev 12:497–505

Iacobuzio-Donahue CA, Ashfaq R, Maitra A et al (2003) Highly expressed genes in pancreatic ductal adenocarcinomas: a comprehensive characterization and comparison of the transcription profiles obtained from three major technologies. Cancer Res 63:8614–8622

Ismail OZ, Bhayana V (2017) Lipase or amylase for the diagnosis of acute pancreatitis? Clin Biochem 50(18):1275–1280

Jansen PL (2004) Non-alcoholic steatohepatitis. Eur J Gastroenterol Hepatol 16(11):1079–1085

Jedeszko C, Sloane BF (2004) Cysteine cathepsins in human cancer. Biol Chem 385:1017–1027

Jefferies JL (2013) Barth syndrome. Am J Med Genet C Semin Med Genet 163C:198–205

Jemal A, Tiwari RC, Murray T, Ghafoor A et al (2004) Cancer statistics. CA Cancer J Clin 54:8–29

Ji F, Chen YL, Jin EY, Wang WL, Yang ZL, Li YM (2005) Relationship between matrixmetalloproteinase-2 mRNA expression and clinicopathological and urokinase type plasminogen activator system parameters and prognosis in human gastric cancer. World J Gastroenterol 11:3222–3226

Kang X, Dong C, Shen B et al (2017) Serum thioredoxinreductase 1 as a diagnostic marker for non-small cell lung cancer. J Clin Oncol 35(15_suppl):e23069–e23069

Kaul N, Siveski-Iliskovic N, Thomas TP, Hill M, Khaper N, Singal PK (1995) Probucol improves antioxidant activity and modulates development of diabetic cardiomyopathy. Nutrition 11:551–554

Kennedy H, Haack TB, Hartill V et al (2016) Sudden cardiac death due to defficiency of the mitochondrial inorganic pyrophosphatase PPA2. Am J Hum Genet 99:674–682

Keshaviah A, Dellapasqua S, Rotmensz N et al (2007) CA15-3 and alkaline phosphatase as predictors for breast cancer recurrence: a combined analysis of seven International Breast Cancer Study Group trials. Ann Oncologia 18(4):701–708

Khan SQ, Kelly D, Quinn P, Davies JE, Ng LL (2007) Myeloperoxidase aids prognostication together with N-terminal pro-B-type natriuretic peptide in high-risk patients with acute ST elevation myocardial infarction. Heart 93:826–831

Kim H, Scorilas A, Katsaros D et al (2001) Human kallikrein gene 5 (KLK5) expressions is an indicator of poor prognosis in ovarian cancer. Br J Cancer 84:643–650

Kos J, Werle B, Lah T, Brunner N (2000) Cysteine proteinases and their inhibitors in extracellular fluids: markers for diagnosis and prognosis in cancer. Int J Biol Markers 15:84–89

Kumar JK, Aronsson AC, Pilko G et al (2016) A clinical evaluation of the TK 210 ELISA in sera from breast cancer patients demonstrates high sensitivity and specificity in all stages of disease. Tumour Biol 37:11937–11945

Lawrence JB, Oxvig C, Overgaard MT et al (1999) The insulin-like growth factor (IGF)-dependent IGF binding protein-4 protease secreted by human fibroblasts is pregnancy-associated plasma protein-A. Proc Natl Acad Sci U S A 96:3149–3153

Li J, Cheng ZJ, Liu Y et al (2015) Serum thioredoxin is a diagnostic marker for hepatocellular carcinoma. Oncotarget 6:9551–9563

Liang SL, Chan DW (2007) Enzymes and related proteins as cancer biomarkers: a proteomic approach. Clin Chim Acta 381:93–97

Liggett SB, Cresci S, Kelly RJ, Syed FM, Matkovich SJ, Hahn HS, Diwan A, Martini JS, Sparks L, Parekh RR, Spertus JA, Koch WJ, Kardia SL, Dorn GW 2nd (2008) A GRK5 polymorphism that inhibits beta-adrenergic receptor signaling is protective in heart failure. Nat Med 14 (5):510–517

Lincoln DT, Al-Yatama F, Mohammed FM et al (2010) Thioredoxin and thioredoxinreductase expression in thyroid cancer depends on tumour aggressiveness. Anticancer Res 30:767–775

Löhr JM (2016) Clinical and laboratory diagnosis of chronic pancreatitis. Pancreapedia: the exocrine pancreas knowledge base. doi:https://doi.org/10.3998/panc.2016.14

Lopes LR, Elliott PM (2013) Genetics of heart failure. Biochim Biophys Acta 832(12):2451–2461

Memarzadeh S, Kozak KR, Chang L et al (2002) Urokinase plasminogen activator receptor: prognostic biomarker for endometrial cancer. Proc Natl Acad Sci U S A 99:10647–10652

Mohamed MM, Sloane BF (2006) Cysteine cathepsins: multifunctional enzymes in cancer. Nat Rev Cancer 6:764–775

Munnich A, Rötig A, Rio M (2012) Defects of the respiratory chain. Inborn metabolic diseases diagnosis and treatment. Springer, Berlin, pp 223–238

Mythili S, Malathi N (2015) Diagnostic markers of acute myocardial infarction. Biomed Rep 3 (6):743–748

Newsome PN, Cramb R, Davison SM, Dillon JF, Foulerton M, Godfrey EM, Hall R, Harrower U, Hudson M, Langford A, Mackie A (2018) Guidelines on the management of abnormal liver blood tests. Gut 67(1):6–19

Oh HC, Kwon CI, El Hajj II, Easler JJ, Watkins J, Fogel EL, McHenry L, Sherman S, Zimmerman MK, Lehman GA (2017) The diagnostic value of pepsin detection in saliva for gastro-esophageal reflux disease: a preliminary study from China. BMC Gastroenterol 17(1):107

Oxvig C, Sand O, Kristensen T, Kristensen L, Sottrup-Jensen L (1994) Isolation and characterization of circulating complex between human pregnancy-associated plasma protein-A and proform of eosinophil major basic protein. Biochim Biophys Acta 1201:415–423

Palermo C, Johanna AJ (2008) Cysteine cathepsin proteases as pharmacologicaltargets in cancer. Trends Pharmacol Sci 29(1):22–28

Paliouras M, Borgono C, Diamandis EP (2007) Human tissue kallikreins: the cancer biomarker family. Cancer Lett 249:61–79

Parrett M, Harris R, Joarder F, Ross M, Clausen K, Robertson F (1997) Cyclooxygenase-2 gene expression in human breast cancer. Int J Oncol 10:503–507

Pitkanen S, Robinson BH (1996) Mitochondrial complex I deficiency leads to increased production of superoxide radicals and induction of superoxide dismutase. J Clin Invest 98:345–351

Pratt DS, Kaplan MM (2000) Evaluation of abnormal liver-enzyme results in asymptomatic patients. New Eng J Med 342(17):1266–1271

Ramachandran J, Sajith KG, Priya S, Dutta AK, Balasubramanian KA (2015) Serum cholinesterase is an excellent biomarker of liver cirrhosis. Trop Gastroenterol 35(1):15–20

Rao SR, Snaith AE, Marino D et al (2017) Tumour-derived alkaline phosphatase regulates tumour growth, epithelial plasticity and disease-free survival in metastatic prostate cancer. Br J Cancer 116:227–236

Romero-Alvira D, Roche E, Placer L (1996) Cardiomyopathies and oxidative stress. Med Hypotheses 47:137–144

Rosalki SB, Tarlow D, Rau D (1971) Plasma gamma-glutamyltranspeptidase elevation in patients receiving enzyme-inducing drugs. Lancet 298(7720):376–377

Rucci N, Teti A (2010) Osteomimicry: how tumor cells try to deceive the bone. Front Biosci (Schol Ed) 2:907–915

Saif MW, Alexander D, Wicox CM (2005) Serum alkaline phosphatase level as a prognostic tool in colorectal cancer: a study of 105 patients. J Appl Res 5:88–95

Selenius M, Rundlöf AK, Olm E, Fernandes AP, Björnstedt M (2010) Selenium and the selenoprotein thioredoxin reductase in the prevention, treatment and diagnostics of cancer. Antioxid Redox Signal 12:867–880

Shimizu K, Yukawa T, Okita R, Saisho S, Maeda A, Nojima Y, Nakata M (2015) Cyclooxygenase-2 expression is a prognostic biomarker for non-small cell lung cancer patients treated with adjuvant platinum-based chemotherapy. World J Surg Oncol 13:21

Shiomi H, Eguchi Y, Tani T, Kodama M, Hattori T (2000) Cellular distribution and clinical value of urokinase-type plasminogen activator, its receptor, and plasminogen activator inhibitor-2 in esophageal squamous cell carcinoma. Am J Pathol 156:567–575

Sifrim D (2015) The role of salivary pepsin in the diagnosis of reflux. Gastroenterol Hepatol 11 (6):417

Singal PK, Kirshenbaum LA (1990) A relative deficit in antioxidant reserve may contribute in cardiac failure. Can J Cardiol 6:47–49

Singh N, Dhalla AK, Seneviratne C, Singal PK (1995) Oxidative stress and heart failure. Mol Cell Biochem 147:77–81

Sobolewski C, Cerella C, Dicato M, Ghibelli L, Diederich M (2010) The role of cyclooxygenase-2 in cell proliferation and cell death in human malignancies. Int J Cell Biol 2010:215158

Stefanini M (1985) Enzymes, isozymes, and enzyme variants in the diagnosis of cancer. A short review. Cancer 55:1931–1936

Thapa BR, Walia A (2007) Liver function tests and their interpretation. Indian J Pediatr 74 (7):663–671

Tomlins SA, Mehra R, Rhodes DR et al (2007) Integrative molecular concept modeling of prostate cancer progression. Nat Genet 39:41–51
Tracy C, Warren JS, Szulik M, Wang L, Garcia J, Makaju A, Russell K, Miller M, Franklin S (2018) The Smyd family of methyltransferases: role in cardiac and skeletal muscle physiology and pathology. Curr Opin Physiol 1:140–152
Turk B (2006) Targeting proteases: successes, failures and future prospects. Nat Rev Drug Discov 5:785–799
Turk D, Guncar G (2003) Lysosomal cysteine proteases (cathepsins): promising drugtargets. Acta Crystallogr D Biol Crystallogr 59:203–213
Ullah MF, Aatif M (2009) The footprints of cancer development: cancer biomarkers. Cancer Treat Rev 3:193–200
Wang D, Dubois RN (2010) The role of COX-2 in intestinal inflammation and colorectal cancer. Oncogene 29:781–788
Wang J, Zhang S, Jin Y, Qin G, Yu L, Zhang J (2007) Elevated levels of platelet-monocyte aggregates and related circulating biomarkers in patients with acute coronary syndrome. Int J Cardiol 115:361–365
Wang Y, Jiang X, Dong S et al (2016) Serum TK1 is a more reliable marker than CEA and AFP for cancer screening in a study of 56,286 people. Cancer Biomark 16:529–536
Warren JS, Tracy CM, Miller MR et al (2018) Histone methyltransferase Smyd1 regulates mitochondrial energetics in the heart. Proc Natl Acad Sci U S A 115(33):E7871–E7880
Whitfield JB (2001) Gamma glutamyltransferase. Crit Rev Clin Lab Sci 38(4):263–355
World Health Organization (WHO) working group, corp-author. The establishment of ischaemic heart disease registers (report of the fifth working group) WHO Reg Publ Eur Ser, Copenhagen. 1972;821(suppl 5)
Wortmann SB, Duran M, Anikster Y, Barth PG, Sperl W, Zschocke J et al (2013) Inborn errors of metabolism with 3-methylglutaconic aciduria as discriminative feature: proper classification and nomenclature. J Inherit Metab Dis 36:923–928
Wu AH, Wang XM, Gornet TG, OrdonezLlanos J (1992) Creatine kinase MB isoforms in patients with skeletal muscle injury: Ramifications for early detection of acute myocardial infarction. Clin Chem 38:2396–2400
Yang B, Jia L, Guo Q, Ren H, Hu Y, Xie T (2016) Clinicopathological and prognostic significance of cyclooxygenase-2 expression in head and neck cancer: a meta-analysis. Oncotarget 7:47265–47277
Yousef GM, Stephan C, Scorilas A et al (2003) Differential expression of the human kallikrein gene 14 (KLK14) in normal and cancerous prostatic tissues. Prostate 56:287–292
Zarghi A, Arfaei S (2011) Selective COX-2 inhibitors: a review of their structure-activity relationships. Iran J Pharm Res 10(4):655–683
Zhu X, Wang X, Chen X (2017) The diagnostic value of serum microRNA-183 and TK1 as biomarkers for colorectal cancer diagnosis. Open Life Sci 12:243–247

Enzymes in Food and Feed Industries: Where Tradition Meets Innovation

Pedro Fernandes

1 Introduction: Scope and Historical Background

The use of enzymes in the production of food for humans and animals is well established (Fraatz et al. 2014). Within such scope, enzymes can fulfill different roles, typically identified as processing aids or additives (Husain 2010; Ravindran 2013; Fraatz et al. 2014; Chandrasekaran et al. 2015; Habte-Tsion et al. 2018).

(a) As processing aids: it involves the transformation of a raw material into the intended product, e.g. hydrolysis of starch into glucose towards glucose syrups, or the targeted chemical modification of a particular component in food preparations for technical purposes, e.g. hydrolysis of lactose in milk towards lactose-free milk. Enzymes used as processing aids have no function in the final product and are either inactivated or absent in the final food and have application in the production of foods in different sectors, such as baking, brewing, dairy, fish, meat and vegetables processing, fruit juice production, lipid modification or winemaking. Most food enzymes are used as processing aids, although a few have also been classified as additives, such as invertase and lysozyme

(b) As additives: enzymes are used to enhance the nutritional properties, to preserve/improve taste and appearance of the food or to increase its shelf-life. Enzymes as additives are vastly used to improve the nutritional features of feed, namely to enhance the availability of essential nutrients, complementing the range of enzyme activities already displayed by the endogenous enzymes, which can be present in feed and in the digestive tract of animals. Feed enzymes have been

P. Fernandes (✉)
Department of Bioengineering and IBB, Institute for Bioengineering and Biosciences, Instituto Superior Técnico, Universidade de Lisboa, Lisbon, Portugal

Faculty of Engineering and Dreams, Universidade Lusófona de Humanidades e Tecnologias, Lisbon, Portugal
e-mail: pedro.fernandes@tecnico.ulisboa.pt

Q. Husain, M. F. Ullah (eds.), *Biocatalysis*,
https://doi.org/10.1007/978-3-030-25023-2_12

considered the most researched topic among feed additives (Ravindran 2013; Fraatz et al. 2014; Fallahi et al. 2018).

Regulatory agencies from different countries have approved the use of enzymes for the envisaged application, although the designation of the role played is not yet uniform (Nunes et al. 2018).

Currently supported by an increasing scientific knowledge and with their commercial use under regulation, the action of enzymes in food and feed is a field where the technical application came clearly ahead of mastering the scientific background (Linko 1989; Fraatz et al. 2014, Schulz 2015; Fernandes 2018; Nunes et al. 2018). Thus, and although unaware of their (bio) chemical nature, mankind has relied on enzymes for food processing/production from around 9000 BC henceforth, a pattern that started with: the use of yeast for bread making, a crude application of rennet for the separation of curd from whey for cheese making by storing fresh milk in the intestine of goats and the production of alcoholic beverages through the fermentation of sugar-rich solutions (Snodgrass 2013; Rasmussen 2015). However, only in the eighteenth century experimental research dedicated to enzyme activity surfaced, with the work of René-Antoine de Réaumuron, the action of the digestive tract of falcons on meat (Cornish-Bowden et al. 2005; Bedford and O'Neill, 2018). In the nineteenth century Payen and Persoz isolated an active extract, diastase that hydrolyzed starch to dextrin and sugar, paving the way for the US patent granted to Jokichi Takamine, which involved the production of takadiastase from *Aspergillus oryzae*, an enzyme source still vastly used (Buchholz and Poulsen 2000; Bedford and O'Neill 2018). Between these two events, other enzymes also currently use in food production/processing were isolated, namely pepsin, emulsin (β-glucosidase), lipase, invertase and trypsin (Cornish-Bowden et al. 2005), Kuhne coined the term enzyme (Fraatz et al. 2014; Bedford and O'Neill 2018) and Christian Hansen started a company featuring a standard enzyme preparation of rennet, a mixture of chymosin (rennin) and pepsin, for cheese making (Buchholz and Poulsen 2000). Still, it took until 1926 for the chemical nature of enzymes to be unveiled (Cornish-Bowden et al. 2005). Shortly after, by the 1930s, enzymes started to be used in the commercial production of fruit juices (van den Berg et al. 2010; Bedford and O'Neill 2018) but it took a little longer, until the 1960s, for the use of microbial enzymes in the food industry to clearly emerge, as amylases and glucoamylase were introduced in the starch industry to almost fully hydrolyze starch to glucose, thus replacing acid hydrolysis (Husain 2017a). This milestone, later complemented with the introduction of xylose (glucose) isomerase to isomerize glucose into fructose, turned the starch industry one of the larger users of enzymes, second only to detergent manufacture (Hobbs 2009; Bedford and O'Neill, 2018; Sheldon and Woodley 2018). Enzymes are currently involved in virtually all sectors of the food industry and novel applications and improvements are frequently presented. Alongside with the production of foodstuffs, enzymes also modify their functional properties, structure and texture, increase the shelf life, enhance flavors, odors and mouth feel (Simpson et al. 2012; Sandri et al. 2013; Fernandes 2016; Patel et al. 2016; Ray et al. 2016; Takei et al. 2016; Husain 2017b; Raveendran et al. 2018; Singh and Kumar 2018).

Somehow curiously, since the early research efforts to probe the role of enzymes on food processing by Réaumur and Spallanzani relied on the study of the effect of the digestive tract of birds in meat digestion, only by the 1920s emerged the first study on the role offered enzymes in animal diet. This work reported improved bird performance as an outcome of the use of an enzyme extract with multiple activities, but primarily amylase, from *A. orizae* in poultry feeding (Clickner and Follwell 1926; Choct 2006; Bedford and O'Neill 2018). Albeit some research was made henceforth, mostly involving amylolytic enzyme preparations to barley-based broiler feeds, there were no major developments until the 1950 and particularly the 1960s. In the latter decade, the relevance of non-starch polysaccharides (NSP) hydrolytic enzymes, e.g. β-glucanase, as tool to improve the digestibility of feeds was much appreciated (Husain 2017c). By hydrolyzing NSP into smaller oligosaccharides or monosaccharides, these enzymes decrease the viscosity of the digest, which in the presence of soluble NSP is such that inhibits the uptake of nutrients and favors the growth of pathogens (Ravindran 2013; O'Neill et al. 2014). Pioneering efforts were made in the 1970s to evaluate the use of phytase, which stands as the only enzyme specifically developed for the feed industry, to improve phosphorus availability from plant sources (Ravindran 2013; Bedford and O'Neill 2018). Research also suggested that previous positive results associated with the use of a given enzyme as feed, e.g. amylase, could have been due to contaminating and unacknowledged activity, e.g. β-glucanase, xylanase. These developments prompted the first commercial application in the mid-1980s and also further research on the role of NSP, typically present in feed and that are not hydrolyzed by endogenous digestive enzymes in monogastric animals, and of NSP hydrolytic enzymes (NSPases) in animal nutrition (Choct 2006; O'Neill et al. 2014; Bedford and O'Neill 2018). It also became evident that the extent of hydrolysis of NSP, which ultimately leads to monosaccharides, should not be sole criterion for the selection of NSPases, as some products resulting from partial hydrolysis have a positive effect in animal performance, e.g. prebiotic role, and some monosaccharides, above given concentrations proved deleterious (Bindelle et al. 2011; O'Neill et al. 2014; Strube et al. 2015). Thus, the use of multiple enzymatic activities that can convey additive, sub-additive, or synergistic effects should be carefully assessed and evaluated considering multiple parameters, e.g. growth performance, digestibility. Still, and although overall acknowledged as positive, the real impact of NSPases has proved difficult to establish in full. This limitation has been ascribed to a partial identification only of the enzymatic activities present in most of the works published. This can be a drawback in establishing cause-effect relationships, since NSPases can be provided: as a single component, with a minor and unacknowledged, yet relevant, side activity; as a blend of enzymes with distinct activities, where the presence of unknown activities can be multiplied; or as enzyme cocktails, a crude formulation resulting of a fermentation where multiple activities are present but hardly ever fully quality controlled by the manufacturer (Ravindran 2013; O'Neill et al. 2014). Finally, the nutrients are present in the raw materials in complex forms, not as individual entities, a feature that also conditions the selection of a suitable selection of enzyme activities (Ravindran 2013; O'Neill et al. 2014; Wealleans et al. 2017; Yaghobfar and Kalantar 2017; Yuan et al. 2017).

Of the established six classes of enzymes, five have relevant roles in commercial food and feed applications, with hydrolases as the most dominant class, with widespread use of enzymes representative of different subclasses, such as: α- and β-amylases (catalyze the endo- or exo-hydrolysis of α-1,4-glycosidic bonds, respectively in starch); chymosin (a proteolytic enzyme that hydrolyzes κ-casein between Phe105 and Met106); glucoamylase (exo-acting enzyme that catalyze the hydrolysis of starch from the non-reducing end, by acting on α-1,4-glycosidic bonds, although with minor activity on α-1,6-glycosidic bonds, releasing β-glucose); lipase (promotes the hydrolysis of ester bonds by acting on triacylglycerols, although it may have a synthetic role); invertase (hydrolyzes sucrose); naringinase (hydrolyzes the bond between terminal L(+)-rhamnose and the aglycone of rhamnose-containing glycosides, hence allowing the hydrolysis of naringin, the main bitter flavanone glycoside in citrus fruits); and papain (a cysteine protease often used as meat tenderizer). On the other tip of the scale, lyases are only represented by acetolactate decarboxylase (Patel et al. 2016; Fernandes 2018). Still, this enzyme is used in brewery, where acetolactate decarboxylase speeds up beer fermentation/maturation by shunting the formation of diacetyl and its unwanted buttery flavor from α-acetolactate and producing acetoin directly, with no impact on the organoleptic features of the beer (Krogerus and Gibson 2013; Choi et al. 2015). Isomerases used to be solely represented by glucose (xylose) isomerase, albeit in a large-scale process in the starch industry, where glucose is partially isomerized to fructose by immobilized glucose isomerase for the annual production of about 10^6 tons high-fructose syrups, a process that requires an amount of enzyme exceeding 500 tons (DiCosimo et al. 2013; Sheldon and Woodley 2018). Other isomerases have however gained relevance in recent years, partly due to the growing concern of the general public on the negative impact of excessive use of conventional, caloric sweeteners on public health. This trend favored the production of low caloric sugars, such as D-allulose (70% as sweet as sucrose) or D-tagatose (92% as sweet as sucrose), that nevertheless have physical chemical properties similar to those of common sugars, e.g. fructose, glucose or sucrose, and can thus replace these (Chattopadhyay et al. 2014; Lamothe et al. 2019). D-Allulose, commonly known as D-psicose, is produced industrially by the reversible isomerization of D-fructose to D-allulose using either D-allulose 3-epimerase or D-tagatose 3-epimerase (Nandakumar and Wakayama 2015; Venkitasubramanian 2016; Yoshihara et al. 2017; Lamothe et al. 2019; Zhang et al. 2018). D-Tagatoseis preferably produced industrially by the reversible isomerization of galactose using L-arabinose isomerase (Hussein et al. 2016; Mei et al. 2016; Vera and Illanes 2016; Jayamuthunagai et al. 2017). The production of the slowly digestible disaccharide isomaltulose, 50% as sweet as sucrose, but with a low glycemic index, relatively small caloric no cariogenic nature and with a prebiotic role, is produced on industrial scale from sucrose, using sucrose isomerase, which rearranges the glycosidic bond between glucose and fructose in sucrose molecules into an α-1,6 bond (Hellmers et al. 2018; Lamothe et al. 2019). Transferases are mostly represented by: transglutaminases, enzymes that act as texturizing agents as they promote the acyl-transfer reaction between primary amines (viz. the ε-amino group of lysine) and the γ-carboxyamide group of glutamine in proteins, thus promoting intra- and intermolecular bonds, the latter leading to cross-linking between peptide chains (Kieliszek and Misiewicz 2014; Patel et al. 2016); cyclodextrin glycosyltransferases, multifunctional enzymes able to catalyze three transglycosylation reactions (disproportionation, cyclization, and coupling) as well as hydrolysis, and are used in the production of cyclodextrins from starch (van der Veen et al. 2000; Patel et al. 2016); and fructosyl transferases, that transfer a fructosyl group to a molecule of sucrose or a fructooligosaccharide, and are used in the production of prebiotics (Michel et al. 2016; Patel et al. 2016). Oxidoreductases are mostly represented by: glucose oxidase, a flavoprotein that in the presence of oxygen converts glucose to gluconolactone, which spontaneously hydrolyzes to gluconic acid, while releasing hydrogen peroxide, and that is typically used to improve shelf-life, among other roles (Jan et al. 2001; Patel et al. 2016; Raveendran et al. 2018); lipoxygenase, associated to the dioxygenation of polyunsaturated fatty acids in lipids that present a cis-1,4-pentadiene, and are used for aroma generation (Patel et al. 2016; Raveendran et al. 2018); and laccases, the major subgroup of multicopper enzymes that catalyze the oxidation of a vastgroup of compounds coupled to the four-electron reduction of molecular oxygen to water, and are used for color modification and flavor enhancement (Patel et al. 2016; Mate and Alcalde 2017; Raveendran et al. 2018). Ligases, the only class of enzymes missing a major role in commercial food and feed applications, has nevertheless an enzyme, L-amino acid ligase that has been successfully evaluated as catalyst for the synthesis of dipeptides L-leucyl-L-serine and L-glutaminyl- L-threonine, a precursor of L-glutamyl- L-threonine. Both L-leucyl-L-serine and L-glutamyl- L-threonine enhance saltiness, hence they may be incorporated in foodstuff while simultaneously reducing the amount of salt, with positive impact in health, as the risk of high blood pressure associated with excessive salt consumption is minimized (Arai et al. 2013).

2 Enzymes: Required Features, Sources, Strategies for Improvement

The features required for enzymes used in food and feed are quite diverse and can be relatively casuistic, depending on the specificity of the application. Enzymes displaying high activity and thermal stability in the mesophilic/thermophilic range are typically aimed at, due to the operational conditions required in several processes, due to thermodynamic constraints, such as those associated with the reactions in the relevant starch and sweetener applications (Synowiecki et al. 2014; Chandrasekaran et al. 2015; Guerrand 2018). On the other end of the temperature scale, enzymes featuring high activity at low temperature are also sought after for application in food and feed, such as dough processing in bakery, cheese making, dairy, meat tenderization, fruit juice production or seafood processing and wine making, where some steps, even if performed at room temperature, may impart unwanted by-products/product degradation (Fernandes 2016; Kuddus 2018). Provided they display similar relative activities/turnover and catalytic efficiency, the more heat stable will be preferred for glucose isomerization to fructose. Currently the largest industrial scale application of an immobilized enzyme, glucose isomerases is utilized on a commercial basis at 60 °C, and this only allows for 50% fructose. Relatively poor heat stability above that temperature prevents operation at higher temperatures, which would allow for equilibrium to further shift towards product formation, hence glucose isomerases (or their formulations) with high thermal stability are looked after (DiCosimo et al. 2013; Li et al. 2017; Pellis et al. 2018). On the other hand, considering enzymatic meat tenderization, where the hydrolysis of myofibrilar proteins and connective tissue (mostly collagen) is typically envisaged, the use of bromelain over papain may be preferred. The former is less heat stable than the latter, hence inactivation will be more effective upon cooking, thus preventing the risk of excessive tenderization and concomitant production of unpalatable soft texture (Cutiño-Avila et al. 2014; Dong et al. 2016; Takei et al. 2016; Bhat et al. 2018). Cold active lactases are of interest in dairy industry as they can enable lactose hydrolysis during shipping and storage of milk, therefore shortening the production process and minimize the risk of both microflora contamination with mesophiles and of the formation of browning products that typically occur when lactose hydrolysis is carried out at higher temperatures (Stougaard and Schmidt 2012; Alikkunju et al. 2016).

Plants, mammalian tissues, and terrestrial microorganisms are the traditional sources of enzymes (Simpson et al. 2012; Fallahi et al. 2018). Enzymes from the later are obtained upon fermentation of selected sources, whereas enzymes from the remaining are typically obtained by extraction (Simpson et al. 2012). The sources of the enzymes in the framework of food and feed application must be of non-pathogenic and non-toxigenic nature (Subin and Bhat 2015). Representative examples of enzymes isolated from these different sources and used for commercial applications are provided in Table 1.

Currently the major source of enzymes for use in the food and feed industries are terrestrial microorganisms. Microbial-based production is more cost effective than mammalian- or plant-based approaches, due to faster growth, lower complexity/cost of the broths and of the production process, which is more amenable to large-scale operation, higher productivity and improved prospect of genetic manipulation of the former. Enzymes from microbial sources provide a wide diversity of optimal parameters, e.g. pH and temperature, for catalytic activity/stability. For example, bacterial α-amylases can operate under temperatures and pH ranging from 80 to 105 °C and 6.0 to 7.0, respectively, while fungal α-amylases can operate under temperatures and pH ranging from 55 to 70 °C and 3.0 to 5.0. Accordingly, microbial enzymes enable to operate in a wide range of environments with different physical and chemical properties, such as those featured in food processing, which can be relatively extreme, e.g. temperatures close to 100 °C endured by α-amylase in starch liquefaction, where animal and plant enzymes are rapidly inactivated. Moreover, microbial enzyme production is not affected by seasonality (deSouza and de

Table 1 Illustrative examples on the use of exogenous enzymes from different sources in applications with commercial interest in food and feed (Simpson et al. 2012; Hellmuth and van den Brink 2013; Ravindran 2013; O'Neill et al. 2014; Suresh et al. 2015; Mishra et al. 2016; Mate and Alcalde 2017; Fallahi et al. 2018)

Source	Enzyme	Relevant applications
Plant	α- and β-amylases	Bakery, brewing
	β-Glucanases	Beer making by reducing haze and easing filtration, and in feed production
	Bromelain, ficin and papain	Meat tenderization and hydrolysis of proteins in milk, animals and plants. Production of low allergy infant food
	Chymosin (key component of rennet) and pepsin (minor component of rennet)	Coagulation of milk in cheese making (mostly Portugal and Spain)
	Lipoxygenase	Bleach flour and strengthen dough
	Pectinase	Fruit juice production and wine making
Mammalian	Catalase, lactoperoxidase and lysozyme	Food preservation due to antimicrobial features
	Chymotrypsin	Extraction of oils rich in ω-3 polyunsaturated fatty acids from fish
	Chymosin (key component of rennet) and pepsin (minor component of rennet)	Coagulation of milk in cheese making
	Collagenase and pepsin	Processing of roe
	Lipase	Flavor enhancement in cheese making
Microbial	α-Acetolactate decarboxylase	Brewing, decarboxylation of diacetyl into acetoin
	D-Allulose (psicose) 3-epimerase	Sweeteners industry, production of D-psicose from D-fructose
	α- and β-amylases	Bakery, brewing, juice production low calorie beers, starch industry: liquefaction of starch, production of maltose syrups. Feed production
	Amyloglucosidase	Starch industry: saccharification (production of glucose syrups)
	L-Arabinose isomerase	Sweeteners industry, production of D-tagatose from D-galactose
	L-Asparaginase	Baked and fried foods, e.g. French fries, cereals, bread, and roasted coffee: prevention of acrylamide formation during processing of starchy foods at high temperatures
	Catalase and glucose oxidase	Bakery (removal of glucose from egg-white), dough strengthening, preservation of seafood; mayonnaise production (catalase)

(continued)

Table 1 (continued)

Source	Enzyme	Relevant applications
	Cellulases and hemicellulases (namely xylanases)	Bakery (dough conditioning), brewing, fruit juice, wine making and feed production: degradation of cell wall, hydrolysis of non-starch polysaccharides and improved extraction of fermentable sugars, maceration of fruits and vegetables, reduction of viscosity and improved digestibility or enhanced filterability
	Chymosin (key component of rennet) and pepsin (minor component of rennet). Mixed (commercial) proteases: Subtilisin (Alcalase®), thermolysin like protease (Neutrase®), Flavourzyme® (a mixture of endopeptidases, aminopeptidases, and dipeptidyl peptidases), Protamex® (mixture of endo- and exopeptidases)	Coagulation of milk in cheese making (typically lower chymosin to pepsin ratio than animal rennet) Hydrolysis of proteins in milk, animals and plants. Production of bioactive peptides, assistance in the extraction of oils and lipids from fish, improvement of the traditional fermentation process in the production of fermented fish products, production of low allergy infant food, prevention of haze formation in beer
	Cyclodextrin glycosyl-transferases and fructosyl transferases	Production of cyclodextrins, useful as carriers and stabilizers for flavors, colors, and sweeteners and as water-solubilizers for fatty acids and some vitamins. Production of short chain fructooligosaccharides (prebiotics)
	α-Galactosidase	Feed production: increased digestibility of grains
	β-Glucanases	Beer making, feed production: reduced viscosity increased digestibility
	Glucose (xylose) isomerase	Starch industry, production of high fructose syrups: glucose isomerization
	Invertase	Confectionery, chocolates: sucrose hydrolysis
	Inulinases (endo- and exo-inulinases)	Sugar industry (production of fructose: exo-inulinase or endo- and exo-inulinase for total hydrolysis of inulin) and nutraceuticals (oligofructose or fructooligosaccharides: endo-inulinase for partial hydrolysis of inulin)
	Laccase	Oxidation of polyphenols, oxygen scavenging, gelling effects. Bakery (improved dough machinability), brewery fruit juice production and wine making (stabilization)

(continued)

Table 1 (continued)

Source	Enzyme	Relevant applications
	Lactase (β-galactosidase)	Dairy: lactose hydrolysis, whey hydrolysis
	Lipase	Bakery: improved properties of dough; chocolate: synthesis of cocoa butter substitutes; Dairy: cheese making (improved cheese ripening, customized flavor), manufacture of human milk fat substitutes; fish: improved titer of ω-3 polyunsaturated fatty acids, isolation of oils and fats from seafood products
	Naringinase	Fruit juice: removal of bitter-tasting compounds in citrus juices
	Pectinases	Brewing, fruit juices production and wine making
	Phytase	Feed production: release of phosphate from phytate and enhanced digestibility
	Pullulanase	Starch industry: starch debranching and improved saccharification
	Phospholipase	Cheese making: synthesis of a lysophospholipid (emulsifier) preventing loss of fat into the whey and increasing cheese yields
	Sucrose isomerase	Functional foods and prebiotics: production of isomaltulose from sucrose
	D-tagatose 3-epimerase	Sweeteners industry, production of D-psicose from D-fructose
	Tannase	Manufacture of coffee-flavored drinks and of instant tea; clarification of beer and fruit juices, improvement in the flavor of grape wine, production of animal feed; hydrolysis of the ester bond of tannins
	Thermolysin	Sweetener industry: production of aspartame, a high intensity sweetener
	Transglutaminase	Crosslinking of proteins and modification of viscoelastic properties. Dairy: production of yoghurt; meat and fish, production of restructured products at low temperature. The potential application to improve shelf-life of seafood by minimizing microbial growth and suppressing the formation of free amino acids, and decreasing the risk of biogenic amine production, was recently highlighted

Oliveira Magalhães 2010; Vengadaramana 2013; Subin and Bhat 2015; Fallahi et al. 2018; Poeta et al. 2018). Enzymes able to cope with the relatively extreme conditions required in some processes can be obtained from microorganisms (extremophiles) that thrive in extreme environments. Extreme environments include environments with temperatures within 55 °C to 121 °C (hot springs/thermal vents) or −2 °C to 20 °C (glaciers/deep sea), pH above 8 or below 4 (chemical industries and wastewater effluents), but also pressures well above atmospheric pressure (deep sea), high salinity such as NaCl or KCl (lakes) and low nutrient availability. Most identified extremophiles have been found in the domains Archaea and Bacteria (Sarmiento et al. 2015; Dumorné et al. 2017). Many of those extreme conditions are found in marine environments, hence marine microorganisms have been gaining relevance as sources for enzymes for application in food and feed (Chinnathambi and Muthusamy 2016). Classic examples of enzymes from extremophiles with use in food and feed include: glucose isomerases from thermophiles *Thermotoga napolitana* and *T. maritima* (Starnes et al. 1993) and α-amylase from thermophilic *Bacillus licheniformis* (Krishnan and Chandra 1983), both involved in the production of high fructose corn syrup (HFCS) from starch; moreover, several cold-active pectinases, obtained from *Aspergillus* spp. and with application in beer, fruit juice and wine production that are commercially available, namely Novoshape® from Novozymes, Pectinase 62 L from Biocatalysts and Lallzyme® from Lallem and Sarmiento et al. (2015). More recent examples of enzymes from extremophiles with potential application in food and feed are given in Table 2.

Currently, the large-scale production of food and feed enzymes strongly rely on recombinant microorganisms. This trend has been implemented mainly because the typical source of the enzyme intended was such that the natural conditions required for production were not compatible with large-scale application. Typical examples of such limitation are plant or animal cells, such as in the case of chymosin, papain or transglutaminase, slow-growing or even pathogenic microorganisms. The introduction of enzymes produced by extremophiles further added to the need of using recombinant, typically mesophilic microorganisms, as the original sources often proved difficult to grow under commonly used conditions. Additionally, genetically engineered microorganisms allows for improved productivity and processability (Subin and Bhat 2015; Mishra et al. 2016; Fallahi et al. 2018). Due to security concerns, only a small number of bacterial and fungal origin, mostly *A. niger*, *A. oryzae*, *B. licheniformis* and *B. subtilis* are generally recognized as safe (GRAS) based on FDA regulations, and can be used as hosts (Adrio and Demain 2014). Recombinant DNA technology enabled the production of enzymes tailored to the specific conditions for food and feed processing, viz. production of highly thermo-tolerant amylases with low requirement for metal ions. Effective gene cloning allowed the production of these enzymes and ultimately their application in multi-step processes, such as starch hydrolysis to HFCS with minimal pH adjustments and addition/removal of ion metals, which added to cost and complexity of the process. Alongside with recombinant DNA technology, protein engineering, either by rational design or by random mutagenesis has also been used to obtain enzymes with improved features, such as improved activity and/or stability, such as

Table 2 Some recent representative examples of extremozymes with potential use in food and feed

Enzyme	Source	Comments	References
Glucose isomerase	*Thermus oshimai*	The hyperthermophilic enzyme was expressed in *E. coli* and allowed a conversion yield of 52% for the isomerization of 400 g/L D-glucose to D-fructose at 85 °C and pH 8.0. Moreover, the activity at pH 6.5 is over 90% of the maximum. This suggest the possibility of operation in mild acidic environment, thus reducing the risk of by-product formation	Jia et al. (2017)
L-Arabinose isomerase	*Geobacillus kaustophilus*	The thermophilic enzyme was expressed in *E. coli*. The holo form of the enzyme displayed optimal activity at 50–60 °C and pH 7.5 although it was highly active at 70 °C and pH 7.0, The latter operational conditions can favor the production D-tagatose from D-galactose. Additionally, the work provided information for better understanding enzyme substrate interactions	Choi et al. (2016)
L-Asparaginase	*Pyrococcus yayanosii* CH1	The thermophilic enzyme was cloned and expressed in *B. subtilis*. The residual activity of the enzyme exceeded 80% for temperatures within 85 and 100 °C and pH 6.5–10, with activity peaking at 95 °C and pH 8.0. Moreover, it retained approximately 55–35% of the initial activity when	Li et al. (2018)

(continued)

Table 2 (continued)

Enzyme	Source	Comments	References
		incubated within temperatures of 70–95 °C for 2 h, displaying potential for application in processing of starch-based foods for acrylamide removal	
	Thermococcus zilligii AN1 TziAN1_1	The residual activity of this thermophilic enzyme exceeded 90% within 80–90 °C, peaking at the latter. Moreover, the residual activity was close to 80% at 95 °C. The residual activity of the enzyme exceeded 80% in pH within 6.5–10. The thermal stability of the enzyme was clearly established, since it displayed half-lives of 557, 533, and 102 min for temperatures of 80, 85 and 90 °C, respectively. When French fries were processed with the enzyme at 80 °C for 15 min, the concentration of acrylamide was reduced by 85%, thus highlighting the role played by the enzyme in the removal of carcinogenic acrylamide from starch-based cooked foods	Zuo et al. (2015)
Cyclodextrin glycosyltransferase	*Geobacillusthermoglucosidans* CHB1	The enzyme displayed over 80% of activity within 65–75 °C, with optimal pH 5.5. The activity was stimulated, up to 1.4-fold, in the presence of Ni^{2+} and Mg^{2+} (1–10 mM) and was highly stable up to 50 °C. The enzyme was active on different substrates,	Jia et al. (2018)

(continued)

Table 2 (continued)

Enzyme	Source	Comments	References
		namely soluble starch, potato starch and maltodextrin (3% solutions) to produce α-, γ- and β-cyclodextrins, although the latter was predominant and soluble starch was the preferred substrate (60.3% conversion yield in 21 h)	
	Thermoanaerobacter sp. 5 K	The enzyme retained over 90% of activity within 75–95 °C, with optimal pH 7.0. The enzyme displayed no activity decay under incubation at 80 °C for 4 h and displayed residual activity over 40% within pH 5.0–9.0. Based on β-CD formation from 2% soluble starch, the enzyme displayed a specific cyclization activity of 403 U/ $mg_{protein}$, where U refers to the amount of enzyme producing 1 mg β-CD in 1 min	Avci et al. (2014)
Phytase	*Geobacillus* sp. TF16	The enzyme retained over 70% of activity within 80–95 °C with optimal temperature at 85 °C and pH 4.0. The residual activity of the phytase activity exceeded 75% and 50% under incubation 85 °C for 24 h and for 48 h, respectively. The high thermal stability and thermophilic nature of this phytase makes it appealing for pretreatment of monogastric animal feeds at high temperatures	Dokuzparmak et al. (2017)

(continued)

Table 2 (continued)

Enzyme	Source	Comments	References
		(>80 °C), as to improve the availability of inositol, phosphoric acid and minerals. This pattern is further reinforced as residual activities of roughly 80 and 60% were observed upon phytase incubation in the presence of proteases for 0.5 and 1 h, Proteolysis endurance being a relevant asset for enzymes used as feed supplement	
Collagenase	*Pseudoalteromonas* sp. SM9913	The cold-active protease was shown to be highly selective for collagen degradation at 4 °C, unlike commercially used papain and bromelain. The enzyme was thus able to significantly reduce the beef meat shear force while maintaining the fresh color and moisture of the meat. Overall these features suggest a promising enzyme for commercial application	Zhao et al. (2012)
Lactase (β-galactosidase)	*Pseudoalteromonas haloplanktis*	The cold-active enzyme was used for lactose hydrolysis in whey. A conversion yield of 96% in 24 h at 23 °C out of an initial lactose concentration in whey of 122 g/L was highlighted as the most promising result. D-Galactose formed is foreseen as substrate for D-tagatose production	van de Voorde et al. (2014)
	Thalassospira sp. 3SC-21	The residual activity of the cold-active enzymes was above	Ghosh et al. (2012)

(continued)

Table 2 (continued)

Enzyme	Source	Comments	References
		90% within 15–25 °C and was of 45% at 10 °C, and was highly stable up to 25 °C, although activity steeply decreased henceforth to total deactivation at 40 °C, suggesting its feasibility to remove lactose from refrigerated milk and whey. Additionally, relative activity and stability was above 80% within pH of 5.5 to 7.5.	
Pectinase	*Cryptococcus saitoi* GM-4 and *Rhodotorula dairenensis* GM-15	The cold-adapted pectinases from both strains displayed pectinolytic activity at 12 °C and grape pH (3.5), namely 0.858 U/mL and 1.104 U/mL for *Cr. saitoi* and *R. dairenensis*, respectively, where U is the amount of enzyme releasing 1 μmol reducing sugars/min. Additionally, to simulate real operation, the effect of 15% (v/v) ethanol and 120 mg/L SO_2 on enzyme activity at 12 and 28 °C and either individually or combined was assessed. Overall, the results suggested that pectinase from *R. dairenensis* GM-15 could have application in the production of red wine at 26–28 °C (the typical temperatures for the process), although at 12 °C the enzyme still retained noticeable residual activity	Merín et al. (2015)

α-amilase, glucose isomerase or phytase (Patel et al. 2016; Fallahi et al. 2018; Fernandes 2018). Finally, immobilization of enzymes has been also used to improve enzyme performance. This more relevant application of this approach involves the immobilization of glucose isomerase in the production of HFCS, that ultimately allows productivities that exceed 104 $kg_{product}/kg_{biocatalyst}$ (DiCosimo et al. 2013; Sheldon and Woodley 2018). Additionally, immobilization is also used in some processes for the production of lactose free milk. One of these, which processes 8000 L/day with a conversion yield within 70–81%, relies on the immobilization of β-galactosidase in cellulose triacetate (Thum et al. 2014).

3 Conclusions

Enzymes have been for long a key asset in food and feed applications, where it has evolved from largely empirical to carefully designed and rationally implemented methodologies. Additionally, due to their highly selective and environmentally friendly nature, enzymes have replaced chemical catalysis in several processes. As the knowledge on the mechanisms of catalysis and stability increase and a wider range of enzymes and enzyme activities are identified, either as the result of screening or of genetic manipulation/protein engineering, not only traditional processes and products are improving but also new products and processes have been introduced. These developments result in better nutrition for both humans and animals. Further breakthroughs are foreseen as more powerful computational methods coupled to experimental techniques such as X-ray crystallography allow to gain more insight on structure-activity and structure-stability relationships and developments in metagenomics, combined with traditional screening methods anchored in high throughput approaches. These provide access to a wider array of genetic manipulation and rapid, highly parallelized evaluation of catalytic activities. It is also expected that more detailed knowledge on the factors that can improve the nutritional value in animal feed and in human diets, if performed in an integrated manner, as these two fields ultimately overlap. Overall, these matters configure the introduction of enzymes with increased performance and range of applications, which can be designed to better tailor process requirements and to produce goods needed to comply with the demand for healthier nutrition.

References

Adrio JL, Demain AL (2014) Microbial enzymes: tools for biotechnological processes. Biomol Ther 4(1):117–139

Alikkunju AP, Sainjan N, Silvester R, Joseph A, Rahiman M, Antony AC, Kumaran RC, Hatha M (2016) Screening and characterization of cold-active β-galactosidase producing psychrotrophic *Enterobacter ludwigii* from the sediments of arctic fjord. Appl Biochem Biotechnol 180 (3):477–490

Arai T, Arimura Y, Ishikura S, Kino K (2013) L-amino acid ligase from *Pseudomonas syringae* producing tabtoxin can be used for enzymatic synthesis of various functional peptides. Appl Environ Microbiol 79(16):5023–5029

Avci A, Nichols NN, Saha BC, Frazer SE, Cotta MA, Dönmez S (2014) A thermostable cyclodextrin glycosyltransferase from *Thermoanaerobacter* sp. 5K. Curr Biotechnol 3:305–312

Bedford MR, O'Neill HVM (2018) Introduction. In: Nunes CS, Kumar V (eds) Enzymes in human and animal nutrition. Academic, London, pp xxvii–xxvxl

Bhat ZF, Morton JD, Mason SL, Bekhit A, El-D A (2018) Applied and emerging methods for meat tenderization: a comparative perspective. Compr Rev Food Sci Food Saf 17:841–859

Bindelle J, Pieper R, Montoya CA, Van Kessel AG, Leterme P (2011) Nonstarch polysaccharide-degrading enzymes alter the microbial community and the fermentation patterns of barley cultivars and wheat products in an in vitro model of the porcine gastrointestinal tract. FEMS Microbiol Ecol 76(3):553–563

Buchholz KB, Poulsen PB (2000) Introduction. In: Straathof AJ, Adlercreutz P (eds) Applied biocatalysis, 2nd edn. Taylor & Francis, Amsterdam, pp 1–15

Chandrasekaran M, Basheer SM, Chellappan S, Krishna JG, Beena PS (2015) Enzymes in food and beverage production: an overview. In: Chandrasekaran M (ed) Enzymes in food and beverage processing. CRC, Boca Raton, FL, pp 117–137

Chattopadhyay S, Raychaudhuri U, Chakraborty R (2014) Artificial sweeteners – a review. J Food Sci Technol 51(4):611–621

Chinnathambi A, Muthusamy C (2016) Marine microbial enzymes in food application. In: Ray RC, Rosell CM (eds) Microbial enzyme technology in food applications. CRC, Boca Raton, FL, pp 396–419

Choct M (2006) Enzymes for the feed industry: past, present and future. Worlds Poult Sci J 62:5–16

Choi EJ, Ahn HW, Kim WJ (2015) Effect of α-acetolactate decarboxylase on diacetyl content of beer. Food Sci Biotechnol 24(4):1373–1380

Choi JM, Lee YJ, Cao TP, Shin SM, Park MK, Lee HS, di Luccio E, Kim SB, Lee SJ, Lee SJ, Lee SH, Lee DW (2016) Structure of the thermophilic L-Arabinose isomerase from *Geobacilluskaustophilus* reveals metal-mediated intersubunit interactions for activity and thermostability. Arch Biochem Biophys 596:51–62

Clickner FH, Follwell EH (1926) Application of protozyme (*Aspergillus orizae*) to poultry feeding. Poult Sci 5:241–247

Cornish-Bowden A, Jamin M, Saks V (2005) Cinétiqueenzymatique. EDP Sciences, Les Ulis

Cutiño-Avila B, Gil D, Aragón C, Fernández Y, Hernández M, Salas E, Chávez MA, Guisán JM, Díaz J, del Monte-Martínez A (2014) Computer-aided design of bromelain and papain covalent immobilization. Rev Colomb Biotecnol 16(1):19–28

de Souza PM, de Oliveira Magalhães P (2010) Application of microbial α-amylase in industry – a review. Braz J Microbiol 41(4):850–861

DiCosimo R, McAuliffe J, Poulose AJ, Bohlmann G (2013) Industrial use of immobilized enzymes. Chem Soc Rev 42(15):6437–6474

Dokuzparmak E, Sirin Y, Cakmak U, Ertunga NS (2017) Purification and characterization of a novel thermostable phytase from thethermophilic *Geobacillus* sp. TF16. Int J Food Prop 20 (5):1104–1116

Dong K, Samarasinghe YAH, Hua W, Kocherry L, Xu J (2016) Recombinant enzymes in the meat industry and the regulations of recombinant enzymes in food processing. In: Ray RC, Rosell CM (eds) Microbial enzyme technology in food applications. CRC, Boca Raton, FL, pp 363–374

Dumorné K, Córdova DC, Astorga-Eló M, Renganathan P (2017) Extremozymes: a potential source for industrial applications. J Microbiol Biotechnol 27(4):649–659

Fallahi P, Habte-Tsion H-M, Rossi W (2018) Depolymerizating enzymes in human food: bakery, dairy products, and drinks. In: Nunes CS, Kumar V (eds) Enzymes in human and animal nutrition. Academic, London, pp 211–237

Fernandes P (2016) Enzymes in fish and seafood processing. Front Bioeng Biotechnol 4:59

Fernandes P (2018) Enzymatic processing in the food industry. In Reference module in food science. Elsevier. Available via Science Direct http://wwwsciencedirectcom/science/article/pii/B978008100596522341X Assessed 31 Aug 2018

Fraatz MA, Rühl M, Zorn H (2014) Food and feed enzymes. Adv Biochem Eng Biotechnol 143:229–256

Ghosh M, Pulicherla KK, Rekha VP, Raja PK, Sambasiva Rao KR (2012) Cold active β-galactosidase from *Thalassospira* sp. 3SC-21 to use in milk lactose hydrolysis: a novel source from deep waters of Bay-of-Bengal. World J Microbiol Biotechnol 28(9):2859–2869

Guerrand D (2018) Economics of food and feed enzymes: status and prospectives. In: Nunes CS, Kumar V (eds) Enzymes in human and animal nutrition. Academic, London, pp 487–514

Habte-Tsion H-B, Kumar V, Rossi W (2018) Perspectives of nonstarch polysaccharide enzymes in nutrition. In: Nunes CS, Kumar V (eds) Enzymes in human and animal nutrition. Academic, London, pp 239–254

Hellmers F, Takors R, Thum O (2018) Robust enzyme immobilizates for industrial isomalt production. Mol Catal 445:293–298

Hellmuth K, van den Brink JM (2013) Microbial production of enzymes used in food applications. In: McNeil B, Archer D, Giavasis I, Harvey L (eds) microbial production of food ingredients, enzymes and nutraceuticals. Woodhead Publishing, Cambridge, pp 262–287

Hobbs L (2009) Sweeteners from starch: production, properties and uses. In: BeMiller JN, Whistler R (eds) Starch: chemistry and technology, 3rd edn. Academic, New York, pp 797–832

Husain Q (2010) β Galactosidases and their potential applications: a review. Crit Rev Biotechnol 30 (1):41–62

Husain Q (2017a) Nanomaterials as novel support for the immobilization of amylolytic enzymes and their applications: a review. DeGruyter Biocatal 3(1):37–53

Husain Q (2017b) Nanosupport bound lipases their stability and applications. Biointerface Res Appl Chem 7(6):2194–2216

Husain Q (2017c) Nanomaterials immobilized cellulolytic enzymes and their industrial applications: a literature review. JSM Biochem Mol Biol 4(3):1029

Hussein AH, Lisowska BK, Leak DJ (2016) The genus *Geobacillus* and their biotechnological potential. Adv Appl Microbiol 92:1–48

Jan J, Husain Q, Saleemuddin M (2001) Preparation of stable, highly active and immobilized glucose oxidase using the anti-enzyme antibodies and $F(ab)_2$. Biotechnol Appl Biochem 34:13–17

Jayamuthunagai J, Gautam P, Srisowmeya G, Chakravarthy M (2017) Biocatalytic production of D-tagatose: a potential rare sugar with versatile applications. Crit Rev Food Sci Nutr 57 (16):3430–3437

Jia DX, Zhou L, Zheng YG (2017) Properties of a novel thermostable glucose isomerase mined from *Thermus oshimai* and its application to preparation of high fructose corn syrup. Enzym Microb Technol 99:1–8

Jia X, Ye X, Chen J, Lin X, Vasseur L, You M (2018) Purification and biochemical characterization of a cyclodextrin glycosyltransferase from *Geobacillusthermoglucosidans* CHB1. Starch – Stärke 70:1700016

Kieliszek M, Misiewicz A (2014) Microbial transglutaminase and its application in the food industry. A review. Folia Microbiol (Praha) 59(3):241–250

Krishnan T, Chandra AK (1983) Purification and characterization of α-amylase from *Bacillus licheniformis* CUMC305. Appl Environ Microbiol 46(2):430–437

Krogerus K, Gibson BR (2013) 125th anniversary review: diacetyl and its control during brewery fermentation. J Inst Brew 119(3):86–97

Kuddus M (2018) Cold-active enzymes in food biotechnology: an updated mini review. J Appl Biol Biotechnol 6(3):58–63

Lamothe LM, Lê KA, Samra RA, Roger O, Green H, Macé K (2019) The scientific basis for healthful carbohydrate profile. Crit Rev Food Sci Nutr 59(7):1058–1070

Li H, Yang S, Saravanamurugan S, Riisager A (2017) Glucose isomerization by enzymes and chemo-catalysts: status and current advances. ACS Catal 7:3010–3029
Li X, Zhang X, Xu S, Zhang H, Xu M, Yang T, Wang L, Qian H, Zhang H, Fang H, Osire T, Rao Z, Yang S (2018) Simultaneous cell disruption and semi-quantitative activity assays for high-throughput screening of thermostable L-asparaginases. Sci Rep 8(1):7915
Linko M (1989) Enzymes in the forefront of food and feed industries. Food Biotechnol 3(1):1–9
Mate DM, Alcalde M (2017) Laccase: a multi-purpose biocatalyst at the forefront of biotechnology. Microb Biotechnol 10(6):1457–1467
Mei W, Wang L, Zang Y, Zheng Z, Ouyang J (2016) Characterization of an L-arabinose isomerase from *Bacillus coagulans* NL01 and its application for D-tagatose production. BMC Biotechnol 16(1):55
Merín MG, Martín MC, Rantsiou K, Cocolin L, de Ambrosini VI (2015) Characterization of pectinase activity for enology from yeasts occurring in Argentine Bonarda grape. Braz J Microbiol 46(3):815–823
Michel MR, Rodríguez-Jasso RM, Aguilar CN, Gonzalez-Herrera SM, Flores-Gallegos AC, Rodríguez-Herrera R (2016) Fructosyltransferase sources, production, and applications for prebiotics production. In: Rao V, Rao L (eds) Probiotics and prebiotics in human nutrition and health. IntechOpen, London, pp 169–189
Mishra SS, Ray RC, Rosell CM, Panda D (2016) Microbial enzymes in food applications: history of progress. In: Ray RC, Rosell CM (eds) Microbial enzyme technology in food applications. CRC, Boca Raton, FL, pp 3–18
Nandakumar R, Wakayama M (2015) Enzymes in flavors and food additives. In: Chandrasekaran M (ed) Enzymes in food and beverage processing. CRC, Boca Raton, FL, pp 305–324
Nunes CS, Kunamneni A, Kumar V, Habte-Tsion H-M (2018) Registration of food and feed additives (enzymes) in the United States, Canada, and China. In: Nunes CS, Kumar V (eds) Enzymes in human and animal nutrition. Academic, London, pp 457–480
O'Neill HVM, Smith JA, Bedford MR (2014) Multicarbohydrase enzymes for nonruminants. Asian-Australas J Anim Sci 27(2):290–301
Patel A, Singhania R, Pandey A (2016) Novel enzymatic processes applied to the food industry. Curr Opin Food Sci 7:64–72
Pellis A, Cantone S, Ebert C, Gardossi L (2018) Evolving biocatalysis to meet bioeconomy challenges and opportunities. N Biotechnol 40(Pt A):154–169
Poeta P, Dias AA, Igrejas G, Silva V, Bezerra R, Nunes CS (2018) Selection, engineering, and expression of microbial enzymes. In: Nunes CS, Kumar V (eds) Enzymes in human and animal nutrition. Academic, London, pp 1–29
Rasmussen SC (2015) Introduction – the role of chemical technology in early civilizations. In: Rasmussen RC (ed) ACS symposium series 1211 – chemical technology in antiquity. American Chemical Society, Washington, pp 1–15
Raveendran S, Parameswaran B, Ummalyma SB, Abraham A, Mathew AK, Madhavan A, Rebello S, Pandey A (2018) Applications of microbial enzymes in food industry. Food Technol Biotechnol 56(1):16–30
Ravindran V (2013) Feed enzymes: the science, practice, and metabolic realities. J Appl Poult Res 22:628–636
Ray L, Pramanik S, Bera D (2016) Enzymes-an existing and promising tool of food processing industry. Recent Pat Biotechnol 10(1):58–71
Sandri IG, Piemolini-Barreto LT, Fontana RC (2013) Enzymes in food processing. In: Panesar PS, Marwaha SS (eds) Biotechnology in agriculture and food processing: opportunities and challenges, 1st edn. CRC, Boca Raton, FL, pp 329–353
Sarmiento F, Peralta R, Blamey JM (2015) Cold and hot extremozymes: industrial relevance and current trends. Front Bioeng Biotechnol 3:148
Schulz S (2015) New EU labeling law: omission of food additives and enzymes from the list of ingredients under regulation (EC) no. 1169/2011. Eur Food Feed Law Rev (EFFL) 1:14–19

Sheldon RA, Woodley JM (2018) Role of biocatalysis in sustainable chemistry. Chem Rev 118 (2):801–838
Simpson BK, Rui X, Klomklao S (2012) Enzymes in food processing. In: Simpson BK (ed) Food biochemistry and food processing. Wiley-Blackwell, Oxford, pp 181–206
Singh P, Kumar J (2018) Microbial enzymes in food biotechnology. In: Kuddus M (ed) Enzymes in food biotechnology – production, applications, and future prospects, 1st edn. Academic, New York, pp 19–28
Snodgrass ME (2013) World Food: an encyclopedia of history, culture, and social influence from hunter-gatherers to the age of globalization. Sharpe, New York
Starnes RL, Kelly RM, Brown SH (1993). Method for glucose isomerization using xylose isomerase purified from *Thermotoga maritima* and *Thermotoga neapolitana*. Patent US5268280
Stougaard P, Schmidt M (2012) Cold-active beta-galactosidase, a method of producing same and use of such enzyme. Patent US8288143B2
Strube ML, Jensen TK, Meyer AS, Boye M (2015) In situ prebiotics: enzymatic release of galacto-rhamnogalacturonan from potato pulp in vivo in the gastrointestinal tract of the weaning piglet. AMB Express 5(1):66
Subin SR, Bhat SG (2015) Enzymes: concepts, nomenclature, mechanism of action and kinetics, characteristics and sources of food-grade enzymes. In: Chandrasekaran M (ed) Enzymes in food and beverage processing. CRC, Boca Raton, FL, pp 3–38
Suresh PV, Nidheesh T, Pal GK (2015) Enzymes in seafood processing. In: Chandrasekaran M (ed) Enzymes in food and beverage processing. CRC, Boca Raton, FL, pp 354–377
Synowiecki J, Panek A, Pietrow O (2014) industrial starch processing. In: Grunwald P (ed) Industrial biocatalysis. Pan Stanford, New York, pp 985–1006
Takei R, Hayashi M, Umene S, Kobayashi Y, Masunaga H (2016) Texture and microstructure of enzyme-treated chicken breast meat for people with difficulties in mastication. J Texture Stud 47 (3):231–238
Thum O, Hellmers F, Ansorge-Schumacher M (2014) Robust enzyme preparations for industrial application. In: Gruneald P (ed) Industrial biocatalysis. Pan Stanford Publishing Pte, CRC, Boca Raton, FL, pp 553–582
van de Voorde I, Goiris K, Syryn E, Van den Bussche C, Aerts G (2014) Evaluation of the cold-active *Pseudoalteromonas haloplanktis* β-galactosidase enzyme for lactose hydrolysis in whey permeate as primary step of D-tagatose production. Process Biochem 49:2134–2140
van den Berg MA, Roubos JA, Parenicová L (2010) Enzymes in fruit and vegetable processing – future trends in enzyme discovery, design, production, and application. In: Bayindirli A (ed) Enzymes in fruit and vegetable processing: chemistry and engineering applications. CRC, Boca Raton, FL, pp 342–358
van der Veen BA, Uitdehaag JC, Penninga D, van Alebeek GJ, Smith LM, Dijkstra BW, Dijkhuizen L (2000) Rational design of cyclodextrin glycosyltransferase from Bacillus circulans strain 251 to increase alpha-cyclodextrin production. J Mol Biol 296:1027–1038
Vengadaramana A (2013) Industrial important microbial alpha-amylase on starch-converting process. Sch Acad J Pharm 2(3):209–221
Venkitasubramanian P (2016) Fructose to allulose conversion. Patent WO2016160573A1
Vera C, Illanes A (2016) Lactose-derived nondigestible oligosaccharides and other high added-value products. In: Illanes A, Guerrero C, Vera C, Wilson L, Conejeros R, Scott F (eds) Lactose-derived prebiotics – a process prospective, 1st edn. Elsevier, San Diego, pp 87–110
Wealleans AL, Walsh MC, Romero LF, Ravindran V (2017) Comparative effects of two multi-enzyme combinations and a *Bacillus* probiotic on growth performance, digestibility of energy and nutrients, disappearance of non-starch polysaccharides, and gut microflora in broiler chickens. Poult Sci 96(12):4287–4297
Yaghobfar A, Kalantar M (2017) Effect of non-starch polysaccharide (NSP) of wheat and barley supplemented with exogenous enzyme blend on growth performance, gut microbial, pancreatic enzyme activities, expression of glucose transporter (SGLT1) and mucin producer (MUC2) genes of broiler chickens. Braz J Poultry Sci 19(4):629–638

Yoshihara A, Kozakai T, Shintani T, Matsutani R, Ohtani K, Iida T, Akimitsu K, Izumori K, Gullapalli PK (2017) Purification and characterization of D-allulose 3-epimerase derived from *Arthrobacter globiformis* M30, a GRAS microorganism. J Biosci Bioeng 123(2):170–176

Yuan L, Wang M, Zhang X, Wang Z (2017) Effects of protease and non-starch polysaccharide enzyme on performance, digestive function, activity and gene expression of endogenous enzyme of broilers. PLoS One 12(3):e0173941

Zhang W, Zhang Y, Huang J, Chen Z, Zhang T, Guang C, Mu W (2018) Thermostability improvement of the D-allulose 3-epimerase from *Dorea* sp. CAG317 by site-directed mutagenesis at the interface regions. J Agric Food Chem 66(22):5593–5601

Zhao GY, Zhou MY, Zhao HL, Chen XL, Xie BB, Zhang XY, He HL, Zhou BC, Zhang YZ (2012) Tenderization effect of cold-adapted collagenolytic protease MCP-01 on beef meat at low temperature and its mechanism. Food Chem 134(4):1738–1744

Zuo S, Zhang T, Jiang B, Mu W (2015) Reduction of acrylamide level through blanching with treatment by an extremely thermostable L-asparaginase during French fries processing. Extremophiles 19(4):841–851

Microbial Enzymes in Food Processing

Praveen Kumar Mehta and Shelly Sehgal

1 Introduction

Enzymes are biological macromolecules (typically proteins), produced by living organisms that accelerate the rate of chemical reactions required for life without permanent alteration to themselves. Enzymes are highly specific, incredibly efficient and operate under mild conditions of temperature and pH. Enzymes are substantial for the sustainability of life in all life forms. They are very efficient catalysts for biochemical reactions and can be obtained from various sources such as animals, plants, and microorganisms.

However, the enzymes derived from the microbial origin are more active and stable than plant and animal enzymes. Moreover, they are considered as potentially interesting candidates for industrial uses because of the ease of culturing substantial quantities in a short span of time by fermentation on incredibly diverse, easy to produce, a wide range of inexpensive and readily available carbon and nitrogen sources.

Microbes represent an important source of the biocatalysts to bring about specific biochemical reactions. Microorganisms are of great importance and proven to be a suitable and leading source of enzymes (Demain and Adrio 2008). History of microbial enzymes reveal that since ancient times, these enzymes have been exploited by humans to production of food products such as cheese making, brewing of beer, bread baking, winemaking and vinegar production etc. (Kirk et al. 2002).

More recently, with the advent of metagenomics and low-cost high-throughput sequencing technologies, it has become possible to characterize the variety of microbial species from nature. In addition, recombinant DNA techniques, have made it possible to develop highly efficient expression systems that can lead to further

P. K. Mehta (✉) · S. Sehgal
Centre for Molecular Biology, Central University of Jammu, Jammu, India
e-mail: mehtapk@cujammu.ac.in

Q. Husain, M. F. Ullah (eds.), *Biocatalysis*,
https://doi.org/10.1007/978-3-030-25023-2_13

enhancement of catalytic properties of enzymes. Microbial enzymes in food applications have not only expanded the food industry but also produced economic assets.

Enzymes tremendously contribute to industrial processes by lowering the energy consumption and maximizing efficiency and yield contributing to its sustainability profile. Microbial sources, enzymes are perceived as natural, nontoxic food components and are preferred by consumers over chemical counterparts used in food-processing.

2 History of Enzyme Technology

Enzymes have been exploited by humans knowingly or unknowingly since prehistoric times. Food processing using biological sources is historically a well-established approach. The ability of the yeast to produce alcoholic beverages from barley was known to the Babylonians and Sumerians before 6000 BC.

The use of chymosin and calve's stomach enzymes for producing cheese has been elucidated in Homer's Greek epic poems as early as 800 BC. Though for centuries enzymes produced by yeast have been used by mankind to produce wine from fermented grape juice, they were technically termed as 'enzymes' only in the eighteenth century.

In the nineteenth century later in 1862, Louis Pasteur studied and discovered the role of yeast in the fermentation of sugar to alcohol. His observation led to the conclusion that this fermentation was catalyzed by '*ferments*', which was functioning within the yeast cells.

The history of modern enzyme technology or industrial production of enzymes for use in food processing started in 1874, when Danish pharmacist, Christian D. A. Hansen developed the first conventional sample of rennet from calve's stomach (Nielsen et al. 1994). Furthermore, the enzyme rennet produced from an animal is a well-set standard even today and is still used in cheese manufacturing. However, at present high-value recombinant chymosin is produced using a microbial source that contains the bovine prochymosin gene introduced through recombinant DNA technology. Apparently this was the first enzyme preparation of relatively high purity which was approved for use in food by the US Food and Drug Administration (FDA) using Bovine chymosin expressed in *Escherichia coli* K-12 (Flamm 1991). In 1878, German physiologist Wilhelm Friedrich Kuhne from University of Heidelberg (1837–1900) introduced the term enzyme from Greek term 'ενζυμον' which literally mean 'in leaven' or 'in yeast'. During the twentieth century, enzyme was recognised as protein, but the emergence of advanced scientific techniques, witnessed the development in industrial production of enzymes. The enzyme urease is noteworthy in the history of enzymology as it was purified and crystallized by James B. Sumner of Cornell University In 1926. Use of enzymes in fruit juice clarification started in 1930, followed by production of invert sugar syrup in the starch industry in the 1960s.

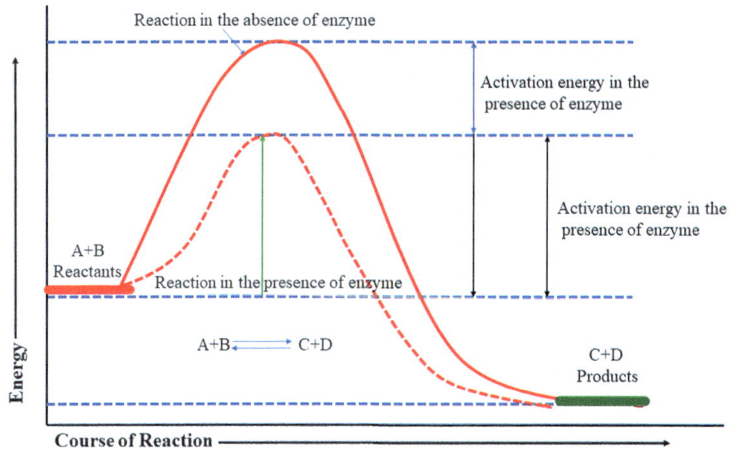

Fig. 1 A schematic diagram showing effect of biocatalyst on free energy changes in a chemical reaction

3 Chemical Nature of Enzymes or Enzyme Classification

Enzymes are natural globular proteins that speed up chemical reactions, and this process is called catalysis. In enzymatic reactions, the molecules upon which enzymes may act are referred to as substrates. Initially, the substrate (S) interacts with the enzyme by binding to the active site of the enzyme (E + S) to form an enzyme-substrate (ES) complex; then the enzyme converts the substrates into different molecules called products.

Enzymes demonstrate extreme specificity for their substrates and therefore catalyze only one or a few types of reactions from among many possibilities. Enzymes facilitate chemical reactions by lowering the activation energy for a reaction (Fig. 1).

Enzymes are remarkably efficient and highly specific biological catalysts produced by a cell that are responsible for carrying out various metabolic processes essential to sustain life (Chapman-Smith and Cronan 1999). Generally, majority of enzymes behave as proteins, although a few important enzymes are catalytic RNA molecules. The latter are known as ribozymes composed of RNA molecules. Enzyme's specificity usually arises from the three-dimensional structure of the enzyme active site. Similar to all other proteins, enzymes are made up of one or more long chain of amino acids linked together by peptide bonds. Each enzyme has a unique sequence of amino acids that causes it to fold into a characteristic shape. Many enzymes require several cofactors (or coenzymes) that is necessary for the enzyme to function properly (Nelson and Cox 2004). These can be metal ions (such as Mn^{2+}, Mg^{2+}, Fe^{2+}, Cu^{2+}) or organic molecules (such as FAD, NAD, haem, biotin, or coenzyme A). Most of the enzymes are involved in the conversion of substrate molecule into a product molecule. Enzymes have a high degree of specificity for their substrates, and they accelerate chemical reactions tremendously. Enzymes are classified on the basis of type of chemical reactions they catalyze and the substrate they act upon. According to the enzyme commission (EC) the enzymes are grouped into six main categories. Examples of enzyme types and some of their applications in the food industry are summarized in Table 1.

Table 1 Industrial microbial enzymes, enzyme classes, types of reactions and their role in food processing industries

S. no.	Class	Reactions	Industrial enzymes	Role
1	Oxidoreductases	Oxidation reactions involve the transfer of electrons or hydrogen or oxygen atoms from one molecule to another.	Catalases, Dehydrogenases, Glucose oxidases, Laccases, Lipoxygenases oxygenases, peroxidases	Cheese processing, Dough strengthening, Clarification of juices, flavor enhancer
2	Transferases	Catalyze the transfer of groups of atoms among molecules.	Fructosyltransferases, Glucosyltransferases, Transglutaminase transketolases, acyltransferases, transaminases	Synthesis of fructose oligomers, Laminated dough strength, dough processing, meat processing
3	Hydrolases	Hydrolytic cleavage of bonds	Amylases, Proteases, Peptidases, Pectinases, Phytases, Pullulanase, amylases, Galactosidases, Glucanases, Glucoamylases, Invertases, Lactases, Cellulase, Xylanases, acylases, lipases	Starch liquefaction and saccharification, Lactose hydrolysis, Mash treatment, dough conditioning, cheese ripening, extraction and clarification of fruit juices
4	Lyases	Non-hydrolytic cleavage by addition or elimination reactions	Pectate lyases, acetolactate decarboxylases, hydratases, dehydratases, fumarase, arginosuccinase	Cell wall degradation, fruit softening, Beer maturation
5	Isomerases	Transfer of group from one position to another within the same molecule	Glucose isomerases	High-fructose corn syrup production
6	Ligases	Join molecules together with covalent bonds followed by input of energy in the form of ATP or similar triphosphates	Not used at present	

4 Microbial Enzymes

Microbial enzymes are those enzymes which are derived from microbial sources and are produced through fermentation of the desired microorganisms. Enzymes, particularly of microbial origin, are the worthiest source of commercial enzymes and acquiring much attention with rapid advances in enzyme technology. Microbial enzymes can be obtained from a wide variety of sources such as algae, fungi, yeasts and bacteria, and their properties differ markedly from their source. The microbial source is preferred over plants and animals for the production of enzymes mainly because of the following points:

1. They are comparatively more stable than corresponding enzymes derived from plant or animals.
2. Large-scale production of enzymes can be achieved in short duration of time due to shorter generation times of microbes as compared to other sources.
3. Enzymes from microbial source can be obtained at desired and regular intervals irrespective of seasonal fluctuations.
4. Enzyme contents obtained from the microbial source are more predictable and controllable.
5. Wide range of inexpensive media can be arranged and utilized to acquire microbial enzyme.
6. Ease of process modification and optimization.
7. Microbial sources have better scope for genetic manipulations for achieving a high yield of the end product as compared to rest.
8. Extraction process and purification of microbial enzymes is quite easier and cost effective.
9. Several enzymes can be efficiently produced in spatial and temporal constraints using diverse environmental conditions.
10. Process development is often more environmentally friendly.

Innumerable strains of bacteria, yeasts and molds are commonly used as sources of enzymes for food processing.

5 Strategies to Improve Properties of Microbial Enzymes

The growing demand for the microbial enzyme in food processing industries has led industrialists and researchers to search for new and improved biocatalysts. The majority of currently used microbial enzymes in the food processing industry are derived from recombinant microorganisms. In the 1900s, classic genetics and protein biochemistry were flourishing. However, it got a major boost with the advances in molecular biology and protein engineering which have been leveraged to achieve substantial stabilization and customization of enzymes. There are several different ways by which enzymes can be modified to achieve an improvized end product. Two

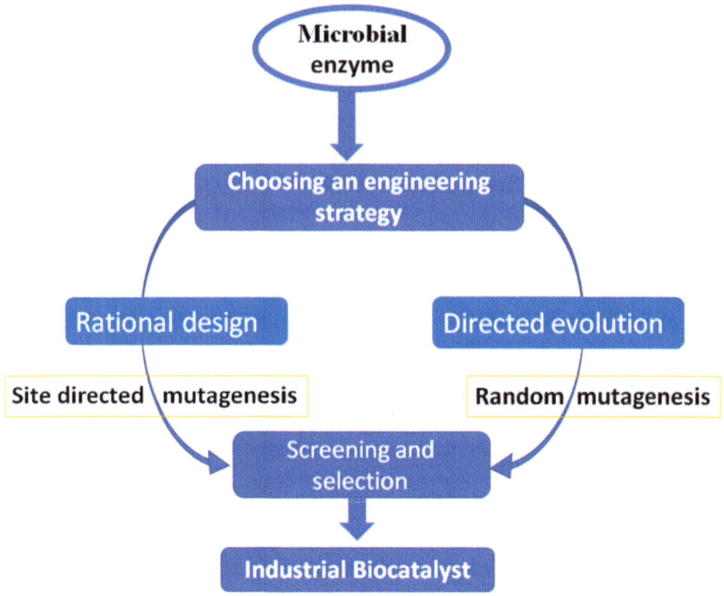

Fig. 2 A schematic diagram illustrating the protein engineering process

strategies which are generally applied in protein engineering are (1) rational protein design and (2) directed evolution. Rational design by site-directed mutagenesis (SDM) is an efficient approach to improve enzyme when the structure, function and catalytic mechanism of the protein of interest are well-known (Fig. 2). In contrast, the directed evolution approach requires only minimal knowledge of its structure and function. It requires only repeated cycles of random mutagenesis and/or gene recombination followed by high throughput screening or selection for positive mutants (Tang and Zhao 2009).

Various recombinant expression systems can be used to produce large scale production of proteins such as *E. coli*, *Bacillus subtilis*, *S. cerevisiae*, *Hansenula polymorpha*, *Pichia pastoris* and species of *Aspergillus* and *Trichoderma*. There are some examples where successful implementation of site-directed mutagenesis method has elevated the overall thermostability of α-amylase after removal of asparagine residues (Declerck et al. 2000). Manipulation of the calcium-binding site of α-amylase such as His-77 to Glu resulted in a four-fold increase in enzyme's half-life (Ghollasi et al. 2013). Likewise, substitution of arginine with proline has resulted in increase in thermostability of the enzyme (Igarashi et al. 1999), or doubly intramolecularly cross-linked lysozyme showed an increase in thermal stability (Ueda et al. 2000). Several enzymes from bacteria and fungi have been engineered to improve their properties to suit their particular applications. Several commercial enzymes for food technology are now derived from particularly selected or genetically modified microorganisms grown in industrial scale fermenters. Table 2 depicts

Table 2 Some examples of microbial technologies and strategies undertaken to improve the performance of microbial enzymes in food processing industries

Enzyme	Role	Targeted improvement	Strategies applied	Reference
α-Amylase	Starch liquefaction	Specific activity increased by 20 times	Error-prone PCR achieved random mutagenesis Directed evolution	Wong et al. (2004)
		Specific activity and pH-activity profile	Random mutagenesis	Bessler et al. (2003)
		Increased thermostability	Directed evolution	Wang et al. (2006)
	Baking	pH-activity profile	Protein engineering through site-directed mutagenesis	Danielsen and Lundqvist (2008)
α-Glucosidase	Hydrolysis of glycosidic linkages	Thermostability	Random mutagenesis using the standard error-prone PCR	Zhou et al. (2015)
β-Galactosidase	Lactose hydrolysis	Increase in the yields of Galacto-oligosaccharides (GOS)	Site-directed mutagenesis	Wu et al. (2013)
L-arabinose isomerase	Tagatose production	pH-activity profile	Protein engineering through directed evolution	Oh et al. (2006)
Glucoamylase	Starch saccharification	Substrate specificity, thermostability and pH optimum	Protein engineering through site-directed mutagenesis	Allen et al. (2003)
	Starch liquefaction	Thermostable	Directed evolution and site-directed mutagenesis	McDaniel et al. (2008)
Lactase	Lactose hydrolysis	Thermostability	Immobilization	Dwevedi and Kayastha (2009)
Pullulanase	Starch debranching	Activity	Protein engineering through directed evolution	England et al. (2013)
		Thermostability and Catalytic Efficiency	Site-directed mutagenesis based on a structure-guided consensus approach	Duan et al. (2013)
Phytase	Animal feed	pH-activity profile	Protein engineering through site-directed mutagenesis	Tomschy et al. (2002)
		Thermostability	Directed evolution was applied to create thermostable phytases	Kim and Lei (2008)

(continued)

Table 2 (continued)

Enzyme	Role	Targeted improvement	Strategies applied	Reference
Xylose isomerase	Isomerization/ epimerization of tetroses, pentoses and hexoses	pH-activity profile	Protein engineering through directed evolution.	Cha and Batt (1998)
		Substrate specificity; three-fold increase in catalytic efficiency	Site-directed mutagenesis	Karimäki et al. (2004)
	Glucose isomerisation	Thermostability; five-fold higher activity at 60 °C	Protein engineering using directed evolution.	Rubin-Pitel and Zhao (2006), Sriprapundh et al. (2003)

Source: Adapted from (Fernandes 2010)

the representative examples of how genetically modified enzyme could be used in food processing.

These classic examples demonstrate how advances in biotechnology, enzyme engineering, and biocatalyst design have emerged as powerful tools for process improvement and product recovery in the food and beverage industries, apart from their other industrial applications.

6 Sources of Microbial Enzymes

Many enzymes from microbial sources such as bacteria, yeasts, and fungi are globally studied and currently being used in various commercial food processing industries. About 80% of all commercial enzymes are derived from fungi and bacteria, and these enzymes are used in food and pharmaceutical applications (Table 3). Furthermore, a limited number of genera *i.e.*, *Aspergillus oryzae*, *Aspergillus niger*, *Penicillium*, *Saccharomyces*, *Lactobacilli* and *Bacillus subtilis* are well established, safe, beneficial and best-known microbial sources for enzymes.

7 Global Demand for Microbial Enzymes

Demand for new enzymes is ever increasing owing to the rising awareness of health and nutritional benefits of food enzymes among the consumers all around the world. This has led to the rising acceptance of food enzymes in various food processing

Table 3 Microbial enzyme sources used in food processing industries

S. No	Microbial source	Enzymes
Bacterial enzymes		
	Bacillus amyloliquefaciens	α-Amylase
	Bacillus licheniformis	α-Amylase, pullulanase
	Bacillus subtilis	α-Acetolactate, decarboxylase, α amylase maltogenic amylase, pullulanase, glucanase, maltase
	Bacillus acidopullulyticus	α-Amylase
	Bacillus lentus	Mannanase
	Escherichia coli	Chymosin
Fungal enzymes		
	Aspergillus niger	Phytase, chymosin, lipase, protease, catalase
	Aspergillus oryzae	Esterase, lipase, aspartic proteinase, glucose oxidase, laccase, lipase
	Fusarium venenatum	Xylanase
	Kluyveromyces marxianus var. lactis	Chymosin
	Pseudomonas fluorescens	α-Amylase
	Penicillium notatum	Glucose oxidase
	Trichoderma reesei	Pectin lyase
	Trichoderma longibrachiatum	Protease, xylanase
Yeast		
	Saccharomyces cerevisiae	Invertase
	Saccharomyces fragilis	Lactase
	candida	Lipase

industries. Food enzymes are essential food additives which transform complex molecules into simpler ones in our diet.

The global food and beverage enzymes market is categorized into carbohydrase, lipase, protease, and others. Carbohydrases, which acquire a major share in industries is further categorized into cellulase, amylase, lactase, invertases, inulinases, galactosidases, glucosidases, fructosyltransferases, pectinases, and glucosyltransferases (Husain 2016). Carbohydrases, which have diverse applications in the food and beverage industry, dominate the world food enzyme market and acquire a major share in industries, accounting for almost 70% of this market followed by proteases and lipases (Husain 2017a, b, c).

Food processing industry is currently the leading consumer of industrial enzymes. Furthermore, microbial enzymes are presently among the well-established products in biotechnology and account for 90% of the worldwide market. According to a recent report, the worldwide food and beverage enzymes market is valued at $1200 million in 2011, $1355.8 million in 2012, and is expected to reach $2306.4 million in

2018 (World Enzymes. Freedonia Group: Cleveland, OH, USA, 2011; Food Enzymes Market: Global Trends and Forecasts to 2018). The demand for food enzymes is expected to reach over $2940 million by 2021, at a CAGR of 7.4% from 2016 to 2021 (https://www.marketsandmarkets.com/Market-Reports/food-enzymes-market-800.html). Furthermore, food and beverage applications will remain the largest industrial enzymes market in the near future 2021. There is a significantly large number of companies which are involved in enzyme manufacturing, but major producers are located mainly in North America, Europe, Latin America and the Asia Pacific. Microbial enzymes account for 90% of the global market, and their producers are located mainly in Europe and Asia. In terms of region, North America dominated the global food enzymes market, due to the developed food and beverage industry in the region. It is followed by Europe and Asia-Pacific. The top three dominant global enzyme producers are Novozymes (Denmark), DuPont (USA) and DSM (The Netherlands), which account for more than 70% of the global enzymes business.

Novozymes is the sole company that has focused much on microbes and enzymes for food processing industries. Today, it holds a dominant share of 48% of the enzyme market and turned out to be the world's largest enzyme company (https://report2016.novozymes.com/our-business/trends).

The other leading companies in food processing industries are Associated British Foods plc (U.K.), Chr. Hansen (Denmark), Connell Bros. (U.S.), Dyadic International, Inc. (U.S.), Enmex (India), Advanced Enzymes (India), Puratos Group (Belgium), Kerry Group (Ireland), Amano Enzyme Inc. (Japan), Advanced Enzymes Technologies (India), Amano Enzyme (Japan), Biocatalysts (U.K.), Biocon (India), Fonterra Co-Op Group Ltd. (U.S.), Fytozimus Biotech (India), Genencor International Inc. (India) and Sternenzym (China).

8 Microbial Enzymes in Food Processing

Food processing industry is one of the leading and most important sectors in the world in terms of production, growth, consumption, and trade. Microbial enzymes are used widely in the food processing industry to improve nutritive value, antioxidant capacity, ripening fruit, developing flavour, appearance, texture strengthening, tenderization and flavor development and nutritive value of processed foods. Every year nearly 30–40% of the food (including fruits and vegetables) is wasted due to various reason in countries like USA, South Africa and India (Sridevi and Ramanujam 2012; Oelofse and Nahman 2013; Gunders 2012). Processing of food using microbial enzyme can assure that our food is safe, prevent spoilage and waste and increase its nutritional value. Consequently, microbial enzymes have found extensive applications in the production and processing of various types of food products. Almost 75% of all industrial enzymes that are widely used in food processing industries are hydrolytic enzymes (e.g., proteases, carbohydrases, lipases, nucleases). Carbohydrases, proteases and lipases dominate the enzyme market, accounting for more than

70% of all enzyme sales. An overview of applications of microbial enzymes in food processing industry are summarised in Table 4.

Enzyme action produces several effects in foods such as clarification, coagulation, color and flavor generation, decolorisation, improved yields, liquefaction, saccharification, tenderization, texture strengthening, maceration and viscosity reduction.

8.1 Microbial Enzymes in Starch Processing

Starch is the most common and main polysaccharide reserve produced by all green plants. It is most abundant biomolecule on earth after cellulose. Also, it is the main component of many agricultural products and forms more than 65% of the dry mass in corn (maize), cereal (wheat, rice) and potato. Starch is primarily composed of two major components *i.e.*, amylose (typically 20–30%) and amylopectin (70–80%) both of which contain α-D-glucopyranosyl units. In amylose 300–3000 glucose units are linked by β-1,4-glycosidic bonds, whereas in amylopectin about one residue in every twenty carbohydrate units are joined by β-1,6-linkages forming branch points (Qureshi and Blaschek 2005; Husain 2017a). It is usually formed by 2000–200,000 glucose units.

Starch is a highly functional natural polymer which possesses many unique properties and is used chemically or enzymatically to process a wide range of derivatives such as starch hydrolysates, glucose syrups, maltodextrin, cyclodextrins, fructose, glucose, and isoglucose.

The enzymatic conversion of the starch comprises of three necessary steps: liquefaction, saccharification, and isomerization. In starch processing industry α-amylase (1,4-α-D-glucan-glucanohydrolase, EC. 3.2.1.1) is one of the main enzymes used in starch liquefaction process during starch hydrolysis, where starch is converted into fructose and glucose syrups.

The second step in the process issaccharification where glucoamylase (Glucan 1,4-α-glucoamylase, EC. 3.2.1.1) hydrolysis starch entirely to glucose along with, maltose and isomaltose. A pullulanase (pullulan 6-glucanohydrolase, EC. 3.2.1.41) is an important de-branching enzyme, which is widely used to hydrolyze the α-1,6 glucosidic linkages in starch and aid in saccharification (Hii et al. 2012). Isomerization is the last step in the process where glucose isomerase, (D-xylose ketol-isomerase EC. 5.3.1.5) catalyzes the reversible isomerization of D-glucose to D-fructose. Amylolytic enzymes have been widely produced by fermentation using a great variety of microbial strains, mainly from bacteria (*Bacillus licheniformis*, *Bacillus flavothermus*, *Bacillus stearothermophilus*, *Bacillus subtilis*, and *Bacillus amyloliquefaciens*) and fungi (*Aspergillus niger*, *Aspergillus phoenicis*, *Aspergillus awamori*, *Penicillium chrysogenum*, *Penicillium notatum*) (Prakash and Jaiswal 2010; Singh et al. 2016b; Husain 2016).

Currently, large quantities of microbial amylases from bacterial and fungal sources are commercially available for industrial applications due to their economic feasibility, consistency, rapid growth, less space requirement for production, ease of control of process optimization and modification (Ellaiah et al. 2002).

Table 4 Application of microbial enzymes in different food processing Industry

Industry	Enzyme	Function/benefits	Microorganisms
Dairy	Acid proteinase	Milk coagulation	*Aspergillus* sp., *Rhizopus oryzae*, *Pleoticusmuelleri*, *Penicillium citrinum*
	Aminopeptidase	Faster cheese ripening, debittering	*Lactobacillus* sp. *Pseudozymahubeiensis*, *Bacillus licheniformis*
	Catalase	Cheese processing	*Aspergillus niger*, *Lactococcus species*
	Lactase (β-galactosidase)	Lactose-reduced milk and whey products	*Escherichia coli*, *Kluveromyces*sp.
	Lipase	Faster cheese ripening, flavorcustomized cheese	*Aspergillus niger*, *A. oryzae*, *Propionibacterium freudenreichii*
	Neutral proteinase	Faster cheese ripening, de-bittering	*Bacillus subtilis*, *A. oryzae*
	Protease	degrade protein and enables aging of cheese	*Aspergillus niger*
	Transglutaminase	Protein cross-linking	*Streptomyces* sp.
Baking	Amylase	Flour adjustment, bread softness	*Aspergillus* sp., *Bacillus* sp.
	Glucose oxidase	Dough strengthening	*Aspergillus niger*, *Penicillium chrysogenum*
	Lipase	Dough stability and conditioning	*Aspergillus niger*
	Maltogenic α-amylase	Enhance the shelf life of bread	*Bacillus stearothermophilus*
	Oxidoreductases	Giving increased gluten strength	*Bacillus subtilis*, *Zymomonasmobilis*
	Pentosanase	Loaf volume increment of bread	*Tricodermareesei Humicolainsolens*
	Proteases	Dough conditioner, enhance dough extensibility	*Aspergillus niger*
	Transglutaminase	Laminated dough strength	*Streptoverticillium* sp., *streptomyces* sp.
	Xylanase	Dough conditioning	*Aspergillus niger*
Beverages	Aminopeptidases	Protein breakdown during mashing	*Lactobacillus brevis*, *L. plantarum*
	Amyloglucosidases	Increasing glucose content	*Aspergillus niger*
	Cellulase	Fruit liquefaction	*Aspergillus niger*, *Trichoderma atroviride*
	Glucose oxidase	Oxygen removal from beer	*Aspergillus niger*
	Limoninase	Debittering	*Aspergillus niger*, *A. oryzae*
	Naringinase	Debittering	*Aspergillus niger*
	Pectinase	Depectinisation	*Aspergillus oryzae*, *Penicillumfuniculosum*

(continued)

Table 4 (continued)

Industry	Enzyme	Function/benefits	Microorganisms
	Protease	Restrict haze formation	*Aspergillus niger*
	Pullulanase	Starch saccharification	*Bacillus* sp., *Klebsiella* sp.
	Tannase	Tea processing	*Aspergillus niger*
	α-Amylase	Starch hydrolysis	*Bacillus*, *Aspergillus*
	β-Amylase	Starch hydrolysis	*Bacillus*, *Streptomyces*, *Rhizopus*
	β-Glucanase	Restrict haze formation, improving wort separation	*Bacillus subtilis*, *Aspergillus* spp.
Juice industry	Amylases	Breaking down starch into glucose	*Aspergillus* spp. *Bacillus* spp. *Microbacterium*
	Cellulases	Pectin hydrolysis	*Aspergillus niger Trichoderma* spp.
	Glucoamylases	Clarifying cloudy juice	*Imperiale*
	Hemicellulose	Lowering viscosity and maintenance of texture	*Aspergillus* spp. *Bacillus subtilis*
	Laccase	Increasing the susceptibility of browning during storage	*Pseudomonas putida*, *Bacillus* sp., *Auerobasidium pullulans*
	Pectinases	Degrading pectins and increasing overall juice production	*Aspergillus* spp., *Penicillium funiculosum*
Starch processing	α-Amylases	Starch hydrolysis	*Aspergillus* spp. *Bacillus* spp. *Microbacterium*
	Amylopullulanases	Starch hydrolysis	*Bacillus* sp., *Staphylothermusmarinus*
	Glucoamylases	Starch hydrolysis	*Aspergillus niger*, *Rhizopus species*
	Glucose isomerases	Catalysing isomerisation of glucose to fructose	*Streptomyces flavogriseus*
	Glycosyltransferases	Increasing the number of branched points to obtain modified starch with improved functional properties such as higher solubility, lower viscosity, and reduced retrogradation	*Streptococcus thermophilus*
	Isoamylases	Starch hydrolysis	*Pseudomonas amyloderamosa*, *Lipmyceskononenkoae*
	Neopullulanases	Starch hydrolysis	*Aspergillus* spp., *Bacillus stearothermophilus*
	Pullulanases	Improvement of saccharification of starch	*Bacillus* spp. *Klebsiella* spp
	β-amylases	Starch hydrolysis	*Bacillus*, *Streptomyces*, *Rhizopus*

(continued)

Table 4 (continued)

Industry	Enzyme	Function/benefits	Microorganisms
Meat and fish processing	Acid proteases	Improve the flavouring, nutritional and functional properties of proteins	*Aspergillus* sp.
	Elastase	Tenderize meat	*Aspergillus fumigatus*, *Pseudomonas aeruginosa*
	Glucose oxidase	Preservation of seafood	*Aspergillus niger*, *Penicillium amagasakiense*
	Glutaminase	Enhance flavor of the meat	*Streptomyces* sp., *Pseudomonas fluorescens*
	Lipase	Hydrolyze triglycerides; Improves flavor in sausages.	*Aspergillus niger*
	Proteases	Removal of scales and skin from fish, production of fish sauce	*Aspergillus niger*, *Bacillus licheniformis*
	Transglutaminase	Improves the structural properties of the processed or cooked meat, meat mince formation	*Streptoverticillium* sp., *streptomyces* sp.
	Tyrosinase	Cross-link meat protein enhances the functional properties of enzymes	*Bacillus megaterium*, *Rhizopus oryzae*, *Trichoderma reesei*
	Urease	Removal of off-odor and fishy taste	*Streptococcus thermophilus*, *Aspergillus niger*

Source: Adapted from (Singh et al. 2016a)

After completion of the process in the starch-processing industries, the syrups and starch derivatives, which comprise of diverse compositions and physical properties, can be used in a wide variety of food and beverage industries. On the basis of type, the market is sectioned into glucose syrup, maltodextrin, cyclodextrin, hydrolysates, and spray-dried starch. Among the types of starch derivatives, glucose syrup acquires the largest market. However, the maltrodextrin sector is growing at the highest compound annual growth rate CAGR of 7.0%.

The global starch derivatives market is one of the burgeoning industries and is projected to grow at a rate of 6.2% from 2014 to 2019 to reach $58.2 billion by 2019. (http://www.marketsandmarkets.com/Market-Reports/starch-derivatives-market-116279237.html).

Globally, the United States dominates the starch industry with 51% of world production. Starch market at industrial level is expected to reach USD 106.73 Billion by 2023, witnessing growth at a CAGR of over 6%, in value terms owing to the rising popularity of ready-to-use convenience foods among the working population who do not have time to cook their meals.

8.2 Enzymes in the Bakery Industry

Microbial enzymes can effectively amend gluten network and are hence useful for modulating flour and dough rheology in addition to flavor enhancement of the baked products. Enzymes lipoxygenases carry out the oxidation of fatty acids and are utilized for bleaching of fat-soluble flour pigments and provide elasticity in the dough (Miguel et al. 2013). They have been obtained from fungus *Ascomycota incertaesedis* such as *Gaeumannomyces*, *Pyricularia*, *Geotrichumcandidum*. The fungal lipoxygenase may be from *Gaeummanomyces graminis*, *Fusarium oxysporum* or *F. proliferatum*, or *Penicillium* sp. (Borch et al. 2006). Microbial proteases (*Bacillus* sp.), α-amylases (*Aspergillus* sp.), xylanases (*Bacillus*, *Aeromonas*, *Cephalosporoim*, *Aspergillus niger*) and lipases (*Acinetobacter*, *Aermonas*, *Bacillus*, *Candida*, *Penicillium*, *Pseudomonas* sp.) are widely used to enhance the homogeneity, softness, colour, longer freshness and crunch to the bakery products (Singh et al. 2016c; Ahmed et al. 2014; Goswami and Rawat 2015; Andualema and Gessesse 2012). The enzymes glucose oxidase and asparaginase often inhibit off flavors in end products due to chemical reactions (e.g. Maillard reaction). Microbial phytases (*Thermomyces lanuginosus*, *Talaromyces thermophilus* and *Sporotrichum thermophile*) cause oxidation of phenol groups and reduce the phytate content and stickiness, thereby increasing dough strength, stability and volume of bread (Dahiya 2016). Transglutaminases (*Streptoverticillium* sp. and *Physarumpolycephalum*) along with laccases (*S. lavendulae*, *S. cyaneus*, *Marinomonas mediterranea*, *Phanerochaete chrysosporium*, *Theiophora terrestris*, *Lenzites*, *betulina*) are used to obtain gluten-free bread and increase the shelf life of the products (Kieliszek and Misiewicz 2014).

8.3 Enzymes in Fish and Sea Food Processing

Both endogenous and added enzymes play a fundamental role in the development of products and processes encompassing seafood and meat processing industries. The paradigm of research has progressed from traditional screening methods to innovative/improved molecular and metagenomics approaches in conjunction with regulatory formulations. Though plenty of plants and mammalian sources of enzymes exist, still microbes remain the preferred source due to ease of cultivation and manipulation (Anbu et al. 2015). Proteases like papain, pepsin, trypsin, subtilisin, bacillolysin, ficin have been widely used for descaling/deskinning, peeling, tenderization, ripening etc. to minimize mechanical damage and to increase shelf-life and yield of the products (Ghaly et al. 2013; Faisal et al. 2015; Fernandes 2016; Husain 2018). Fungi belonging to genus *Penicillin*, *Mucor*, *Fusurium*, *Thermomyces*, *Aspergillus*, *Rhizopus*, *Humicola*, *Trichoderma* and *Thermoascus* remain a powerful source to deliver proteases (Singh et al. 2016c). On the other hand, meat tenderizing enzymes like subtilisin and neutral proteases are produced by bacteria *Bacillus subtilis*, *B. licheniformis*, *B. alcalophilus*, *B. lentus*, etc. Both crude enzymatic preparations (*e.g.* pepsin, trypsin, chymotrypsin, thermolysin) from *Bacillus subtilis*, *Geobacillus stereothermophilus*, *B. thermoproteolyticus*, *T. thalophilus* and commercial formulations (*e.g.*, Alcalase derived from submerged fermentation of *B. licheniformis*, Flavorzyme—a peptidase preparation from *Aspergillus oryzae*, Nutrase from *Bacillus amyloliquefaciens*, etc.) are used for manufacturing the products involving protein hydrolysis *e.g.* fish protein hydrolysate (FPH) and fish sauce. Transglutaminases are another important group of enzymes isolated from *Streptoverticillium* sp., *Streptomyces* sp. and are used for formulations of fish meat mince, texture modification, the formation of collagen and gelatin bonds and minimization of drip after thawing (Zheng et al. 2001; Suresh et al. 2015). Lipases find a great application in this industry for the preparation of ω-3-poly-unsaturated fatty acids (ω-PUFAs) and for isolation of fats and oils from seafood by-products (Chaurasia et al. 2016; Guerrand 2018). Most of the industrial microbial lipases are derived from fungi (*Aspergillus*, *Candida*, *Geotrichum*) and bacteria (*Bacillus*, *Pseudomonas*, *Chromobacterium*, etc.) (Aravindan et al. 2007).

8.4 Enzymes in Meat Industry

Microbial enzymes are routinely used in meat industries in numerous production processes. Faced with the new market trends, production of nutritionally enhanced meat foods is being focused upon by reformulation of raw materials and products and advanced technological processes. For the breakdown of proteins in muscle and hydrolyze myofibrillar and connective tissue proteins of meat and meat products several proteolytic enzymes are utilized. Proteases from both fungal (e.g., *Aspergillus oryzae*) and bacterial (e.g., *Bacillus subtilis*) sources are used for meat tenderization (Arihara 2006). Bacterial proteases like subtilisin and neutral proteases find major applications in the meat industry because of their relative specific activity and low inactivation temperature. Enzymes like alkaline elastase are obtained from the alkalophilic *Bacillus* sp., are routinely used for collagen, elastin, and myofibrillar hydrolysis (Avendano et al. 2016). Transglutaminase is another enzyme of microbial origin produced by *Streptoverticilliummobarence* which in addition to processing also enhances the nutritional value by adding essential amino acids (Maróstica and Pastore 2010). It also aids in binding, texture maintenance, and improvement, emulsification, homogenization of meat and meat products. Together proteases and thermolysin (EC 3.4.24.27) are used to for the ripening of dry sausages (Avendano et al. 2016).

8.5 Enzymes in the Dairy Industry

Microbial enzymes are extensively used for three major applications in dairy industries like cheese production, product ripening, shelf-life extension and function alteration. Flavor development and ripening in cheese are met by controlled hydrolysis of triglycerides. Microbial community structure and hence enzyme secreted by them play an important role for developing different cheese types. *Rhizomucorpusilus*, *R. meihi*, *Endothiaparasitica*, *Aspergillus oryzae*, *Irpexlactis* are the sources of rennet-like proteinases used for cheese manufacture. Milk clotting protease has been isolated from *R. pusilus* (Neelakantan et al. 1999). *Mucor miehei* rennet, and *Cryphonectriaparasitica* rennet is also being widely used in dairy industries. For flavor development in dairy products, lipolysis (enzymatic hydrolysis of triglycerides to fatty acids) of the milk fat is essential. Lipases isolated from *Propionibacterium* sp. contribute to the development of Swiss-type cheese. Esterases produced by *L. helveticus*, *L. delbrueckiisubsp. Bulgaricus*, *L. delbrueckiisubsp. Lactis*, *L. casei*, *L. paracasei*, and *L. plantarum* are lipolytic in nature are widely used in dairy industries for enhancing the cheese flavor (Neelakantan et al. 1999; McSweeny and Sousa 2000; Raveendran and Parameswaran 2018). Enzymes isolated from *Micrococcus* and *Pediococcus* also show weak lipolytic activity. Psychrotrophic bacteria (e.g., *Pseudomonas florescens*) also contribute to lipolysis in cheese made from milk. Studies reveal that K- and β-caseins were most susceptible to proteolysis while the α-casein was less affected enzymes secreted by psychrotrophs (Shammet et al. 1992). *E. faecium*, *E. durans* produce biogenic amines like tyramine used for cheddar cheese production and development (Rea et al. 2004). Lipases are also used for the production of different types of cheese like Camembert cheese using lipase from *Penicillium camemberti* and cheddar cheese using *Aspergillus niger* or *A. oryzae* (Aravindan et al. 2007). Lipase catalysis also prolongs the shelf life of dairy products. Antimicrobial enzymes like bacteriolysins are used to control microbial contamination, improving safety and shelf life of dairy products. Another class of enzymes, *i.e.*, catalases (apparently from the genus *Bacillus*) are also used for the treatment of milk to remove H_2O_2 before cheese production (Tarhan 1995). Galactosidase produced from yeast *Kluyveromyces lactis*, *K. fragilis* is generally used for the hydrolysis of lactose in the milk of whey (Hussein et al. 1989).

8.6 Enzymes in the Vegetable and Fruit Juice Industry

Fruits and vegetables contain approximately 50–70% water creating a turgor pressure and juices are prepared from them by mechanically squeezing or mashing them with little or no application of heat or solvents. Enzymes are used in these industries for (1) maceration (2) treatment of extracted fruit/vegetable juice (3) obtaining better clarification by reducing viscosity (4) increasing product stability (5) colour and flavor enhancement. The cell walls of plant tissues comprise of a methoxylated galacturonic acid polymer called pectin which is present as a complex macromolecular structure and provides integrity and rigidity. Pectinolytic enzymes like protopectinases, lyases and pectin esterases are used for hydrolysis of pectin or dissolution of protopectin (Kumar and Suneetha 2016). Most of these enzymes have been isolated from fungi like *Aspergillus niger*, *Penicillium oxalium*, etc. For clarification of orange, papaya, pear, peach, plum, apricot, apple and blueberry juices pectinase enzyme from *Rhizopus oryzae*, pectin methylesterase from *Aspergillus tubigensis*, pectinase from *Aspergillus niger*, purified polygalacturonase from *Aspergillus awamori* and some strains of *Bacillus* Ar1.2, Ega16, Ega22 and VIT sun-2 are utilized for reducing viscosity and increasing stabilization (Kumar 2015; Garg et al. 2016; Kumar and Suneetha 2016).

9 Conclusion

Role of enzymes is indispensable in our daily lives being the most proficient catalysts with high specificity and stability. Globally these micro-machines, primarily made up of proteins; have been the major research focus and increasing efforts are made for their discovery, engineering and production for industrial applications. Their role is tremendous in competitive and cost-effective processes in food and beverage industries. They have been greatly exploited in the development of innovative bioprocesses and for speeding-up production processes. A vast diversity in the types and actions of enzymes enable them to serve as efficient biological catalysts for innumerable chemical reactions ensuring highest market profitability, yet least logistic load on user and environment.

References

Ahmad Z, Butt MS, Ahmed A, Riaz M, Sabir SM, Farooq U, Rehman FU (2014) Effect of *Aspergillus niger* xylanase on dough characteristics and bread quality attributes. J Food Sci Technol 51(10):2445–2453

Allen MJ, Fang TY, Li Y, Liu HL, Chen HM, Coutinho P, Honzatko R, Ford C (2003) Protein engineering of glucoamylase to increase pH optimum, substrate specificity and thermostability. U.S. Patent No. 6,537,792. Washington, DC: U.S. Patent and Trademark Office

Anbu P, Gopinath SC, Chaulagain BP, Tang TH, Citartan M (2015) Microbial enzymes and their applications in industries and medicine 2014. Biomed Res Int 2015:816419

Andualema B, Gessesse A (2012) Microbial lipases and their industrial applications. Biotechnology 11(3):100–118

Aravindan R, Anbumathi P, Viruthagiri T (2007) Lipase applications in food industry. Indian J Biotechnol 6:141–158

Arihara K (2006) Strategies for designing novel functional meat products. Meat Sci 74(1):219–229

Avendano KA, Anguiano M, Lopez CE, Montanez LE, Sifuentes L, Balagurusamy N (2016) Microbial enzymes applications in food processing. Agro Food Ind Hi Tech 27:63–67

Bessler C, Schmitt J, Maurer KH, Schmid RD (2003) Directed evolution of a bacterial α-amylase: toward enhanced pH-performance and higher specific activity. Protein Sci 12(10):2141–2149

Borch K, Erlandsen L, Jensen M (2006) Treatment of dough with a lipoxygenase and a lipolytic enzyme. U.S. Patent Application No. 10/528,330

Cha JAEHO, Batt CA (1998) Lowering the pH optimum of D-xylose isomerase: the effect of mutations of the negatively charged residues. Mol Cells 8(4):374–382

Chapman-Smith A, Cronan JE Jr (1999) The enzymatic biotinylation of proteins: a post-translational modification of exceptional specificity. Trends Biochem Sci 24(9):359–363

Chaurasia SP, Bhandaria K, Sharmaa A, Dalai AK (2016) A review on lipase catalysed synthesis of DHA rich glyceride from fish oils. Int J Res Sci Innov 3:9–19

Dahiya S (2016) Industrial application of phytases. Int J Appl Res 2:95–98

Danielsen S, Lundqvist H (2008) Bacterial alpha-amylase variants. WO Patent 825, 2008

Declerck N, Machius M, Wiegand G, Huber R, Gaillardin C (2000) Probing structural determinants specifying high thermostability in *Bacillus licheniformis* α-amylase1. J Mol Biol 301(4):1041–1057

Demain AL, Adrio JL (2008) Contributions of microorganisms to industrial biology. Mol Biotechnol 38(1):41

Duan X, Chen J, Wu J (2013) Improving the thermostability and catalytic efficiency of *Bacillus deramificans* pullulanase by site-directed mutagenesis. Appl Environ Microbiol 79(13):4072–4077

Dwevedi A, Kayastha AM (2009) Stabilization of β-galactosidase (from peas) by immobilization onto Amberlite MB-150 beads and its application in lactose hydrolysis. J Agric Food Chem 57 (2):682–688

Ellaiah P, Adinarayana K, Bhavani Y, Padmaja P, Srinivasulu B (2002) Optimization of process parameters for glucoamylase production under solid state fermentation by a newly isolated Aspergillus species. Process Biochem 38(4):615–620

England G, Kolkman M, Miller B, Vroemen C (2013) Pullulanase variants with increased productivity. U.S. Patent No. 8,354,101. Washington, DC: U.S. Patent and Trademark Office

Faisal M, Islami SNE, Islam MN, Kamal M, Khan MNA (2015) Study on microbial and physical changes in fish sauce during fermentation. Res Agric Livestock Fish 2(2):375–383

Fernandes P (2010) Enzymes in food processing: a condensed overview on strategies for better biocatalysts. Enzyme Res 2010:1–19. https://doi.org/10.4061/2010/862537

Fernandes P (2016) Enzymes in fish and seafood processing. Front Bioeng Biotechnol 4:59

Flamm EL (1991) How FDA approved chymosin: a case history. Nat Biotechnol 9(4):349

Food Enzymes Market by Type (Carbohydrase, Protease, Lipase), Application (Beverages, Bakery Products, Dairy Products, Confectionery Products, Processed Foods), Formulation (Lyophilized Powder and Liquid), Source, and Region – Global Forecast to 2023. https://www.marketsandmarkets.com/Market-Reports/food-enzymes-market-800.html

Food Enzymes Market: Global Trends and Forecasts to 2018. http://www.researchandmarkets.com/research/3xgvqh/food_enzymes

Garg G, Singh A, Kaur A, Singh R, Kaur J, Mahajan R (2016) Microbial pectinases: an ecofriendly tool of nature for industries. 3 Biotech 6(1):47

Ghaly AE, Ramakrishnan VV, Brooks MS, Budge SM, Dave D (2013) Fish processing wastes as a potential source of proteins. Amino acids and oils: a critical review. J Microb Biochem Technol 5(4):107–129

Ghollasi M, Ghanbari-Safari M, Khajeh K (2013) Improvement of thermal stability of a mutagenised α-amylase by manipulation of the calcium-binding site. Enzym Microb Technol 53(6–7):406–413

Goswami GK, Rawat S (2015) Microbial Xylanase and their applications – a review. Int J Curr Res Acad Rev 3(6):436–450

Guerrand D (2018) Economics of food and feed enzymes: status and prospectives. In: Enzymes in human and animal nutrition, pp 487–514

Gunders D (2012) Wasted: how America is losing up to 40 percent of its food from farm to fork to landfill. Natural Resources Defense Council, 26

Hii SL, Tan JS, Ling TC, Ariff AB (2012) Pullulanase: role in starch hydrolysis and potential industrial applications. Enzyme Res 2012:921362

Husain Q (2016) Magnetic nanoparticles as a tool for the immobilization/stabilization of hydrolases and their applications: an overview. Biointerfaces Res Appl Chem 6(6):1585–1606

Husain Q (2017a) Nanomaterials as novel support for the immobilization of amylolytic enzymes and their applications: a review. DeGruyter Biocatal 3(1):37–53

Husain Q (2017b) Nanomaterials immobilized cellulolytic enzymes and their industrial applications: a literature review. JSM Biochem Mol Biol 4(3):1029
Husain Q (2017c) Nanosupport bound lipases their stability and applications. Biointerfaces Res Appl Chem 7(6):2194–2216
Husain Q (2018) Nanocarriers immobilized proteases and their industrial applications: an overview. J Nanosci Nanotechnol 18(1):486–499
Hussein L, Elsayed S, Foda S (1989) Reduction of lactose in milk by purified lactase produced by *Kluyveromyces lactis*. J Food Prot 52(1):30–34
Igarashi K, Ozawa T, Ikawa-Kitayama K, Hayashi Y, Araki H, Endo K, Hagihara H, Ozaki K, Kawai S, Ito S (1999) Thermostabilization by proline substitution in an alkaline, liquefying α-amylase from *Bacillus* sp. strain KSM-1378. Biosci Biotechnol Biochem 63(9):1535–1540
Karimäki J, Parkkinen T, Santa H, Pastinen O, Leisola M, Rouvinen J, Turunen O (2004) Engineering the substrate specificity of xylose isomerase. Protein Eng Des Sel 17(12):861–869
Kieliszek M, Misiewicz A (2014) Microbial transglutaminase and its application in the food industry. A review. Folia Microbiol 59(3):241–250
Kim MS, Lei XG (2008) Enhancing thermostability of *Escherichia coli* phytase AppA2 by error-prone PCR. Appl Microbiol Biotechnol 79(1):69–75
Kirk O, Borchert TV, Fuglsang CC (2002) Industrial enzyme applications. Curr Opin Biotechnol 13 (4):345–351
Kumar S (2015) Role of enzymes in fruit juice processing and its quality enhancement. Health 6 (6):114–124
Kumar P, Suneetha V (2016) Microbial pectinases: wonderful enzymes in fruit juice clarification. Int J MediPharm Res 2:119–127
Maróstica MR, Pastore GM (2010) Some nutritional, technological and environmental advances in the use of enzymes in meat products. Enzyme Res 2010:480923
McDaniel A, Fuchs E, Liu Y, Ford C (2008) Directed evolution of *Aspergillus niger* glucoamylase to increase thermostability. Microb Biotechnol 1(6):523–531
McSweeney PL, Sousa MJ (2000) Biochemical pathways for the production of flavour compounds in cheeses during ripening: a review. Lait 80(3):293–324
Miguel ÂSM, Martins-Meyer TS, da Costa Figueiredo ÉV, Lobo BWP, Dellamora-Ortiz GM (2013) Enzymes in bakery: current and future trends. In: Muzzalupo I (ed) Food industry. InTech, UK
Neelakantan S, Mohanty AK, Kaushik JK (1999) Production and use of microbial enzymes for dairy processing. Curr Sci 77(1):143–148
Nelson DL, Cox MM (2004) Lehninger principles of biochemistry, 4th edn. WH Freeman, New York
Nielsen PH, Malmos H, Damhus T, Diderichsen B, Nielsen HK, Simonsen M, SchiV HE, Oestergaard A, Olsen HS, Eigtved P, Nielsen TK (1994) Enzyme applications, industrial. In: Kirk-Othmere Encyclopedia of chemical technology. Wiley, New York
Novozymes (https://report2016.novozymes.com/our-business/trends)
Oelofse SH, Nahman A (2013) Estimating the magnitude of food waste generated in South Africa. Waste Manag Res 31(1):80–86
Oh DK, Oh HJ, Kim HJ, Cheon J, Kim P (2006) Modification of optimal pH in L-arabinose isomerase from *Geobacillus stearothermophilus* for D-galactose isomerization. J Mol Catal B Enzym 43(1–4):108–112
Prakash O, Jaiswal N (2010) Alpha-Amylase: an ideal representative of thermostable enzymes. Appl Biochem Biotechnol 160(8):2401–2414
Qureshi N, Blaschek HP (2005) Butanol production from agricultural biomass. In: Food biotechnology. CRC, Boca Raton, FL, pp 541–565
Raveendran S, Parameswaran B (2018) Applications of microbial enzymes in food industry. Food Technol Biotechnol 56(1):16–30

Rea MC, Franz CMAP, Holzapfel WH, Cogan TM (2004) Development of enterococci and production of tyramine during the manufacture and ripening of Cheddar cheese. Ir J Agric Food Res 43(2):247–258

Rubin-Pitel SB, Zhao H (2006) Recent advances in biocatalysis by directed enzyme evolution. Comb Chem High Throughput Screen 9(4):247–257

Shammet KM, Brown RJ, McMahon DJ (1992) Proteolytic activity of some milk-clotting enzymes on κ-Casein1. J Dairy Sci 75(6):1373–1379

Singh R, Kumar M, Mittal A, Mehta PK (2016a) Microbial enzymes: industrial progress in 21st century. 3 Biotech 6:174

Singh R, Mittal A, Kumar M, Mehta P (2016b) Amylases: a note on current applications. Int Res J Biol Sci 5(11):27–32

Singh R, Mittal A, Kumar M, Mehta PK (2016c) Microbial proteases in commercial applications. J Pharm Chem Biol Sci 4:365–374

Sridevi DV, Ramanujam AR (2012) Performance of mixture of vegetable wastes with high cabohydrate content in anaerobic digestion process. Int J Environ Sci 3(1):181

Sriprapundh D, Vieille C, Zeikus JG (2003) Directed evolution of *Thermotoga neapolitana* xylose isomerase: high activity on glucose at low temperature and low pH. Protein Eng 16(9):683–690

Starch Derivatives Market by Type (Maltodextrin, Cyclodextrin, Glucose Syrup, Spray Dried Starch, Hydrolysates), Application (F & B, Feed, Paper, Cosmetics, Pharmaceutical), Function (Emulsifying, Binding, Stabilizing, Thickening), & by Region – Global Trends & Forecast to 2020. http://www.marketsandmarkets.com/Market-Reports/starch-derivatives-market-116279237.html

Suresh PV, Nidheesh T, Pal GV (2015) Enzymes in seafood processing. In: Chandrasekaran M (ed) Enzymes in food and beverage processing. CRC, Boca Raton, FL, pp 354–377

Tang WL, Zhao H (2009) Industrial biotechnology: tools and applications. Biotechnol J Healthc Nutr Technol 4(12):1725–1739

Tarhan L (1995) Use of immobilised catalase to remove H_2O_2 used in the sterilisation of milk. Process Biochem 30(7):623–628

Tomschy A, Brugger R, Lehmann M, Svendsen A, Vogel K, Kostrewa D, Lassen SF, Burger D, Kronenberger A, van Loon AP, Pasamontes L (2002) Engineering of phytase for improved activity at low pH. Appl Environ Microbiol 68(4):1907–1913

Ueda T, Masumoto K, Ishibashi R, So T, Imoto T (2000) Remarkable thermal stability of doubly intramolecularly cross-linked hen lysozyme. Protein Eng 13(3):193–196

Wang Y, Fuchs E, da Silva R, McDaniel A, Seibel J, Ford C (2006) Improvement of *Aspergillus niger* glucoamylase thermostability by directed evolution. Starch-Stärke 58(10):501–508

Wong DW, Batt SB, Lee CC, Robertson GH (2004) High-activity barley α-amylase by directed evolution. Protein J 23(7):453–460

World Enzymes (2011) World food & beverage enzyme demand by market (million dollars). Freedonia Group, Cleveland, OH. https://www.freedoniagroup.com/industry-study/world-enzymes-2824.htm

Wu Y, Yuan S, Chen S, Wu D, Chen J, Wu J (2013) Enhancing the production of galacto-oligosaccharides by mutagenesis of *Sulfolobus solfataricus* β-galactosidase. Food Chem 138 (2–3):1588–1595

Zheng M, Du G, Guo W, Chen J (2001) A temperature-shift strategy in batch microbial transglutaminase fermentation. Process Biochem 36(6):525–530

Zhou C, Xue Y, Ma Y (2015) Evaluation and directed evolution for thermostability improvement of a GH 13 thermostable α-glucosidase from *Thermus thermophilus* TC11. BMC Biotechnol 15 (1):97

Significance of Enzymes and Their Application in Agriculture

Anna Piotrowska-Długosz

1 Introduction

Almost all enzymes are proteins that act as biological catalysts and speed up the biochemical reactions in living organisms without being used in the process. They can also be extracted from cells and carry out the reactions outside the living organisms (Lever 2015). Such extracellular enzymes play a pivotal role in various life areas and can be applied to catalyze many of commercially significant processes (Robinson 2015). The use of enzymes is not a new idea; they have been exploited throughout the ages. The Egyptians and Sumerians (2000 BC) used specific enzymes (fermentations) in cheese preparation, leather tanning, beer brewing and leavening of bread (Arapoglou et al. 2010). The term *enzyme* came from the Greek words *en*, which means *within* and *zyme* that means *yeast* and was introduced for the first time by Wilhelm Kühne in 1878, while he tested the ability of beer yeast to ferment various sugars. In 1926 enzymes were first known to be protein, while in 1982 the first food application of a product of gene technology, e.g. α-amylase, took place (Arapoglou et al. 2010). Nowadays, enzymes have a wide range of applications in many areas of life. They are used in many branches of industry for different purposes, e.g. in food production and processing, in leather, cloth and textile industry, in cosmetics and detergents production (e.g. α-amylase), beer brewing, dairy industry, meat tenderizing (papain), paper production (xylanases, laccases). They are also used in medicine (alanine transaminase, acid and alkaline phosphatases, aspartate transaminase), pharmacy (streptokinase, cysteine proteinases) and for scientific and analytical purposes (various proteases, dehydrogenases, kinases and phosphatases) (Doubnerová 2012; Wells et al. 2012; Hemalatha et al. 2013; Husain 2016). Finally, environmental and agricultural use of enzymatic proteins is

A. Piotrowska-Długosz (✉)
Laboratory of Soil Science and Biochemistry, Department of Biogeochemistry and Soil Science, Faculty of Agriculture and Biotechnology, UTP University of Science and Technology, Bydgoszcz, Poland

Q. Husain, M. F. Ullah (eds.), *Biocatalysis*,
https://doi.org/10.1007/978-3-030-25023-2_14

also widespread and has been successively increased for many years (Shi 2011). A wide range of enzymes of plant and microbial origin play significant role in waste treatment application (Karam and Nicell 1997; Pandey et al. 2017). Some enzymes can work on peculiar resistant pollutants to dislodge them by precipitation and changing to other compounds. They can also change parameters of some specified pollutants to make them more suitable to treatment or help to convert the waste compounds to value-added products (Karam and Nicell 1997). From an agricultural point of view the most important are soil enzymes and enzymes applied in animal farms as feed additives. In agricultural soils the occurrence of a wide range of enzymes warrant a suitable course of processes at soil-plant-environment interrelations that are crucial to the growth and production of crops for human and animal feed. They are often used in agronomy as indicators of soil health, fertility and productivity (Niculina and Ştefanic 2008; Shi 2011). Better understanding of main properties and functions of soil enzymes is necessary for scientists and other stakeholders engaged in this area.

2 Soil Enzymes: General Consideration, Factors Affecting Soil Enzymatic Activity

Soil is a main constituent of terrestrial ecosystems and the principle matrix essential in agricultural production. The suitable soil functioning is indispensable for supporting the biochemical cycles of main important nutrients, and that is why soil processes influence a range of biotic and abiotic constituents of soil ecosystem (Bünemann et al. 2018). To recognize and understand the functioning of soil and to hold back soil damage that are caused by both natural and anthropogenic factors, including agricultural management practices, it is pivotal to have adequate tools to determine and prognose the soil possible changes. Strategies based on biological indicators, including soil enzymes, are often used to determine the status of soil ecosystem (e.g. Bastida et al. 2008; Baležentiené 2012). Studies of soil enzymes are significant because they play the principal biochemical functions in organic matter formation and decomposition as well as nutrient turnover in a soil ecosystem, the stabilization of soil structure, decomposition of pollutants, therefore playing important role in agriculture (Makoi and Ndakidemi 2008; Burns et al. 2013; Piotrowska-Długosz 2014).

The soil enzymes are mainly produced by soil microorganisms (both living and dead cells), originate from plants (both roots and plant residues) as well as originate from soil animals. Enzymes in soil are classified in three main categories: intracellular (in leaving and proliferating organisms, e.g. soil dehydrogenases), cell-associated (in cell and tissue fragments) and free enzymes occurring in soil solution (Bakshi and Varma 2011). Intracellular enzymes occur in various cellular parts, such as cytoplasm or periplasm or they are connected with cell membrane and cell wall. In turn, extracellular enzymes are produced and secreted by living cells and act outside the parent cells. The small fraction of free enzymes is to be short-living, and is subjected to quick degradation or is immobilized on the surface of soil organic and/or mineral colloids (Nannipieri et al. 2002). The activity of stabilized extracellular enzymatic proteins is independent of current cell growth and immobilization which protect enzymes against degradation and denaturation caused by unprofitable environmental factors and by decomposition by soil proteases (Gianfreda and Scarfi 1991; Safari Sinegani et al. 2005). Although the stabilized enzymes are less active than free enzymes, they may accumulate in soil for the long time and thus the activity of these enzymes is the most important for overall biochemical processes occurring in soil (Quiquampoix et al. 2002; Theng 2012). Because of imperfect methods used in soil enzymology, no unequivocal separation between soil extracellular and intracellular activity has been reached so far (Burns 1982; Ladd 1985; Piotrowska-Długosz 2014). Some of soil enzymes (e.g. β-glucosidase or urease) are able to catalyze reaction both inside and outside the organisms producing them, while other (e.g. cellulases) can act only outside the parent cells, because of the large size of the substrate they catalyze (Das and Varma 2011). In turn, enzymes such as dehydrogenases, nitrate reductase that are involved in the main metabolic processes, are active only endocellularly (Abdelmagid and Tabatabai 1987; Szajdak and Gaca 2010; Moeskops et al. 2010; Zhao et al. 2010; Yuan and Yue 2012; Piotrowska-Długosz and Wilczewski 2014). Enzymes in soil could be constitutive (always secreted and present in cells, aside from the presence and/or addition of any substrates) or inducible (not present constantly in soil, but quickly produced and secreted by the cells after substrate addition) (Das and Varma 2011). Enzyme activities found in soil belong to four classes: oxidoreductases, hydrolases, lyases and transferases; the two first classes being predominant (Gianfreda and Bollag 1996). The most often studied enzymes of agricultural importance are presented in Table 1.

Soil enzymes usually exhibit the properties different than enzymes from other sources (plants, water). They commonly show specified range of stability in extreme soil conditions (e.g. a wide range of soil temperature and reaction) and when subjected to soil protease attack. Soil enzymes usually show different kinetic parameters as compared to enzymes in purified form. Thus, they have lower values of Vmax and Km constant is usually higher, which indicates lower catalytic efficiency and the lower substrate affinity (Dick 2011; Kujur and Pater 2014).

Different soil types contain a set of enzymes varied quantitatively and qualitatively due to different physical, chemical, microbiological and biochemical properties found in these soils. Usually soils with high organic matter, nutrients and clay contents, as well as with high microbial activity reveal greater enzymatic activities compared to soils with low values of these properties. The highly important role of clay-organic matter complexes in maintaining extracellular enzymatic activity in soil is commonly known (Zimmerman and Ahn 2011). Enzymes bound to these complexes retain their activity since they are more resistance to proteolysis and microbial attack and to soil unbeneficial factors, such as changing temperatures and moisture (Nannipieri et al. 1996; Theng 2012). On the other side however, the activity of immobilized enzymes are lower compared to the activity of free enzymes occurring in soil solution (Datta et al. 2017).

Table 1 Some enzymes important in agriculture and their functions

EC number[a]	Accepted name	Agriculture significance and function	Reaction to be carried out	References
I class: oxidoreductases				
1.7.1.1	Nitrate reductase (NADH)	The first enzyme of denitrification process, nitrate reduction II (assimilatory)	Nitrate + NADH + H^+ = nitrite + NAD^+ + H_2O	Szajdak and Gaca (2010)
1.11.1.6	Catalase	Release O_2 from H_2O_2; protects *cells* from oxidative damage by reactive oxygen species.	$2H_2O_2 = 2H_2O + O_2$	Rodríguez-Kábana and Truelove (1982), Uzun and Uyanöz (2011)
1.11.1.7	Peroxidase	Release O_2 from H_2O_2; protects *cells* from oxidative damage by reactive oxygen species. Takes part in lignin decomposition	2 phenolic donor + H_2O_2 = 2 phenoxyl radical of the donor + 2 H_2O	Sinsabaugh (2010), Bach et al. (2013)
1.18.6.1	Nitrogenase	Takes part in N fixation, converts the atmospheric, gaseous dinitrogen (N_2) into ammonia (NH_3)	8 reduced ferredoxin + 8 H^+ + N_2 + 16 ATP + 16 H_2O = 8 oxidized ferredoxin + H_2 + 2 NH_3 + 16 ADP + 16 phosphate	Egamberdieva and Kucharova (2008), Hoffman et al. (2014)
II class: transferases				
2.8.1.1	Thiosulfate sulfurtransferase (rhodanese)	Indicator for S cycle. Cleaves the S–S bond of thiosulphate, forming S and sulphite, and then S is subsequently oxidized into sulphite by sulphur oxygenase, while sulphite is further oxidized into sulphate by sulphite oxidase	thiosulfate $(S_2O3_2{}^-)$ + cyanide (CN) = sulfite (SCN^-) + thiocyanate $(SO_3{}^{2-})$	Szajdak (1996), Siwik-Ziomek et al. (2015)
III class: hydrolases				
3.1.3.1/ 3.1.3.2	Acid/Alkaline phosphatase	Convert organic P compounds into inorganic forms ($HPO_4{}^{-2}$, $H_2PO_4{}^-$), which are directly available to microorganisms and plants	A phosphate monoester + H_2O = an alcohol + phosphate	Wang et al. (2011), Nannipieri et al. (2011)

3.1.6.1	Arylsulfatase	Catalyze the hydrolysis of aromatic sulphate esters (C–O–SO_{3-}) to phenols (R–OH) and sulphate (SO_4^{2-}), thus is important for mobilization of inorganic SO_4^{2-} for plant nutrition.	A phenol sulfate + H_2O = a phenol + sulfate	Siwik-Ziomek et al. (2013), Siwik-Ziomek et al. (2015)
3.1.4.1	Phosphodiesterase I	Indicator for P cycle, hydrolysis of phosphoric esters, revealed to be a good index of the soil P availability to the plant	R_2NaPO_4 + H_2O = ROH + $RNaHPO_4$	Browman and Tabatabai (1978), Nannipieri et al. (2011)
3.6.1.1	Inorganic diphosphatase (pyrophosphatase)	Indicator of phosphorus transformation, ammonium polyphosphate, an inorganic salt of polyphosphoric acid and ammonia, is one of the frequently used phosphoric fertilizers	Diphosphate + H_2O = **2** phosphate (PO_4^{-3})	Wang et al. (2011)
3.5.1.1 and 3.5.1.2.	L-Asparaginase and L-Glutaminase	Act on C–N bonds (other than peptide bonds) on respective amino acids releasing NH_3, important in N mineralization to provide plant available N	Asparagine = aspartic acid + NH_3 Glutamine = glutamic acid + NH_3	Dodor and Tabatabai (2003), Bhattacharyya et al. (2007)
3.5.1.13	Aryl acylamidase	Hydrolyses propanil, which is use as a component of herbicide	Anilide + H_2O = carboxylate + aniline	Zablatowicz et al. (1998)
3.5.1.4	Amidase	Hydrolysis of C–N bonds other than peptide bond in linear amides releasing NH_3, important for N mineralization to provide plant available N form	Monocarboxylacid amide + H_2O = monocarboksyl acid + NH_3	Dodor and Tabatabai (2003)
3.4.11.2	Arylamidase [-α-aminoacyl-peptide hydrolase (microsomal)]	Hydrolysis of a N-terminal amino acid from peptides, amides and arylamides; important in the beginning stages of the soil amino acids mineralization; indicator of soil N mineralization		Dodor and Tabatabai (2007), Muruganandam et al. (2009)

(continued)

Table 1 (continued)

EC number[a]	Accepted name	Agriculture significance and function	Reaction to be carried out	References
3.2.1.4 3.2.1.91	Cellulases (endo-1,4-β-D-glucanase and exo-cellobiohydrolase)	The enzymatic complex involved in degradation of cellulose, the most abundant polysaccharide found in the biosphere, provides readily available C for soil microorganisms, thus increasing soil microbiological activity, and directly soil fertility	Endohydrolysis of (1 → 4)-β-D-glucosidic linkages in cellulose. Hydrolysis of (1 → 4)-β-D-glucosidic linkages in cellulose releasing cellobiose from the non-reducing ends of the chains	Dodor and Tabatabai (2005)
3.2.1.21	β-glucosidase	Is a part of enzymatic complex involved in degradation of cellulose	Hydrolysis of terminal, non-reducing β-D-glucosyl residues with release of β-D-glucose	Dodor and Tabatabai (2005)
3.2.1.26	Invertase	Indicator of carbon transformation, responsible for the breakdown of plant litter in soil; catalyzes the hydrolysis of sucrose to glucose and fructose	$C_{12}H_{22}O_{11} + H_2O = C_6H_{12}O_6 + C_6H_{12}O_6$	Ross (1975), Stemmer et al. (1998)
3.2.1.8	Xylanase	Responsible for decomposition of xylan, a polysaccharide found with cellulose in soil	Hydrolysis of β-1,4-xylan bonds	Rodríguez-Kábana (1982), Stemmer et al. (1998)
3.2.1.1 and 3.2.1.2	α-amylase and β-amylase	The amylase system synergistically hydrolyze starch (glycogen and other poly- and oligosaccharides). The products of this reaction are dextrins, oligosaccharides, maltose and finally monosaccharides, like glucose. Important in the transformation of plant residues entering the soil.	α-amylase: endohydrolysis of (1 → 4)-α-D-glucosidic linkages in polysaccharides containing three or more (1 → 4)-α-linked D-glucose units; β-amylase: hydrolysis of (1 → 4)-α-D-glucosidic linkages in polysaccharides so as to remove successive maltose units from the non-reducing ends of the chains	Pancholy and Rice (1972), EC Webb (1992), Deng and Popova (2011)
3.4	Proteolytic enzymes	They hydrolyze proteins and peptides and liberating amino acids. The proteins degradation (proteolysis) is believed to be a limiting step of N mineralization in soil		Vranová et al. (2013), Asha and Palaniswamy (2018)

3.5.1.5	Urease	The urease activity is crucial in regulating the N supply to plants after urea fertilization. Important in more effective way of managing N fertilizers. Used as an index of N transformation in soil.	Urea + $H_2O = CO_2 + 2\ NH_3$	Glibert et al. (2006), Balota and Chaves (2010)

[a]The enzymes are classified according to the Nomenclature Committee of the International Union of Biochemistry and Molecular Biology (NC-IUBMB); ROH—Hydroksylated compound

The soil enzymatic activity could be significantly affected by the set of natural and anthropogenic factors. Since their response to the various factors is more evident and earlier than in the case of other soil properties (e.g. physicochemical ones), enzymes was found to be the useful indicator of soil possible changes (Bastida et al. 2008; Shi 2011; Utobo and Tewari 2015). Natural factors such as soil moisture and temperature related to seasonal changes, topographical distribution, in situ location, physicochemical properties, the content of available forms of nutrients, soil texture (clay minerals), organic matter content and microbial biomass mostly influence soil enzymes by changes in their production and their constancy in natural conditions. The soil physical-chemical features as well as high content of organic and mineral colloids (clay and humus) are committed in the stabilization and protection of extracellular enzymes by their immobilization (Demkina et al. 2017).

Although the seasonal and temporal changes of soil enzymatic activity is often studied, the direct and clear relationship between soil enzymes and seasonal variation was not specified since is dependent on many factors, such as temperature changing and seasonal precipitation patterns (Wolińska and Stępniewska 2011; Rao et al. 2017; Veeraragavan et al. 2018). Additionally, different vegetation cover can influence the seasonal changes significantly (Yin et al. 2014; Khan et al. 2014). Marked seasonally related activities of soil enzymes were demonstrated e.g. by Paz-Ferreiro et al. (2011) and Zhou and Zhang (2014). Some studies showed that maximum enzymatic activity occurred in spring (Rastin et al. 1988; Piotrowska and Długosz 2012), in summer or in autumn (Tripathi et al. 2007; Yuan and Yue 2012), while the lowest was often found in winter (Yuan and Yue 2012).

Changing temperature could have either positive or negative impact on the character and kinetics of individual extracellular enzymes. Urease and β-glucosidase activity in Mediterranean soils were positively correlated with soil temperatures in winter which was related to the highest moisture, while the negative relationship between these properties was found in summer (Sardans et al. 2012). In the study of Steinweg et al. (2013) enzymatic activity increased with increase in temperature up to some optimum, and further increase can lead to enzyme denaturation. Temperature increasing in a grassland soils studied by Gong et al. (2015) enhanced the phosphatase activity by almost 36%, while inhibited the cellulase activity about 30%. Changes in seasonal precipitation may increase the duration of wet and dry events in the soil. There is commonly known that the distribution of enzymatic proteins, their substrates, and reaction products in the soil is relative significantly to soil type (soil texture) and moisture (Acosta-Martinez et al. 2007; Zhang et al. 2011). When the soil moisture is low, enzymatic activity is low and long droughts usually inhibit enzymes secretion and in consequence their activities. However, decreased enzymatic activity in dry soils, accompanied by permanent production of enzymatic proteins, even in small quantities, may lead to maintenance of total enzymatic pool size during a dry period (Burns et al. 2013). Increasing of soil moisture leads to accelerated organic matter transformation and its availability because of cell lysis and disruption of soil aggregates. This may, in turn, activate the microbial biomass and its transformation thereby contributes to temporary increase in overall soil enzymatic activity (Spohn et al. 2013).

As was mentioned above, various natural and anthropogenic factors may affect directly or indirectly the soil enzymatic activity, which allow them to be commonly used as good indicators of possible changes of soil status.

3 Soil Enzymes as Indicators of Agricultural Practice and Agriculture Related Soil Contamination

The significance of soil enzymes in agriculture could be considered on two levels. Firstly, they are necessary for organic matter transformation (both formation and decomposition) and nutrient cycling, strongly influencing soil fertility and productivity (Balota and Dias Chaves 2010; Sherene 2017). Secondly, since they are highly sensitive towards various agricultural practices and quickly respond to various environmental conditions, their activities are suitable indicators of soil status and are willingly applied to determine the influence of management practices on general soil status, with special attention to soil biological functioning (Benitez et al. 2006; Melero et al. 2007). Enzymatic proteins respond to soil management practices much earlier than other soil status indices (e.g. physical and/or chemical properties) alteration is visible. Determination of microbiological biomass and different enzyme groups have been widely used to monitor the soil status and to determine the influence of various factors related to diverse land use and agricultural management (Gonzalez et al. 2007).

Agricultural practices, such as application of inorganic fertilizers, organic amendments and bio-fertilizers, tillage, cropping systems and vegetation cover, irrigation, mulching, use of pesticides, urease inhibitors, as well as environmental pollution related to the above human impact (e.g. heavy metals, PAHs) influence the soil structure and its chemical composition, which results in changes in abundance of soil microbes, the species structure and their metabolic activity. In turn, changes in soil microbial status entail changes in the overall soil enzymatic activity by the increasing or decreasing the individual enzyme production and the total enzymatic pool (Gianfreda and Ruggiero 2006).

Among various agricultural practices, organic and mineral fertilization could have a major impact on soil properties including soil enzymes. Various results have been obtained, dependent on soil type, cropping system, the type and the dose of the fertilizer and its application time, climatic conditions and character of enzymatic protein (Gianfreda and Ruggiero 2006). Generally, recent amending with organic materials like farmyard manure, compost or plant residues usually increase soil organic matter content, improve soil physico-chemical features, and increase the activity of soil enzymes, such as urease activity and acid phosphatase as an indirect result of microbial proliferation or/and enzyme induction (Mandal et al. 2007; Mohammadi 2011; Pajares et al. 2011). The activities of some enzymes (urease, protease, phosphatase, dehydrogenase, nitrate reductase) after 6 years were found to be significantly higher in soil treated with 12 tons ha^{-1} of compost as compared to the control soil. The double dose of the compost did not induce any further increase in enzymatic activity, and even its decrease, like in the case of protease activity (Crecchio et al. 2004). Regarding some organic materials, such as liquid pig manure or MSW compost, it is not recommended to use their higher doses, since they can decrease of some soil enzymes and the microbial counts due to inhibitory effects of heavy metals (Balota et al. 2011; King et al. 2015). The influence of applied organic materials on soil properties also depends on the stability of fresh versus mature material organic amendments. The use of fresh material compared to mature compost led to rapid increase of soil microbial biomass and enzymatic activity in the first stage of the experiment that was followed by a decrease due to the decrease of easily degradable substrates available to microorganisms (Saviozzi et al. 2002).

Contradictory results have been obtained in the studies devoted to the effect of inorganic N fertilization on soil enzymatic activities. According to some authors increasing amount of N fertilization can increase the activity of enzymes, such as cellulases, urease, phosphatases (Saiya-Cork et al. 2002; Sinsabaugh et al. 2005; Guo et al. 2011) or reduce the activity of catalase and invertase (Yang et al. 2008), peroxidase, urease, proteases, phosphatase and invertase (Giacometti et al. 2013; Husain 2017; Liao et al. 2017). More frequently however, enzymatic activity has been activated after the combined application of organic and inorganic N fertilization (e.g. Piotrowska and Wilczewski 2012; Shi et al. 2018). Some earlier studies (Klose et al. 1999; Klose and Tabatabai 2000) found a lack of significant effect of N fertilization on soil arylsulphatase and urease. Nitrogen fertilization can affect soil enzymes production and functions in two ways. The direct effect is based on the influence on the production of soil enzymes by microorganisms, which is in turn related to changes in the microbial community composition and hence the enzyme secretion (Iyyemperumal and Shi 2008; Merino et al. 2016). Differently, N fertilization, particularly in mineral forms, may have an indirect impact on the soil enzymatic activity by changing the set of soil physico-chemical properties, like soil reaction, porosity and nutrient concentration (Gianfreda and Ruggiero 2006).

Systematic studies have been performed to determine the influence of tillage or no-tillage on soil enzymatic activity e.g. (Roldán et al. 2005; Ulrich et al. 2010; Huang et al. 2012). Tillage affects some soil properties such as nutrient content and their availability, soil organic matter (OM) components and distribution by increasing decomposition leading to very quick loss of its content (Madejon et al. 2007; Kibet et al. 2016). Indeed, some researchers have found decrease in enzymatic activity because of the reduction in OM content what was caused by the mixing of soil horizons by plowing (e.g., Dick 1994). According to some other authors, no-tillage practice activated soil enzymes in comparison to conventional tillage system (Roldán et al. 2003; Mangalassery et al. 2015). Indeed, Mangalassery et al. (2015) showed that soil enzymes such as dehydrogenase, xylanase, cellulase, phenol oxidase, β-glucosidase, and peroxidase were significantly higher in zero-tilled soils than in tilled soils. Similarly, Mina et al. (2008) showed that soil dehydrogenase, alkaline phosphatase and protease activities were in most cases more active under a zero-till practice as compared to conventional tillage system. Most often, the activity of enzymes in the surface horizon of a zero-tilled soil has been higher as compared to those of the same layer of conventionally tilled soils, as long as the contrary trends occurred for the deepest soil layers (e.g. Balota et al. 2004). These trends could be explain by the fact that microbial activity of surface no-tilled soils is significantly higher than in conventional tillage, and this in turn is due to the fact that OM is more evenly disposed in soil under traditional tillage in comparison with that in soils under diminished tillage in which organic residues, being substrates for soil microbes, are mainly focused on the soil surface (Ekenler and Tabatabai 2003).

Earlier it was found that plowing increased enzymatic activity in agricultural soils due to the break of soil aggregates which causes the exposure of new enzymatic active centers (e.g., Khan 1996). Indeed, Ulrich et al. (2010), noted that the arginine ammonification revealed relatively high activity to the depth of 0.3 m of soil profile because of the high aeration in the ploughed soil. Likewise, Seifert et al. (2001) showed that conventional tillage speed up the oxidation of organic matter, and thus increasing a soil microbial activity. For example, two commonly used indicators of microbial activity, namely dehydrogenase activity and FDAH (fluorescein diacetate hydrolysis), were significantly higher in soil conventionally tilled than in soil with plant residues incorporating onto its surface, due to more labile carbon substrates, essential for hold up microbiological activity. In turn, in the study of Acosta-Martinez et al. (2011) conventional tillage practice, as compared to no-till, had no significant influence on the enzymatic activity in some field studies carried out in different places in USA in a loam soil. In the same way, soil dehydrogenase, urease and phosphatase activities were no significantly affected by the tillage practice in soil under 4-year long spring barley cultivation (Corchran et al. 1989).

The soil enzymatic activity can be also influenced by the type of crop cultivated. Generally, it is considered that the monoculture system cause the worsening of physico-chemical soil properties and nutrient losses and hence possible negative effect on soil microorganisms and enzymes, while crop rotation usually positively affects soil biological properties (Gajda and Martyniuk 2005; Gianfreda and Ruggiero 2006). In traditional crop rotations there is the highest and more diversified input of plant residues which positively influenced soil microbial activity leading to higher microbial biomass and enzymatic activity as compared to monoculture system (McDaniel et al. 2014; Samuel and Ciobanu 2018). Factors affecting soil enzymatic activity connected with cropping systems are crop type, the plant growth stage, chemical character of plant residues and root system extension, which affects the rhizosphere (Fang et al. 2013). Application of winter cover crop plant residue in a double-cropping rice (*Oryza sativa* L.) system in China increased enzyme activities, whereat each enzyme reached the highest activity in different growth stages, e.g. β-glucosidase was the most active in the tillering stage, alkaline phosphatase at the booting stage and arylsulfatase and arylamidase at the maturity stage (Hai-Ming et al. 2014).

Enzymes are also used as indicators of soil pollution caused by agricultural alteration such as pesticides, organic wastes (these including some pollutants, e.g. heavy metals) or excessive mineral fertilizers application (Gianfreda and Ruggiero 2006). Application of agrochemicals to increase crop productivity may cause a variety of negative environmental changes including significant inhibition of soil biological activity, especially when they are used in doses higher than these recommended (Wyszkowska and Kucharski 2004). Modern agriculture is dependent on many different groups of pesticides (insecticides, herbicides and fungicides) that are frequently applied together (in combination) or successionally to prevent effectively the number of pests and increase crop yields (Riah et al. 2014). Pesticides are often applied several times during one crop season, and most of them enter soil causing pollution of soil and water and finally influence human health by entering into the food chain, which has raised noticeable public concern (Hussain et al. 2009; Deborah et al. 2013). Ramudu et al. (2011) stated that only about 0.1% of used pesticide reaches the target organism, while the rest is deposited on the surrounding soil, mainly the 2–5 cm of top soil with concomitant side effects on non-target microorganisms and affects soil metabolism, including soil enzymatic activity. As has already been mentioned, using enzymatic tests is the reliable method to assess changes caused by pesticides and usually negative changes soil enzymatic activity has been found, especially when they have been used in the long term and at the doses higher than these recommended (Sannino and Gianfreda 2001; Bielińska and Pranagal 2007; Riah et al. 2014). A set of pesticides applied at the recommended or even higher doses has little (both positive and negative) or no impact on soil enzymatic activity. Thus, insecticide endosulfan was found to stimulate dehydrogenase activity when it was applied in the dose 100–200 higher than the standard rate of application (Kalyani et al. 2010; Defo et al. 2011). Sukul (2006) found phosphatase activity to be stimulated by fungicide metalaxyl, while other fungicides had no effect or even inhibited its activity (Tejada et al. 2011; Yan et al. 2011). Additionally it was found that the same enzyme reacts differently toward the same pesticide, and thus the same insecticide inhibit acid phosphatase and stimulate alkaline phosphatase activity, and vice versa (Defo et al. 2011; Jastrzębska 2011), what was explained by different structure and character of these enzymes related to their source (microorganism or plants) and their sensitivity to pesticides applications (Klose et al. 2006).

As regards the influence of pesticides on enzymes, the dose and the time of incubation/interaction was found to have significant effect on the activity. In the study of Deborah et al. (2013) the differentiated effects of imidacloprid (insecticide) and triadimefon (fungicide) applied at the concentrations of 0.2, 0.5 and 0.7 kg ha^{-1} on soil enzymatic activity under tomato cultivation was obtained. The highest rate of pesticides decreased the amylase activity, while the field rate (0.5 kg ha^{-1}) stimulated this activity. In turn, cellulase activity was inhibited at all doses used compared to the control. Earlier, the effect of fenamiphos, a widely used OP pesticide, on important soil microbial activities such as dehydrogenase, urease and potential nitrification in four soils from Australia and Ecuador was studied by Caceres et al. (2009). The results showed that fenamiphos in general was not toxic to dehydrogenase as well as urease up to 100 mg kg^{-1} soil. Deborah et al. (2013) studied also the influence of incubation time of pesticides on soil enzymes and after 24 h of incubation, the highest decrease was found for invertase activity at all applied rates. After 48 h, the activity was revived to some extent and imidacloprid showed enhanced activity at 0.5 μg g^{-1} (field rate). Earlier, soil enzymatic activity was significantly inhibited by the 90-day long presence of chlorothalonil in soil as compared to pesticide-free soils (Singh et al. 2002).

4 Soil Enzymes as Possible Indicators of Soil Fertility and Soil Nutrient Dynamics

There are different terms to be used to describe the state of the soil environment, such as soil quality, soil fertility, soil productivity, and soil health. In the context of agriculture, the term soil fertility is most often used. Soil fertility determine the ability of soil to provide adequate quantity of nutrients in a suitable equilibrium for plants growth and development on the assumption that other soil factors, such as light, moisture, temperature and the physical properties are favorable (Feller et al. 2012; Dwivedi 2017). A fertile soil is able to produce a high yield of good quality. From a sustainable agricultural point of view, soil fertility is the capacity of soil to serve as a proper matrix for sustainable plants growth and development (Adjei-Nsiah et al. 2007). In turn, soil productivity is the capability of soil to produce high yield of crops under a specified set of management practices. Soil productivity is the effect of selected agents such as soil fertility, proper soil management practices and suitable climatic factors (Dwivedi 2017). A given soil can reveal high fertility, i.e. it has high content of available forms of nutrients but its productivity can be low. For instance, the unbeneficial physical conditions in waterlogged soils have been responsible for low crop yield although the soil was found to be highly fertile (Tishchenko et al. 2013). In turn, soil quality, is defined as the capacity of a soil to function within a given set of conditions (land management and ecosystem limitations), to maintain biological activity, sustain environmental quality and improve plant, animal and human health (Laishram et al. 2012).

Soil enzymes are often considered to be a good index of soil quality and fertility and can be used for determination of ecosystem responses to management and overall environmental changes as well as sustainability of agricultural ecosystems (Nannipieri et al. 2002; Gianfreda and Ruggiero 2006). Justification for the use of soil enzymatic activities as soil status indicators is based on the facts that they (1) are usually related to other soil properties and processes such as the rate of organic matter decomposition, the level of physico-chemical properties and nutrient turnover, (2) they respond to environmental conditions more quickly than other soil properties, e.g. physical or/and chemical features, (3) assays of enzymes determination are simple, rapid, cheap and can be carried out routinely in most laboratories (Laudicina et al. 2012).

In reference to the above, the main role of soil enzymes in agricultural is based on their participation in nutrients transformation and releasing their forms available to plants and microorganisms. Thus, enzymes play a key role in transformation of C (e.g. celullases, glucosidases, galactosidases), N (e.g. proteases, urease, amidase, nitrate reductase), P (acid and alkaline phosphomonoesterases, phosphodiesterases) and S (rhodanase, arylsulphatase). The activity of these enzymes is often considered to be a good indicator of soil nutrient availability (Sherene 2017). Some of enzymes are produced by microorganisms and plants only in the presence of organic compounds that are to be decomposed or in the case of limited amount of available nutrients (positive feedback). A positive correlation between the soil enzymatic activity and nutrient cycling has been mentioned for some agricultural sites. For example, in a long-term cropping systems with N fertilization, the activity of arylamidase and N-acetylglucosaminidase was well related to the N mineralization rate (Muruganandam et al. 2009; Cenini et al. 2016). In turn, in the case of negative feedback, the enzyme production can be low when the end-products of the reaction (nutrients) are abundant or substrates to the enzymatic reaction (e.g. organic compounds) are limited (Allison and Vitousek 2005). Some authors have stated that the enzymes activity of the P- and N-cycling is negatively related to the P and N available forms (Balota and Dias Chaves 2010; Orczewska et al. 2012; Sherene 2017).

The growing interest in research on soil enzymology undertaken in the 1950s gave a lot of information about the structure and functions of enzymes in soil with in relation to soil agricultural practice and environment. The scientists believed that the knowledge on extracellular enzymes would allow use them as useful indicators of the soil total biological activity level. Consequently it was hoped to formulate a *soil fertility index* that would be suitable for practical purposes in agriculture. In 1960s the use of a single enzyme, such as invertase, proteases, catalase, phosphatase and asparaginase, was a common practice that was used to determine soil fertility (Skujins 1978). With time, however, a lot of contradictory and ambiguous data and conclusions have been obtained, showing that the use of soil enzymatic activity as a soil fertility indicator was considerably restricted. Some enzymes, such as dehydrogenase, did not show any significant correlation with other soil properties (Moore and Russel 1972). Some authors have found no close correlation between soil enzymatic activity and plant yields (e.g. Herrero et al. 1998), but other researchers have found such kind of relationship and that is why they considered enzymes to be a good indicator of soil fertility (Zhang et al. 2010; Wang et al. 2011; De Castro Lopes et al. 2012). However, in managed soils such correlations are dubious since other factors may interfere between enzymatic activity and plant productivity. This is found in agricultural soils in which the fertilization and irrigation (external delivery of nutrients and water) can significantly increase plant growth and development without a simultaneous reaction in soil microbiological and enzymatic activity (Dick 1994). The application of single enzymes as indicators of soil fertility and productivity has been often criticized (Nannipieri 1994; Nannipieri et al. 2012):

1. Since soil enzymes catalyze a specific reaction (e.g. urease), they cannot be responsible for the general soil microbial activity, which consist of many various biochemical processes. The production of a specific enzymatic protein can be inhibited or activated by a defined chemical compounds, without having any influence on the overall soil microbial activity as well as soil fertility and productivity.
2. Most of enzymes are substrate specific that is why they cannot indicate the overall soil nutrient status. For example the enzyme urease is frequently used as an index of organic nitrogen mineralization though this enzyme catalyzes the hydrolysis of

urea and urea, as is commonly known, is not an significant component of soil organic N transformation, especially if urea is not applied as a fertilizer.

3. The methodologies that are used for soil enzyme measurements in different laboratories are not the same and therefore it is difficult to compare the data of soil enzymatic research. There is no methods standardization and differences between substrates, assay conditions, incubation times and detection methods contribute to differences in the enzyme readings (Burns et al. 2013). Additionally, in different laboratories soil samples are subjected to various pretreatments, such as sample collection and their storage, before they are analyzed (Gil-Sotres et al. 2005).

Due to the above criticism some attempts have been made to integrate different enzyme activities in single and complex indexes that can be used in the assessment of agricultural soils (Tables 2 and 3). Because of the complexity of soil environment, a useful soil quality and fertility index must be the integrative combination of a number of properties into an easy and quantitative measure (García-Ruiz et al. 2008). Furthermore, because soil enzymes reveal different source, function and position in soil and answer distinctly to environmental agents, it would be beneficial to represent the information they give into one single numeric expression. Much effort has been made to combine various enzyme activities in single or complex indexes that can be utilized in the determination of agricultural soils status (Tables 2 and 3). Thus, first two proposed indexes, the Biological Index of soil Fertility and Enzyme Activity Number (Beck 1984; Stefanic et al. 1984) have been widely tested and discussed by many authors so far and some limitations of both indexes was indicated (e.g. Nannipieri 1994; Riffaldi et al. 2002). Thus e.g., the EAN can be used only in neutral and alkaline soil because of alkaline phosphatase used in this formula (Nannipieri 1994). An increase in EAN and BIF in untilled management field compared to tilled area was found by Riffaldi et al. (2002). As stated by Nannipieri (1994) both these indices were less useful to access the influence of various factors in the range of soil under different management systems than indicator proposed later.

More preferable choice of enzymatic activities was performed out by Sinsabaugh et al. (1992) who assessed six enzymatic activities that degrade lignocelluloses material working in a cascade. On the basis of principal component analysis (PCA), the authors proposed so called *lignocelluloses factor* (LF), consisting of the subsequent enzymes: β-endoglucanase (or endocellulase), β-glucosidase, β-xylosidase, β-exoglucanase, phenolxidase and peroxidase (Table 2). The LF factor was markedly related to the amount of weight-loss of the plant residues over time. Some authors have considered this approach as one of the best among factors using enzymatic activities as indicators of soil fertility and quality because of its precise choice of enzyme activities (Nannipieri et al. 2002). The LF factor was only criticized for the fact that it considers only the enzymes engaged in the carbon cycle, excluding those of the nitrogen, phosphorus and sulfur chnages. Thus, the LF factor can be considered as a good index of a soil's ability to transform lignocellulosic material, but not as an index of the overall ability of the soil to break down organic compounds.

Table 2 Indexes of soil status based on enzymatic activity and other selected soil properties

Authors	Index	Formula	Purpose/application
Beck (1984)	Enzymatic Activity Number (EAN)	$EAN = 0.2\left(0.15\ DHA + CAT + 1.25.10^{-5}\ PHA + 4.10^{-2}PROT + 6.10^{-4}AML\right)$	Assessment of the influence of soil management (e.g. municipal refuse) on the quality of cultivated and forest soils as well as pastures
Stefanic et al. (1984)	Biological Index of Fertility (BIF)	$BIF = (DHA + kCAT)/2$	Often used in order to determine soil changes as affected by various management practices and degrading factors, e.g. by soil salinity, heavy metals, varied organic and mineral fertilization, crop rotation
Sinsabaugh et al. (1992)	Lignocelluloses factor (LF)	$LF = \left(\beta\text{-GLU}, \beta\text{-enGL or enCEL}, \beta\text{-exG}, \beta\text{-XYL}, \text{PHOX and PR}\right)$	Used as a good indicator of soil's capacity to degrade lignocelluloses material in various soils. The LF index was significantly related to the amount of weight-loss of the plant remains over time
Myśków et al. (1996)	The Biological Index of Soil Fertility (F)	$F = \sqrt{M^2 + H^2 + T^2}$	Valid for assessment of the influence of organic and mineral fertilization on the quality of light and medium agricultural soils. The rightness of the proposed index was confirmed by the positive relationship between its values and yields of maize and potatoes
Gajda et al. (2000)	Index of Soil Biological Activity (SBI)	$SBI = \left[(PHAac \times 10) + (PHAal \times 10) + DHA + MB\right]/100$	Used for evaluation of soil fertility in arable soils. The SBI index was significantly correlated with the yield of potatoes and spring barley
Piotrowska (2002)	Biochemical Index of Soil Fertility (B)	$B = Corg + Ntot + DHA + PHAal + PROT + AML$	Assessment of the influence of soil management, e.g. varied organic and mineral fertilizers, on the status of arable soils
Puglisi et al. (2006)	Soil Alteration Indexes Al_1, Al_2 Al_3	$Al_1 = -21.30 \times ARYL + 35.2 \times \beta\text{-GLU} - 10.20 \times PHA - 0.52 \times UR - 4.53 \times INV + 14.3 \times DHA + 0.003 \times PHOX$ $Al_2 = 36.18 \times \beta\text{-GLU} - 8.72 \times PHA - 0.48 \times UR - 4.19 \times INV$ $Al_3 = 7.87 \times \beta\text{-GLU} - 8.22 \times PHA - 0.49 \times UR$	Evaluation of the quality of three agricultural sites contaminated with industrial and municipal wastes, organic fertilization or irrigation under different crops

Bastida et al. (2006)	Microbial Degradation Index (MID)	$\mathrm{MID} = [0.89(1/1 + (\mathrm{DHA}/4.87) - 2.5)] + [0.86(1/1 + (\mathrm{WSC}/11.09) - 2.5)] + [0.84(1/1 + (\mathrm{UR}/1.79) - 2.5)] + [0.72(1/1 + (\mathrm{RESP}/18.01) - 2.5)]$	Valid for determination of the degree of degradation of semiarid soils. Due to high sensitivity of included parameters the MID allowed to establish four levels of soil degradation (very high, high, low and null)

DHA dehydrogenases, *CAT* catalase, *PHA* phosphatase, *PROT* proteases, *AML* amylases, *k* proportionality coefficient (0.01), *β-GLU* β glucosidase, *β-enGL* β-1,4-endoglucanase, *enCEL* endocellulase, *β-exGL* β-1,4-exoglucanase, *β-XYL* β-xylosidase, *PHOX* phenoloxidase, *PR* peroxidase, *M* soil biological activity (interchangeably: bacteria count, bacteria/fungi ratio, phosphatase activity, dehydrogenase activity), *H* organic carbon content, *T* soil sorption capacity, *PHAac* acid phosphatase, *PHAal* alkaline phosphatse, *MB* microbial biomass content, *Corg* organic carbon, *Ntot* total nitrogen, *ARL* arylsulphatase, *UR* urease, *INV* invertase, *WSC* water-soluble carbohydrates, *RESP* soil respiration

Table 3 Empirical indices formulated on the basis of some enzymatic activities and other soil properties

Dependent variable	Independent variables	Mathematical expression	Purpose/application	Bibliography
Organic matter content	MBC, β-GLU, ARYL	Organic matter content $\left(\mathrm{g\ kg^{-1}}\right) = 5.35 + 0.017\ (\mathrm{MBC}, \mu\mathrm{g\ g^{-1}}), + 0.06\ \left(\beta\text{-GLU}, \mu\mathrm{M}\ p\text{-NP}\ \mathrm{g^{-1}h^{-1}}\right) + 1.08\ \left(\mathrm{ARYL}, \mu\mathrm{M}p\text{-NP}\ \mathrm{g^{-1}h^{-1}}\right)$	Assessment of 21 soils submitted to a high degree of erosion	García and Hernández (1997)
Total nitrogen	MBC, PMN, PHA, β-GLU, UR	$\text{Total N}(\%\mathrm{w/w}) = 0.38 \times 10^{-3}\left({}^{1}\mathrm{MBC}, \mu\mathrm{g\ g^{-1}}\right) + 1.4 \times 10^{-3}\left(\mathrm{PMN, mg\ kg^{-1}}\right) + 13.6 \times 10^{-3}\left(\mathrm{PHA}, \mu\mathrm{M\ product\ g^{-1}h^{-1}}\right) + 8.9 \times 10^{-3}\left(\beta\text{-GLU}, \mu\mathrm{M\ product\ g^{-1}h^{-1}}\right) + 1.6 \times 10^{-3}\left(\mathrm{UR}, \mu\mathrm{M\ product\ g^{-1}h^{-1}}\right)$	Determination of soils under climax vegetation	Trasar-Cepeda et al. (1998)
Total carbon	C-H_2Oextr, CEL, DHA, RESP	Biological Quality Index (BQI);Total C $= -2.92 + 0.037 \times$ C-H_2Oextr $- 0.096 \times$ CEL $+ 0.081$ DHA $+ 0.009 \times$ RESP	Assessment of variation in relation to the eco-system degradation	Arms et al. (2007)
Organic carbon	ARYL, DHA, PHA, β-GLU	Organic C $=$ -0.4008 $\times$ ARYL $+ 0.4153 \times$ DHA $+ 0.40033 \times$ PHA $+ 0.4916 \times$ β-GLU $= 0.8675\ (P < 0.0001)$	Evaluation of soil in different states of degradation	de la Paz Jimenez et al. (2002)
Total nitrogen	P, MWC, PHA, UR, MBC, β-GLU	Total N $\left(\mathrm{g\ kg^{-1}}\right) = 0.448\ \left(\mathrm{P\ mg\ kg^{-1}}\right) + 0.017\ (\mathrm{MWC\%}) + 0.410\ \left(\mathrm{PHA}\ \mu\mathrm{M}p\text{-NP}\ \mathrm{g^{-1}h^{-1}}\right) - 0.567\ \mathrm{UR}\ \left(\mu M\ \mathrm{NH^{+}{}_{4}g^{-1}h^{-1}}\right) + 0.001\ \left(\mathrm{MBC\ mg\ kg^{-1}}\right) + 0.410\ \left(\beta\text{-GLU}\ \mu\mathrm{M}\ p\text{-NP}\ \mathrm{g^{-1}h^{-1}}\right) - 0.980$	Valid for Mollisol. Evaluation of forests soils under natural vegetation without human intervention	Zarnoza et al. (2007)
Organic carbon	P, β-GLU, UR	Organic C $\left(\mathrm{g\ kg^{-1}}\right) = 4247\ \left(\mathrm{P\ mg\ kg^{-1}}\right) + 8183\ \left(\beta\text{-GLU}\ \mu M\ p\text{-NP}\ \mathrm{g^{-1}h^{-1}}\right) - 7949\ \left(\mathrm{UR}\,\mu M\ \mathrm{NH^{+}{}_{4}\ g^{-1}h^{-1}}\right) + 17{,}333$	Valid for Entisol. Evaluation of forests soils under natural vegetation without human intervention	Zarnoza et al. (2007)

Total carbon	MBC, CAT, UR, PHA	Total carbon (%) $= 0.764 + (2.304 \times 10^{-3}$ ^{1}MBC mg kg^{-1}) $+ (0.936$ CAT mM $H_2O_2 g^{-1} h^{-1}) + (0.017$ UR mM N-$NH_4^+ g^{-1} h^{-1}) + (0.206$ PHA mM pNP $g^{-1} h^{-1})$	Used for evaluation of native grassland soils in Galicia and to assess the biochemical equilibrium of different grassland soils under contrasting management systems	Paz-Ferreiro et al. (2007, 2011)

MBC microbial biomass carbon, *β-GLU* β-glucosidase, *ARYL* arylsulphatase activity, *PMN* potentially mineralizable nitrogen, *PHA* phosphatase, *UR* urease; *C-H_2Oextr* water extractable carbon, *CEL* cellulase, *DHA* dehydrogenase, *P* phosphorus, *RESP* soil respiration, *MWC* maximum water capacity, *CAT* catalase

Also Puglisi et al. (2006) proposed indicators consisting of enzyme activities only. The authors presented three indicators (AI1, AI2, AI3) based on various enzymatic activities that have been used to assess the soil degradation caused by agricultural practices, such as application of organic fertilization and crop density (Table 2). The above-mentioned indexes have been prepared after selection (reduction) of the data with the canonical discriminant analysis (CDA) technique. By using the AI3 index the authors were able to differentiate soils that had been subjected to intense agriculture practices, irrigation with salty waste waters, pollution by a tanning industry wastes and heavy metals (Puglisi et al. 2006).

The above-characterized indicators, developed by using the enzymatic activities solely, did not take into account the concentration of the most important nutrients in soil, which are significant for plant growth and development. That is why Kang et al. (2005) developed the indicator that consisted of three separate sub-indicators: a soil nutrient indicator, a soil microbiological indicator and a crop-related indicator. The microbiological indicator (MI*i*) was based on the following properties: soil phosphatase and dehydrogenase activities, microbial biomass carbon (MBC) and nitrogen (MBN), soil respiration, potentially mineralizable nitrogen (PMN) and bacterial count. The authors used then the sub-indexes to calculate so called *Sustainability Index* in soil under wheat fertilized with different manures, and stated that the soil fertility increased in proportion to the applied manures.

Summarizing, in order to develop the global, widely used soil fertility indicator, a encouraging way is the use of complex indicators consisting of both biochemical (enzymatic) as well as different chemical and physical features. The addition of various properties into the complex indicator lead to better reflection of the complexness of a soil environment, although for the condition in which they were developed (Bastida et al. 2008). The major difficulty with the actually accessible indexes is that they generally have not been tried in sites besides those for which they were proposed. Thus, they can be used rather regionally but not globally (Bastida et al. 2008). As stated Burns et al. (2013), the elaboration of a reliable soil fertility indicator based on enzyme activities is today one of the most important research tasks in soil enzymology. The universal index for the determination of soil fertility as affected by various soil management practices, different climatic conditions in varied geographic areas is difficult to express because of the great changeability of microbiological and biochemical features themselves and the influence of different site-specific agents that influence soil biological properties. The attempt to form a universal index of soil fertility ought to be performed at the global level, considering different factors found in the study sites, like climatic parameters, various soil types and their management as well as different vegetation cover.

5 Enzymes and Their Use in Farm Animal Feed

Another area of enzymes application in agriculture is their use in farm animal nutrition, mainly in pigs and poultry feeding (Ravindran 2013). All animals need enzymes to digest feed. Although they produce some enzymes themselves, pigs and poultry cannot digest even to 25% of feed they receive as fodder (Barletta 2010). The absence of certain enzymes since the feed constituents contain indigestible anti-nutritional compounds that inhibit the digestive process and/or they do not have some peculiar enzymes that hydrolyze some compounds that are contained in the livestock has led to the wastage of few essential nutrients; the attempt to compensate for the appropriate nutritional elements has increased the cost of feed. However, the inclusion of feed enzymes is an effective substitute to achieve optimum feed efficiency (Bedford 2018).

The beginning of feed enzyme technology in Europe dates back to 1980s. Europe is the largest market for feed enzymes, with poultry being the largest segment globally. Asia-Pacific is the fastest-growing region, with a share of 33.4%, mainly due to the increased awareness among the farmers about the benefits of enzymes and the high demand for quality meat and meat products. Countries like India, Vietnam, and Thailand are likely to provide the future growth potential for the market studied. Presently, there are many active companies producing feed enzymes, such as Alltech Inc., BASF, DuPont-Danisco, and others. Leading companies are focused on acquiring small-scale enzyme manufacturing units for the expansion of their feed enzymes business in local as well as foreign markets). The global animal feed enzymes market was valued at US$ 842.9 Million in 2016, and is expected to reach US$ 1487 Million by 2025, expanding at a Compound Annual Growth Rate (CAGR) of 7.4% from 2017 to 2025 (https://www.marketsandmarkets.com).

Nowadays enzymes added to the feed breaking down fibre, starch, proteins and phytate are most often used in animal feeding. The major enzyme categories used for commercial manufacturing of feed enzymes are carbohydrase and phytase (Ravindran 2013). Carbohydrases are the most widely used enzymes, accounting for around 41.67% of the global feed enzyme market by volume. Among the carbohydrases fibre- and starch-degrading enzymes are most often used (Asmare and Mekuriaw 2014). The break-down of fibre, which could be soluble and insoluble, is significant because it acts as an anti-nutrient agent in various manners. Several nutritive compounds like protein and starch are caught within not soluble fibrous cell walls. Since animals such as swine and poultry do not have suitable enzymes to be able to digest the fiber occurring in the cell walls, they are unable to exploit these bound nutrients. Afterwards, soluble fibers dissolve in the animal's digestive track, creating sticky gels that keep nutrients and decrease the speed of digestion and movement of fodder through the digestive track (Ravindran 2013). Moreover, fiber can keep water and hold nutrients dissolved in water. Ultimately, fiber forms bulk in the intestine, which can diminish the passage of feed, decreasing feed uptake and further animal growth. β-glucanase and xylanase are the two most important fibre-decomposing enzymes supplemented to animal feeding (Singh et al.

2018). Arabinoxylans, present in large amount in cereals and their derivatives, are broke down by xylanases, In turn, β-glucanases decompose β-glucans that are especially present in cereals and their derivatives (O'Shea et al. 2010). An animal feed is less supplemented with other fiber degrading enzymes, such as α-galactosidase, β-mannanase and pectinase (Yiğit et al. 2014)

Starch, the most often occurring storage polysaccharide in plants, is the commonly used carbohydrate in the animal feeding. The rate of starch availability in plant-originated fodder will change with the rate of *resistant starch* (RS), size of starch granules, its structure and encapsulation (Nissar et al. 2017). Variability in plant genes, factors affecting plant grow and development, conditions during starch harvesting and storage as well as feed processing are influencing changeability in starch digestibility and availability. Amylases, family of the starch-degrading enzymes that decompose starch in grains, their derivatives and some plant proteins (Dhital et al. 2017). Amylolytic enzymes by accelerating starch degradability and availability, allow animals to get more nutrients from the fodder, which will in turn improve animal production efficiency. Supplementing with amylases is especially important in young pig feeding, because of their immature digestive track where low feed uptake post-weaning is connected with a low level of produced amylase. Additionally, amylases make also possible the use of less processed grain in the diet, what allow the breeders to reduce the cost associated with feeding, without decreasing animal growth after weaning (Isaksen et al. 2010).

Protein-degrading enzymes (proteases) applied in swine and poultry feeding are able to decompose storage proteins in different plant components and proteinaceous anti-nutrients in plant proteins that interfere with normal physiological activities of animals (Isaksen et al. 2010). Grains, especially of leguminous plants like soybean, have a high content of storage proteins that can be bound to starch. Proteolytic enzymes are responsible for breaking down the proteins, releasing bound starch that can then be decomposed by amylolytic enzymes. Trypsin and lectins are two most often occurring proteinaceous anti-nutrients compounds that are inhibitors of enzymes taking part in digestion (Emire et al. 2013). Trypsin inhibitors (TI) are determined in raw plant (e.g. soybeans) proteins. They can hamper digestion by decreasing the activity of trypsin, which is the enzyme excreted by the pancreas and takes part in decomposing of protein in the small intestine (Erdaw and Beyene 2018). Lectins, in turn, are carbohydrates binding (glyco) proteins and they also are considered to decrease proteins digestibility (Vasconcelos and Oliveira 2004). The heat processing is the common practice to decrease the amount of TI and lectins in soybean-based feed. The excessive heating however diminishes the availability of amino acids, e.g. lysine. Thus only optimally processed soybean-based feed will contain required amount of amino acids and residual (not demanding) levels of TI and lectins. By reducing the levels of lectins and trypsin inhibitors, proteases are important for increasing protein digestibility (Erdaw and Beyene 2018).

Phosphorus (P) is commonly known to be important in some metabolic processes in animals as well as for their bone development. Majority of P in plant-origin food could be found as phytate (phytic acid), which is break down by the enzyme phytase (Greiner and Konietzny 2010; Humer et al. 2015). Corn, which is often used as fodder in poultry diet, consists of a large content of phytate. In the plant-origin fodder, the phytate is complexed with minerals such as phosphorus, calcium and with starch and proteins, making them difficult or even unassimilable for animals. Poultry and pigs are not able to secrete the phytase, which is why the addition of the enzyme to the fodder is highly advisable. The enzyme releases minerals and other phytate-bound compounds, which in turn can then be digested and assimilated by the animal to enhance the efficiency of meat, milk and egg production (Greiner and Konietzny 2010). Phytases also decrease the risk of pollution of waters from excessive phosphorus produced by pigs and poultry (Vats et al. 2007).

Summarizing, application of feed enzymes has some benefits:

- improving the use of feed and to reduce the associated cost—by decomposition of anti-nutrient components, leading to better digestion of the feed, resulting in increase of meat, caw milk or eggs production per kilogram of feed applied and their better quality
- reducing pollution of the environment. by improvement of digestion and absorption of nutrients from feed, decreasing the amount of manure produced as well as reduction of N and P secretion
- enhancing feed consistency—decreasing the nutritional changes in fodder, leading to more balanced feed for more steady animal growth and eggs and milk production
- help to keep the health of digestive truck—by amelioration of feed digestibility, fewer nutrients are available in the animal's intestine for the potential growth of pathogenic, causing disease microorganisms.

6 Conclusions and Future Challenges

Based on the reviewed literature, it can be concluded that interest in the application of enzymes to agricultural purposes is still very high. Special attention has been paid to soil enzymes since they are considered to play a significant role in soil organic matter transformation and nutrient cycling. Understanding the presence and activity of enzymes in soil may have important implications on ecosystem disturbances and can help to understand the transformation of organic matter and nutrients in sustainable soil management and sustaining agricultural productivity. Soil enzymes are willingly used in agronomy as reliable indicators of soil health, fertility and productivity as affected by differentiated natural and anthropogenic factors since they are more sensitive to any changes than other soil variables. Future research should investigate the global, widely used soil fertility/productivity index by using a complex mathematical expression consisting of various enzymatic and physicochemical properties. Such approach will lead to better reflection of the soil environment complexness, not only for the condition in which it has been formed. The development of a reliable soil fertility indicator based on enzyme activities is today one of the most important research tasks in soil enzymology.

The technology of feed enzyme is a fast-growing area of research in recent years. Exogenous enzymes are increasingly added to animal feed to help their own digestive enzymes and to decompose anti-nutritive fraction occurring in the feed. An appropriate supplementation of feed with exogenous enzymes allows to improving the use of feed and to reduce the associated cost. Future research should focus on finding and testing enzymes that are better adapted to the conditions, which exist in the digestive tract of animals. These approaches will further improve the effectiveness of animal production after feed enzymes application. Additionally, application of feed enzymes can reduce pollution of the environment. by improvement of digestion and absorption of nutrients from feed, decreasing the amount of manure produced as well as reduction of N and P secretion.

References

Abdelmagid HM, Tabatabai MA (1987) Nitrate reductase activity of soils. Soil Biol Biochem 19 (4):421–427

Acosta-Martinez V, Cruz L, Sotomayor-Ramírez D, Pérez-Algería L (2007) Enzyme activities as affected by soil properties and land use in tropical watershed. Appl Soil Ecol 35:35–45

Acosta-Martinez V, Mikha MM, Sistani KR, Stahlman PW, Benjamin JG, Vigil MF, Erickson R (2011) Multi-location study of soil enzyme activities as affected by types and rates of manure application and tillage practices. Agriculture 1:4–21

Adjei-Nsiah S, Kuyper TW, Leeuwis C, Abekoe MK, Giller KE (2007) Evaluating sustainable and profitable cropping sequences with cassava and four legume crops: effects on soil fertility and maize yields in the forest/savannah transitional agro-ecological zone of Ghana. Field Crops Res 103:87–97

Agricultural Enzymes Market – Global Forecast to 2022. https://www.marketsandmarkets.com/Market-Reports/feed-enzyme-market-1157.html

Allison SD, Vitousek PM (2005) Responses of extracellular enzymes to simple and complex nutrient inputs. Soil Biol Biochem 37:937–944

Arapoglou DG, Labropoulos AL, Varzakas TH (2010) Enzymes applied in food technology. In: Yildiz F (ed) Advances in food biochemistry. CRC, Taylor and Francis Group, Boca Ratan, London, New York, pp 101–126

Arms CM, Santana B, Mora JL, Notario JS, Arbelo CD, Rodriguez A (2007) A biological quality index for volcanic Andisols and Andisols (Canary Islands, Spain): variation related to the ecosystem development. Sci Total Environ 378:238–244

Asha B, Palaniswamy M (2018) Optimization of alkaline protease production by *Bacillus cereus* FT 1 isolated from soil. J Appl Pharm Sci 8(02):119–127

Asmare B, Mekuriaw Y (2014) Exogenous enzymes in nutrient digestion and health of non-ruminant: carbohydrases enzyme. Res J Agric Environ Manage 3(12):605–611

Bach CE, Warnock DD, Van Horn DJ, Weintraub MN, Sinsabaugh RL et al (2013) Measuring phenol oxidase and peroxidase activities with pyrogallol, L-DOPA, and ABTS: effect of assay conditions and soil type. Soil Biol Biochem 67:183–191

Bakshi M, Varma A (2011) Soil enzymes: state of art. In: Shukla S, Varma A (eds) Soil enzymology, soil biology, vol 22. Springer, Berlin, pp 1–23

Baležentienė L (2012) Hydrolases related to C and N cycles and soil fertility amendment: responses to different management styles and agro-ecosystems. Pol J Environ Stud 21(5):1153–1159

Balota EL, Dias Chaves JC (2010) Enzymatic activity and mineralization of carbon and nitrogen in soil cultivated with coffee and green manures. Rev Bras Ciênc Solo 43(5):1573–1583

Balota EL, Kanashiro M, Colozzi Filho A, Andrade DS, Dick RP (2004) Soil enzyme activities under long-term tillage and crop rotation systems in subtropical agro-ecosystems. Braz J Microbiol 35:300–306

Balota EL, Machineski O, Truber PV (2011) Soil enzyme activities under pig slurry addition and different tillage systems. Acta Sci Agron 33(4):729–737

Barletta A (2010) Introduction: current marked and expected development. In: Bedford MR, Partidge GG (eds) Enzymes in farm animal nutrition, 2nd edn. CAB International, Bodmin, pp 1–11

Bastida F, Moreno JL, Hernández T, García C (2006) Microbiological degradation index of soils in a semiarid climate. Soil Biol Biochem 38:3463–3473

Bastida F, Zsolnay A, Hernández T, García C (2008) Past, present and future of soil quality indices: a biological perspective. Geoderma 147:159–171

Beck TH (1984) Methods and application of soil microbiological analysis at the Landensanstalt fur Bodenkultur und Pflanzenbau (LBB) for determination of some aspects of soil fertility. In: Nemes MP, Kiss S, Papacostea P, Stefanic C, Rusan M (eds) Proceedings of the fifth symposium on soil biology. Rumania, Rumanian National Society of Soil Science, Bucharest, pp 13–20

Bedford MR (2018) The evolution and application of enzymes in the animal feed industry: the role of data interpretation. Br Poult Sci 59(5):486–493

Benitez E, Nogales R, Campos M, Ruano F (2006) Biochemical variability of olive-orchard soils under different management systems. Appl Soil Ecol 32:221–231

Bhattacharyya P, Chakrabarti K, Tripathy S, Chakrabarti A, Kim K, Kim SH (2007) L-asparaginase and L-glutaminase activities in submerged rice soil amended with municipal solid waste compost and decomposed cow manure. J Environ Sci Health B 42(5):593–598

Bielińska EJ, Pranagal J (2007) Enzymatic activity of soil contaminated with triazine herbicides. Pol J Environ Stud 16(2):295–300

Browman MG, Tabatabai MA (1978) Phosphodiesterase activity of soils. Soil Sci Am J 42(2):284–290

Bünemann EK, Bongiamo G, Bai Z, Creamer RE, De Deyn G, de Goede R, Fleskens L, Geissen V, Kuyper TW, Mäder P, Pulleman M, Sukkel W, van Groenigen JW, Brussaard L (2018) Soil quality – a critical review. Soil Biol Biochem 120:105–125

Burns RG (1982) Enzyme activity in soil: location and a possible role in microbial ecology. Soil Biol Biochem 14:423–427

Burns RG, DeForest JL, Marxsen J, Sinsabaugh RL, Stromberger ME, Wallenstein MD, Weintraub MN, Zoppini A (2013) Soil enzymes in a changing environment: current knowledge and future directions. Soil Biol Biochem 58:216–234

Caceres TP, He WX, Megharaj M, Naidu R (2009) Effect of insecticide fenamiphos on soil microbial activities in Australian and Ecuadorean soils. J Environ Sci Health B 44:13–17

Cenini VL, Fornara DA, McMullan G, Ternan N, Corlan R, Crawley MJ, Clément JC, Lavorel S (2016) Linkages between extracellular enzyme activities and the carbon and nitrogen content of grassland soils. Soil Biol Biochem 96:198–206

Corchran VL, Elliott LF, Lewis CE (1989) Soil microbial biomass and enzyme activity in subarctic agricultural and forest soils. Biol Fertil Soils 7:283–288

Crecchio C, Curci M, Pizzigallo MDR, Riccuti P, Ruggiero P (2004) Effects of municipal solid waste compost amendments on soil enzyme activities and bacterial genetic diversity. Soil Biol Biochem 36:1595–1605

Das SK, Varma A (2011) Role of enzymes in maintaining soil health. In: Shukla G, Varma A (eds) Soil enzymology, soil biology, vol 22. Springer, Berlin, pp 25–42

Datta R, Anand S, Moulick A, Baraniya D, Pathan SI, Rejsek K, Vranova V, Sharma M, Sharma D, Kelkar A, Formanek P (2017) How enzymes are adsorbed on soil solid phase and factors limiting its activity: a review. Int Agrophys 3:287–302

De Castro Lopes AA, Gomes de Suosa DM, Montandon Chaer G, dos Reis FB, Goedert WJ, de Carvalho MI (2012) Interpretation of microbial soil indicators as function of crop yield and organic carbon. Soil Sci Soc Am J 77:461–472

De la Paz-Jiménez M, De la Horra AM, Pruzzo L, Palma RM (2002) Soil quality: a new index base microbiological and biochemical parameters. Biol Fertil Soils 35:302–306
Deborah BV, Mohiddin MJ, Madhuri RJ (2013) Interaction effects of selected pesticides on soil enzymes. Toxicol Int 209(3):195–200
Defo MA, Njine T, Nola M, Beboua FS (2011) Microcosm study of the long term effect of endosulfan on enzyme and microbial activities on two agricultural soils of Yaounde-Cameroon. Afr J Agric Res 6:2039–2050
Demkina EV, Shanenko EF, Nikolaev YA, El-Registan GI (2017) Model of the regulation of activity of immobilized enzymes (amylases) in soil. Microbiology 86(2):231–240
Deng S, Popova I (2011) Carbohydrate hydrolases. In: Dick RP (ed) Methods of soil enzymology, vol 9. Soil Science Society of America, Madison, WI, pp 185–207
Dhital S, Warren FJ, Butterworth PJ, Ellis PR, Gidley MJ (2017) Mechanisms of starch digestion by α-amylase—structural basis for kinetic properties. Crit Rev Food Sci Nutr 57(5):875–892
Dick RP (1994) Soil enzyme activity as an indicator of soil quality. In: Doran JW, Leman DC, Bezdicek DF, Steward BA (eds) Defining soil quality for a sustainable environment. SSSA, Special Publication 35, Madison, WI, pp 107–124
Dick RP (ed) (2011) Methods of soil enzymology. Soil Science Society of America, Madison, WI
Dodor DE, Tabatabai MA (2003) Aminohydrolases in soil as affected by cropping systems. Appl Soil Ecol 24:73–90
Dodor DE, Tabatabai MA (2005) Glycosidases in soils as affected by cropping systems. J Plant Nutr Soil Sci 168:749–758
Dodor DE, Tabatabai MA (2007) Arylamidase activity as an index of nitrogen mineralization in soils. Commun Soil Sci Plant Anal 36(15–16):2197–2207
Doubnerová V (2012) Utilization of enzymes in biochemistry and analytical biochemistry. Biochem Anal Biochem 1(4):1–2
Dwivedi RS (2017) Soil fertility. In: Remote sensing of soils. Springer, Berlin, pp 457–495
Egamberdieva D, Kucharova Z (2008) Cropping effects on microbial population and nitrogenase activity in saline arid soil. Turk J Biol 32:85–90
Ekenler M, Tabatabai MA (2003) Effects of liming and tillage systems on soil microbial biomass and glycosidases in soil. Biol Fertil Soils 39:51–61
Emire SA, Jha Y, Mekam F (2013) Role of anti-nutrient factors in food industry. Beverage Food World:23–28
Erdaw MM, Beyene WT (2018) Anti-nutrients reduce poultry productivity: influence of trypsin inhibitors on pancreas. Asian J Pouls Sci 12(1):1–12
Fang S, Liu D, Tian Y, Deng S, Shang X (2013) Tree species composition influences enzyme activities and microbial biomass in the rhizosphere: a rhizobox approach. PLoS One 8(4):e61461
Feller C, Blanchart E, Bernoux M, Lal R, Manlay R (2012) Soil fertility concepts over the past two centuries: the importance attributed to soil organic matter in developed and developing countries. Arch Agron Soil Sci 58, S1, S3–S21, pp 19
Gajda A, Martyniuk S (2005) Microbial biomass C and N and activity of enzymes in soil under winter wheat grown in different crop management systems. Pol J Environ Stud 14(2):159–163
Gajda AM, Martyniuk S, Stachyra AM, Wróblewska B, Zięba S (2000) Relations between microbiological and biochemical properties of soil under different agrotechnical conditions and its productivity. Pol J Soil Sci 33(2):55–60
García C, Hernández T (1997) Biological and biochemical indicators in derelict soils subject to erosion. Soil Biol Biochem 29:171–177
García-Ruiz R, Ochoa V, Hijanosa MB, Carreira JA (2008) Suitability of enzyme activities for the monitoring of soil quality improvement in organic agricultural systems. Soil Biol Biochem 40:2137–2145
Giacometti C, Deyman MS, Cavani L, Marzadori C, Ciavatta C, Kandeler E (2013) Chemical and biochemical soil quality indicators and their potential to differentiate fertilization regimes in temperate agroecosystems. Appl Soil Ecol 64:32–48
Gianfreda L, Bollag JM (1996) Influence of natural and anthropogenic factors on enzyme activity in soil. In: Stotzky G, Bollag LM (eds) Soil biochemistry, vol 9. Marcel Dekker, New York, Basel, pp 123–193

Gianfreda L, Ruggiero P (2006) Enzyme activities in soil. In: Nannipieri P, Smalla K (eds) Nucleic acids and proteins in soil. Springer, Berlin, pp 258–311
Gianfreda L, Scarfi MR (1991) Enzyme stabilization: state of art. Mol Cell Biochem 100:97–128
Gil-Sotres F, Trasar-Cepeda C, Leirós MC, Seoane S (2005) Different approaches to evaluating soil quality using biochemical properties. Soil Biol Biochem 37:877–887
Glibert P, Harrison J, Heil C, Seitzinger S (2006) Escalating worldwide use of urea – a global change contributing to coastal eutrophication. Biogeochemistry 77:441–463
Gong S, Zhang T, Guo R, Hongbin C, Lianxuan S, Guo J, Sun W (2015) Response of soil enzyme activity to warming and nitrogen addition in a meadow steppe. Soil Res 53(3):242–252
Gonzalez MG, Gallardo JF, Gomez E, Masciandaro G, Ceccanti B, Pajares S (2007) Potential universal applicability of soil bioindicators: evaluation in three temperate ecosystems. Cienc Suelo 25(2):151–158
Greiner R, Konietzny U (2010) Phytases: biochemistry, enzymology and characteristics relevant to animal feed use. In: Bedford MR, Partidge GG (eds) Enzymes in farm animal nutrition, 2nd edn. CAB International, Wallingford, pp 96–128
Guo P, Wang C, Feng X, Su M, Zhu W, Tian X (2011) Mixed inorganic and organic nitrogen addition enhanced extracellular enzymatic activities in a subtropical forest soil in East China. Water Air Soil Pollut 216(1):229–237
Hai-Ming T, Xiao-Ping X, Wen-Guang T, Ye-Chun L, Ke W, Guang-Li Y (2014) Effects of winter cover crops residue returning on soil enzyme activities and soil microbial community in double-cropping rice fields. PLoS One 9(6):e100443
Hemalatha T, Uma Maheswari T, Krithiga G, Sankaranarayanan K, Puvanakrishnan R (2013) Enzymes in clinical medicine: an overview. Indian J Exp Biol 51:777–788
Herrero O, Canet R, Albiach R, Pomares F (1998) Enzymatical activities and content of mineral nitrogen in soil after the application of two rates of different organic products. Agrochimica 42 (6):296–303
Hoffman BM, Lukoyanov D, Yang Z-Y, Dean DR, Seefeldt LC (2014) Mechanism of nitrogen fixation by nitrogenase: the next stage. Chem Rev 114:4041–4062
Huang M, Zou Y, Jiang P, Xia B, Feng Y, Cheng Z, Mo Y (2012) Effect of tillage on soil and crop properties of wet-seeded flooded rice. Field Crops Res 129:28–38
Humer E, Schwarz C, Schedle K (2015) Phytate in pig and poultry nutrition. J Anim Physiol Anim Nutr 99:605–625
Husain Q (2016) Magnetic nanoparticles as a tool for the immobilization/stabilization of hydrolases and their applications: an overview. Biointerfaces Res Appl Chem 6(6):1585–1606
Husain Q (2017) High Yield immobilization and stabilization of oxidoreductases using magnetic nanosupports and their potential applications: an update. Curr Catal 6(3):168–187
Hussain S, Siddique T, Saleem M, Arshad M, Khalid A (2009) Impact of pesticides on soil microbial diversity, enzymes, and biochemical reactions. Adv Agron 102(1):159–200
Isaksen MF, Cowieson AJ, Kragh KM (2010) Starch- and protein-degrading enzymes: biochemistry, enzymology and characteristics relevant to animal feed use. In: Bedford MR, Partidge GG (eds) Enzymes in farm animal nutrition, 2nd edn. CAB International, Wallingford, pp 85–95
Iyyemperumal K, Shi W (2008) Soil enzyme activities in two forage systems following application of different rates of swine lagoon effluent or ammonium nitrate. Appl Soil Ecol 38:128–136
Jastrzębska E (2011) The effect of chlorpyrifos and teflubenzuron on the enzymatic activity of soil. Pol J Environ Stud 20:903–910
Kalyani SS, Sharma J, Dureja P, Singh S, Lata (2010) Influence of endosulfan on microbial biomass and soil enzymatic activities of a tropical Alfisol. Bull Environ Contam Toxicol 84:351–356
Kang GS, Beri V, Sidhu BS, Rupela OP (2005) A new index to assess soil quality and sustainability of wheat-based cropping systems. Biol Fertil Soils 41:389–398
Karam J, Nicell JA (1997) Potential applications of enzymes in waste treatment. Rev J Chem Tech Biotechnol 69:141–153
Khan AR (1996) Influence of tillage on soil aeration. J Agron Crop Sci 177:253–259
Khan SJ, Nael M, Safari Sinejani AA, Asadian G, Salari Nik K (2014) The effect of vegetation type on selected soil quality indicators in a semiarid rangeland in Hamedan, Iran. Eur J Soil Sci 4:70–75

Kibet LC, Blanco-Canqui H, Jasa P (2016) Long-term tillage impacts on soil organic matter components and related properties on a Typic Argiudoll. Soil Till Res 155:78–84

King T, Schoenau JJ, Malhi S (2015) Effect of application of liquid swine manure on soil organic carbon and enzyme activities in two contrasting Saskatchewan soils. Sustain Agric Res 4 (1):13–25

Klose S, Tabatabai MA (2000) Urease activity of microbial biomass in soils as affected by cropping systems. Biol Fertil Soils 31:191–199

Klose S, Moore JM, Tabatabai MA (1999) Arylsulfatase activity of microbial biomass in soils as affected by cropping systems. Biol Fertil Soils 29:46–54

Klose S, Acosta-Martinez V, Ajwa HA (2006) Microbial community composition and enzyme activities in a sandy loam soil after fumigation with methyl bromide or alternative biocides. Soil Biol Biochem 38:1243–1254

Kujur M, Pater AK (2014) Kinetics of soil enzyme activities under different ecosystems: an index of soil quality. Chilean J Agric Res 74(10):1–9

Ladd JN (1985) Soil enzymes. In: Vaugham D, Malcolm RE (eds) Soil organic matter and biological activity. Martinus Nijhoff/Dr W Junk Publisher, Dordrecht, pp 175–221

Laishram J, Saxena KG, Maikhuri RK, Rao KS (2012) Soil quality and soil health: a review. Int J Ecol Environ Sci 38(1):19–37

Laudicina VA, Dennis PG, Palazzolo E, Badalucco L (2012) Key biochemical attributes to assess soil ecosystem sustainability. In: Malik A, Grohmann E (eds) Environmental protection strategies for sustainable development, strategies for sustainability. Springer, Dordrecht, pp 193–227

Lever G (2015) Proteins, enzymes and biological catalysis. In: Large-scale Quantum-mechanical enzymology. Springer, Switzerland, pp 9–18

Liao L, Feng Z, Fu J, Liu X, Dong Z, Dong T, Wang Z (2017) Increasing the amount of nitrogen fertilizer decreased the activity of soil enzyme in *cv*. Huangguogan. In: Conference proceedings of 2nd international conference on materials science, resource and environmental engineering (MSREE), AIP Publishing, Melville, NY, pp 1–4

Madejon E, Moreno F, Murillo JM, Pelegrin F (2007) Soil biochemical response to long-term conservation tillage under semi-arid Mediterranean conditions. Soil Till Res 94:346–352

Makoi JH, Ndakidemi PA (2008) Selected soil enzymes: examples of their potential roles in the ecosystem. Afr J Biotechnol 7:181–191

Mandal A, Patra AK, Singh D, Swarup A, Ebhin Masto R (2007) Effect of long-term application of manure and fertilizer on biological and biochemical activities in soil during crop development stages. Bioresour Technol 98:3585–3592

Mangalassery S, Mooney SJ, Sparkes DL, Fraser WT, Sjögersten S (2015) Impacts of zero tillage on soil enzyme activities, microbial characteristics and organic matter functional chemistry in temperate soils. Eur J Soil Biol 68:9–17

McDaniel MD, Tiemann LK, Grandy AS (2014) Does agricultural crop diversity enhance soil microbial biomass and organic matter dynamics? A meta-analysis. Ecol Appl 24(3):560–570

Melero S, Madejón E, Ruiz JC, Herencia JF (2007) Chemical and biochemical properties of a clay soil under dryland agriculture system as affected by organic fertilization. Eur J Agron 26 (3):327–334

Merino C, Godoy R, Matus F (2016) Soil enzymes and biological activity at different levels of organic matter stability. J Soil Sci Plant Nutr 16(1):14–30

Mina BL, Saha S, Kumar N, Srivastova AK, Gupta HS (2008) Changes in soil nutrient content and enzymatic activity under conventional and zero-tillage practices in an Indian sandy clay loam soil. Nutr Cycl Agroecosyst 82:273–281

Moeskops B, Buchan D, Sleutel S, Herawaty L, Husen E, Saraswati R, Setyorini D, De Neve S (2010) Soil microbial communities and activities under intensive organic and conventional vegetable farming in West Java, Indonesia. Appl Soil Ecol 45:112–120

Mohammadi K (2011) Effect of different fertilization methods on soil biological indexes. World Acad Sci Eng Technol 78:407–410

Moore AW, Russel JS (1972) Factors affecting dehydrogenase activity as an index of soil fertility. Plant Soil 37:675–682

Muruganandam S, Israel DW, Robarge WP (2009) Activities of nitrogen-mineralization enzymes associated with soil aggregate size fractions of three tillage systems. Soil Sci Soc Am J 73:751–759

Myśków W, Stachyra A, Zięba S, Masiak D (1996) Biological activity of soil as an index of its fertility. Soil Sci Sci 47(1/2):89–99

Nannipieri P (1994) The potential use of soil enzymes as indicators of productivity, sustainability and pollution. In: Pankhurst CE, Doube BM, Gupta VVSR, Grace PR (eds) Soil biota: management in sustainable farming systems. CSIRO, Adelaide, pp 238–244

Nannipieri P, Sequi P, Fusi P (1996) Humus and enzyme activity. In: Piccolo A (ed) Humic substances in terrestrial ecosystems. Elsevier, Amsterdam, pp 293–328

Nannipieri P, Kandeler E, Rugiero P (2002) Enzyme activities and microbiological and biochemical processes in soil. In: Burs RG, Dick RP (eds) Enzymes in the environment. Activity, ecology and application. Marcel Dekker, New York, pp 1–33

Nannipieri P, Giangoni L, Renella G (2011) Role of phosphatases in soil. In: Bunemann EK et al (eds) Phosphorus in action, soil biology, vol 26. Springer, Berlin, pp 215–243

Nannipieri P, Giagnoni L, Renella G, Puglisi E, Ceccanti B, Masciandaro G, Fornasier F, Moscatelli M, Marinari S (2012) Soil enzymology: classical and molecular approaches. Biol Fertil Soils 48:743–762

Niculina G, Ştefanic G (2008) Agronomical significance of the pedoenzymical tests. Proc Rom Acad Ser B 1–2:123–128

Nissar J, Ahad T, Naik HR, Hussain SZ (2017) Resistant starch – chemistry and nutritional properties. Int J Food Sci Nutr 2(6):95–108

O'Shea CJ, Sweenev T, Lynch MB, Gahdan DA, Callan JJ, O'Doherty JV (2010) Effect of beta-glucans contained in barley- and oat-based diets and exogenous enzyme supplementation on gastrointestinal fermentation of finisher pigs and subsequent manure odor and ammonia emissions. J Anim Sci 88(4):1411–1420

Orczewska A, Piotrowska A, Lemanowicz J (2012) Soil acid phosphomonoesterase activity and phosphorus forms in ancient and postagricultural black alder [*Alnus glutinosa* (L.) Gaertn.] woodlands. Acta Soc Bot Pol 81(2):81–86

Pajares S, Gallardo GF, Masciandaro G, Ceccanti B, Etchevers JD (2011) Enzyme activity as an indicator of soil quality in degraded cultivated acrisols in the Mexican trans volcanic belt. Land Degrad Dev 2(3):373–381

Pancholy SK, Rice EL (1972) Soil enzymes in relation to old field succession: amylase, cellulase, invertase, dehydrogenase, and urease. Soil Sci Soc Am Proc 37:47–50

Pandey K, Singh B, Pandey A, Badruddin IJ, Pandey S, Mishra VK, Jain PA (2017) Application of microbial enzymes in industrial waste water treatment. Int J Curr Microbiol Appl Sci 6 (8):1243–1254

Paz-Ferreiro J, Trasar-Cepeda C, Leirós MC, Seoane S, Gill-Sotres F (2007) Biochemical properties of acid soils under native grassland in a temperate humid zone. NZ J Agric Res 50:537–548

Paz-Ferreiro J, Trasar-Cepeda C, Leirós MC, Seoane S, Gill-Sotres F (2011) Intra-annual variation in biochemical properties and the biochemical equilibrum of different grassland soils under contrasting management and climate. Biol Fertil Soils 47:633–645

Piotrowska A (2002) The usefulness of biochemical indexes for soil fertility evaluation as dependent on crop rotation and differentiated organic-mineral fertilization. PhD thesis, University of Technology and Life Sciences, Bydgoszcz, Poland, pp 128

Piotrowska A, Długosz J (2012) Spatio-temporal variability of microbial biomass content and activities related to some physicochemical properties of *Luvisols*. Geoderma 173–174:199–208

Piotrowska A, Wilczewski E (2012) Effects of catch crops cultivated for green manure and mineral nitrogen fertilization on soil enzyme activities and chemical properties. Geoderma 189–190:72–80

Piotrowska-Długosz A (2014) Enzymes and soil fertility. In: Gianfreda L, Rao MA (eds) Enzymes in agricultural sciences. OMICS Group eBooks, Foster City, pp 44–79

Piotrowska-Długosz A, Wilczewski E (2014) Assessment of soil nitrogen and related enzymes as influenced by the incorporation time of field pea cultivated as a catch crop in *Alfisol*. Environ Monit Assess 186(12):8425–8441

Puglisi E, Del Re AAM, Rao MA, Gianfreda L (2006) Development and validation of numerical indexes integrating enzyme activities of soils. Soil Biol Biochem 38:1673–1681
Quiquampoix H, Servagent-Noinville S, Baron M-H (2002) Enzyme adsorption on mineral surfaces and consequences for the catalytic activity. In: Burns RG, Dick RP (eds) Enzymes in the environment. Activity, ecology and applications. Marcel Dekker, New York, pp 285–306
Ramudu AC, Mohiddin GJ, Srinivasulu M, Madakka M, Rangaswamy V (2011) Impact of fungicides chlorothalonil and propiconazole on microbial activities in groundnut (*Arachis hypogaea* L.) soils. ISRN Microbiol, I, pp 7
Rao CS, Grover M, Kudu S, Desai S (2017) Soil enzymes. In: Lal R (ed) Encyclopedia of soil science. Taylor & Francis Group, New York, pp 8
Rastin N, Rosenplänter K, Hüttermann A (1988) Seasonal variation of enzyme activity and their dependence on certain soil factors in a beech forest soil. Soil Biol Biochem 20(5):637–642
Ravindran V (2013) Feed enzymes: the science, practice, and metabolic realities. J Appl Poult Res 22(3):628–636
Riah W, Laval K, Laroche-Ajzenberg E, Mougin C, Latour X, Trinsoutrot-Gattin I (2014) Effects of pesticides on soil enzymes: a review. Environ Chem Lett 12:257–273
Riffaldi R, Saviozzi A, Levi-Minzi R, Cardelli R (2002) Biochemical properties of Mediterranean soil as affected by long-term crop management systems. Soil Till Res 67:109–114
Robinson PK (2015) Enzymes: principles and biotechnological applications. Essays Biochem 59:1–41
Rodríguez-Kábana R (1982) The effects of crop rotation and fertilization on soil xylanase activity in a soil of the southeastern United States. Plant Soil 64:237–247
Rodríguez-Kábana R, Truelove B (1982) Effects of crop rotation and fertilization on catalase activity in a soil of the southeastern United States. Plant Soil 69:97–104
Roldán A, Caravaca F, Hernández MT, García C, Sánchez-Brito C et al (2003) No-tillage, crop residue addition, and legume cover cropping effects on soil quality characteristics under maize in Patzcuaro watershed (Mexico). Soil Till Res 72:65–73
Roldán A, Salinas-García JR, Alguacil MM, Díaz E, Caravaca F (2005) Soil enzyme activities suggest advantages of conservation tillage practices in sorghum cultivation under subtropical conditions. Geoderma 129:178–185
Ross DJ (1975) Studies on climosequence of soils in tussock grassland. 5. Invertase and amylase activities of topsoil and their relationship with other properties. NZ J Sci 18:511–518
Safari Sinegani AA, Emtiazi G, Shariatmadari H (2005) Sorption and immobilization of cellulose on silicate clay minerals. J Coloid Interface Sci 290:39–44
Saiya-Cork KR, Sinsabaugh RL, Zak DR (2002) The effects of long term nitrogen deposition on extracellular enzyme activity in an *Acer saccharum* forest soil. Soil Biol Biochem 34:1309–1315
Samuel AD, Ciobanu C (2018) Enzyme activities in a preluvosoil as affected by crop rotation and fertilization systems. Rom Agric Res 35:183–191
Sannino F, Gianfreda L (2001) Pesticide influence on soil enzymatic activities. Chemosphere 54:417–425
Sardans J, Peñuelas J, Estiarte M (2012) Changes in soil enzymes related to C and N cycle and in soil C and N content under prolonged warming and drought in a Mediterranean shrubland. Appl Soil Ecol 39:223–235
Saviozzi A, Bufalino P, Levi-Minzi R, Riffaldi R (2002) Biochemical activities in a degraded soil restored by two amendments: a laboratory study. Biol Fertil Soils 35:96–101
Seifert S, Shaw DR, Zablatowicz RM, Wesley RA, Kingery WL (2001) Effect of tillage on microbial characteristics and herbicide degradation in a Sharkey clay soil. Weed Sci 49:685–693
Sherene T (2017) Role of soil enzymes in nutrient transformation: a review. Biol Bull 3(1):109–131
Shi W (2011) Agricultural and ecological significance of soil enzymes: soil carbon sequestration and nutrient cycling. In: Shukla G, Varma A (eds) Soil enzymology, soil biology, vol 22. Springer, Berlin, pp 43–73
Shi B, Zhang J, Wang C, Ma J, Sun W (2018) Responses of hydrolytic enzyme activities in saline-alkaline soil to mixed inorganic and organic nitrogen addition. Sci Rep 8(4543):12

Singh BK, Walker A, Wright DJ (2002) Degradation of chlorpyrifos, fenamiphos, and chlorothalonil alone and in combination and their effects on soil microbial activity. Environ Toxicol Chem 21(12):2600–2605

Singh N, Kuhur S, Priya K, Jaryal R, Yadav R (2018) Phytase: the feed enzyme, an overview. In: Gahlawat SK et al (eds) Advances in animal biotechnology and its applications. Springer Nature, Singapore, pp 269–327

Sinsabaugh RL (2010) Phenol oxidase, peroxidase and organic matter dynamics of soil. Soil Biol Biochem 42:391–404

Sinsabaugh RL, Antibus KR, Linkens EA, McClaugherty AC, Rayburn L, Weiland T (1992) Wood decomposition in a first order watershed: mass loss as a function of exoenzyme activity. Soil Biol Biochem 24:743–749

Sinsabaugh RL, Gallo ME, Lauber C, Waldrop MP, Zak DR (2005) Extracellular enzyme activities and soil organic matter dynamics for northern hardwood forests receiving simulated nitrogen deposition. Biogeochemistry 75:201–215

Siwik-Ziomek A, Lemanowicz J, Koper J (2013) Arylosulphatse activity and the content of total sulphur and its forms under the influence of fertilization with nitrogen and other macroelements. J Elem 3:437–447

Siwik-Ziomek A, Figas A, Rolbiecki R (2015) Influence of irrigation on the sulphur content and soil enzymes activity under *Silphium perfoliatum* L. EJPAU 18(3): #03

Skujins J (1978) History of abiontic soil enzyme research. In: Burns RG (ed) Soil enzymes. Academic, London, pp 1–49

Spohn M, Carminati A, Kuzyakov Y (2013) Soil zymography: a novel in situ method for mapping distribution of enzyme activity in soil. Soil Biol Biochem 58:275–280

Stefanic F, Ellade G, Chirnageanu J (1984) Researches concerning a biological index of soil fertility. In: Nemes MP, Kiss S, Papacostea P, Stefanic C, Rusan M (eds) Proceedings of the fifth symposium on soil biology. Rumanian National Society of Soil Science, Bucharest, pp 35–45

Steinweg JM, Dukes JS, Paul EA, Wallenstein MD (2013) Microbial responses to multifactor climate change: effects on soil enzymes. Front Microbiol 4:146

Stemmer M, Gerzabek MH, Kandeler E (1998) Invertase and xylanase activity of bulk soil and particle-size fraction during maize straw decomposition. Soil Biol Biochem 31:9–19

Sukul P (2006) Enzymatic activities and microbial biomass in soil as influenced by metalaxyl residues. Soil Biol Biochem 38:320–326

Szajdak L (1996) Impact of crop rotation and phonological periods on rhodanese activity and free sulfuric amino acids concentrations in soils under continuous rye cropping and crop rotation. Environ Int 22(5):563–569

Szajdak LW, Gaca W (2010) Nitrate reductase activity in soil under shelterbelt and an adjoining cultivated field. Chem Ecol 26(2):123–143

Tejada M, Gomez I, Garcia-Martinez AM, Osta P, Parrado J (2011) Effects of prochloraz fungicide on soil enzymatic activities and bacterial communities. Ecotox Environ Safe 74:1708–1714

Theng BKG (2012) Proteins and enzymes. In: Development in clay science, Formation and properties of clay-polymer complexes, vol 4. Elsevier, Amsterdam, pp 245–318

Tishchenko SA, Bezuglova OS, Morozov IV (2013) Specificity of the physical properties of soils in locally waterlogged landscapes in the lower reaches of the Don River. Eurasian Soil Sci 46 (3):297–302

Trasar-Cepeda C, Leiros MC, Gil-Sotres F, Seoane S (1998) Toward a biochemical quality index for soils: an expression relating several biological and biochemical properties. Biol Fertil Soils 26:100–106

Tripathi S, Chakraborty A, Chakrabarti K, Bandyopadhyay B (2007) Enzyme activities and microbial biomass in coastal soils of India. Soil Biol Biochem 39:2840–2848

Ulrich S, Tischer S, Hofmann B, Christen O (2010) Biological soil properties in a long-term tillage trial in Germany. J Plant Nutr Soil Sci 173:483–489

Utobo EB, Tewari L (2015) Soil enzymes as bioindicators of soil ecosystem status. Appl Ecol Environ Res 13(1):147–169

Uzun N, Uyanöz R (2011) Determination of urease and catalase activities and CO_2 respiration in different soils obtained from Konya, Turkey. Trends Soil Sci Plant Nutr 2(1):1–6
Vasconcelos IM, Oliveira JTA (2004) Antinutritional properties of plant lectins. Toxicon 44:385–403
Vats P, Bhattacharyya MS, Banerjee UC (2007) Use of phytases (*myo*-inositolhexakisphosphate phosphohydrolases) for combatting environmental pollution: a biological approach. Crit Rev Environ Sci Technol 35(5):469–486
Veeraragavan S, Duraisamy R, Mani S (2018) Seasonal variation of soil enzyme activities in relation to nutrient and carbon cycling in *Senna alata* (L.) Roxb invaded sites of Puducherry region, India. Geol Ecol Lands 2(3):155–168
Vranova V, Rejsek K, Formanek P (2013) Proteolytic activity in soil: a review. Appl Soil Ecol 70:23–32
Wang JB, Chen ZH, Chen LJ, Zhu AN, Wu ZJ (2011) Surface soil phosphorus and phosphatase activities affected by tillage and crop residue input amounts. Plant Soil Environ 57:251–257
Webb EC (1992) Enzyme nomenclature. Recommendations of the nomenclature committee of the international union of biochemistry and molecular biology on the nomenclature and classification of enzymes. Academic, San Diego
Wells AS, Finch GL, Michels PC, Wong JW (2012) Use of enzymes in the manufacture of active pharmaceutical ingredients – a science and safety-based approach to ensure patient safety and drug quality. Org Process Res Dev 16(12):1986–1993
Wolińska A, Stępniewska Z (2011) Microorganisms abundance and dehydrogenase activity as a consequence of soil reoxidation process. In: Miransari M (ed) Soil tillage and microbial activities. Research Singpost, Kerala, pp 111–143
Wyszkowska J, Kucharski J (2004) Biochemical and physicochemical properties of soil contaminated with herbicide Triflurotox 250 EC. Pol J Environ Stud 3:223–231
Yan H, Wang DD, Dong B, Tang FF, Wang BC, Fang H, Yu YL (2011) Dissipation of carbendazim and chloramphenicol alone and in combination and their effects on soil fungal: bacterial ratios and soil enzyme activities. Chemosphere 84:634–641
Yang L, Li T, Li F, Lemcoff JH, Cohen S (2008) Fertilization regulates soil enzymatic activity and fertility dynamics in a cucumber field. Sci Hortic 116(1):21–26
Yiğit NO, Koca SB, Didinen BI, Diler I (2014) Effect of β-mannanase and α-galactosidase supplementation to soybean meal based diets on growth, feed efficiency and nutrient digestibility of Rainbow Trout, *Oncorhynchus mykiss* (Walbaum). Asian Australas J Anim Sci 27(5):700–705
Yin R, Deng H, Wang H, Zhang B (2014) Vegetation type affects soil enzyme activities and microbial functional diversity following re-vegetation of a severely eroded red soil in sub-tropical China. Catena 115:96–103
Yuan B, Yue D (2012) Soil microbial and enzymatic activities across a chronosequence of Chinese pine plantation development on the loess plateau of China. Pedosphere 22:1–12
Zablatowicz RM, Hoagland RE, Wagner SC (1998) 2-Nitroacetanilide as substrate for determination of aryl acylamidase activity in soils. Soil Biol Biochem 30:679–686
Zarnoza R, Mataix-Solera J, Guerrero C, Arcenegui V, Garcia-Orenes F, Mataix-Beneyto J, Morugan A (2007) Evaluation of soil quality using multiple lineal regression based on physical, chemical and biochemical properties. Sci Total Environ 378:233–237
Zhang YL, Chen LJ, Sun CX, Wu ZJ, Chen ZH et al (2010) Soil hydrolase activities and kinetic properties as affected by wheat cropping systems of Northeastern China. Plant Soil Environ 56:526–532
Zhang Y, Chen L, Wu Z, Sun C (2011) Kinetic parameters of soil β-glucosidase response to environmental temperature and moisture regimes. Rev Bras Cienc Solo 35(4):1285–1291
Zhao B, Chen J, Zhang J, Qin S (2010) Soil microbial biomass and activity response to repeated drying-rewetting cycles along a soil fertility gradient modified by long-term fertilization management practices. Geoderma 160:218–224
Zhou X, Zhang Y (2014) Temporal dynamics of soil oxidative enzyme activity across a simulated gradient of nitrogen deposition in the Gurbantunggut Desert, Northwestern China. Geoderma 213:261–267
Zimmerman AR, Ahn MY (2011) Organo-mineral-enzyme interaction and soil enzyme activity. In: Shukla G, Varma A (eds) Soil enzymology, Soil biology, vol 22. Springer, Berlin, pp 271–292

Proteinaceous Trypsin Inhibitors from Plants in Disarming the Insect Pest

Farrukh Jamal, Sangram Singh, Prabhash Kumar Pandey, and Ranjan Singh

1 Introduction

Cells are abundant in proteolytic enzymes. These enzymes are crucial for intracellular proteolysis. The average growth and development of insect pest require the digestive proteolytic enzymes for assimilation of dietary proteins. Consequently, any remarkable interference in these enzymes will serve as an impediment for the normal physiological process. Therefore, using suitable inhibitors to such targets would be an effective strategy to control insect pests. The inhibition of serine proteases has been in focus as these proteases are abundant in a specific order of insects. Although, insect serine proteases are similar to mammalian proteases yet they differ in substrate specificity and behavior in the presence of inhibitors (Gatehouse et al. 1999). Majority of insect fall in the order Lepidoptera and Diptera, whose mid gut proteases are predominantly of serine (trypsin) type (Srinivasan et al. 2006) whereas the Homoptera and Coleoptera employ cysteine proteases for digestion (Schluter et al. 2010). Whatever may be the case, naturally occurring inhibitors from plants with proteinaceous character influence these proteases with different degree of specificity (Christeller et al. 1998). Protease inhibitor(s) can serve as an attractive and alternative strategy for enhancing crop yield (Fan and Wu 2005) due to their inhibitory potential that is significant, selective and protective. As the possibility is surmountable in insect pest management of food crops and approaches eco-friendly, it will be

F. Jamal (✉) · S. Singh
Department of Biochemistry, Dr. Rammanohar Lohia Avadh University, Faizabad, Uttar Pradesh, India

P. K. Pandey
Department of Biochemistry, University of Allahabad, Allahabad, Uttar Pradesh, India

R. Singh
Department of Biotechnology, Coithram College of Professional Studies, Indore, Madhya Pradesh, India

Q. Husain, M. F. Ullah (eds.), *Biocatalysis*,
https://doi.org/10.1007/978-3-030-25023-2_15

interesting to dwell on the characteristics of some recently explored insecticidal proteins and their effectiveness on a popular pest *Helicoverpa armigera.*

2 Protease Inhibitors (PIs)

PIs are ubiquitously distributed in the plant, animal and microbial kingdom (Christeller and Liang 2005). Plant PIs are abundant in storage tissues such as tubers and seeds and expressed elsewhere including aerial parts of plants (De Leo and Gallerani 2002). These are generally small (5–25 kDa), structurally compact molecules, often abundant in disulfide linkage and exhibits thermal, pH stability and resistance to proteolysis (Fan and Wu 2005). With diverse role to play in a cellular system, PIs also play an essential role in the blood coagulation system, complement cascade, apoptosis and the hormone processing pathway (Neurath 1989).

PIs proteins are expressed in specific plant tissues upon herbivory or wounding (Browse and Howe 2008). The defensive action involves interaction and inhibition of cognate proteases either secreted by microorganism or present in insect guts thereby, causing a reduction in the availability of free amino acids necessary for growth and development (Lawrence and Koundal 2002; Chen et al. 2007). To harness the vast potential of such proteins for the increase in the crop yield, gene bank of such defense proteins could be maintained which could be utilized in developing transgenic varieties. Moreover, PIs can be used as essential molecules to design a drug for use in preventing the propagation of causative agents including pests (Johnson and Pellecchia 2006).

Low molecular weight PIs irreversibly modify an amino acid residue of the protease active site as in the case of phenylmethane sulfonyl fluoride and activate the serine proteases (Umezawa 1982). High molecular weight PIs from a variety of organisms including bacteria, animals, and plants (De Leo and Gallerani 2002) undergo tight binding either reversibly or pseudo-irreversibly, causing steric hindrance and thereby preventing substrate access to the active site. Their size varies from 50 residues (e.g., BPTI: Bovine Pancreatic Trypsin Inhibitor) to 400 residues (e.g., alpha-1 proteinase inhibitor).

Naturally occurring plant protease inhibitors are classified by proteases they inhibit, *i.e.,* serine protease inhibitor, cysteine protease inhibitor, aspartic protease inhibitor and metalloprotease inhibitor (De Leo and Gallerani 2002). The classification is also performed by their molecular mass, their protein architecture (monomeric or multimeric), the number of disulfide bridges present and their isoelectric points. They have been grouped into families and subfamilies and into different clans by similarities detectable at the level of amino acid sequence or sequence homology and three-dimensional structure of the protein. MEROPS database assigned PIs to 74 inhibitor families (11–191) and 73 inhibitor clans. The term 'clan' indicates a single evolutionary line of inhibitors defined by the unique type of protein fold or unit. The MEROPS database is a data resource for peptidases (also termed protease, proteinases and proteolytic enzyme) and the proteins that inhibit them. A protein that contained only a single inhibitor unit was termed as a simple inhibitor, and one that included multiple inhibitor units was termed as a compound inhibitor. Among the entire PIs group, serine protease inhibitors have been the largest group of PIs in nature when looking at the number of families (Rawlings et al. 2004; Haq et al. 2004).

2.1 Serine Protease Inhibitors (SPIs)

The SPIs vary from 5 to 25 kDa, with the majority in the range of 8–20 kDa (Hung et al. 2003). Most common SPIs are chymotrypsin inhibitor (EC. 3.4.21.3), trypsin inhibitor (EC. 3.4.21.4) and subtilisin inhibitor (EC. 3.4.21.62). Although they are a generally smaller protein with constituent amino acids ranging from 29 to 190, they exhibit a significant difference in its composition (Ryan 1990). SPIs usually have a high content of cysteine residues (Richardson 1991) that form disulfide bridges (Greenblatt et al. 1989a; Hung et al. 2003) and confer resistance to heat, extreme pH and proteolysis (Richardson 1991). These inhibitors participate in various physiological functions. Their role as defense molecules against plants pests and pathogens has gained prominence in recent year (Lawrence and Koundal 2002).

3 Inhibitors Proteins: Source and Characteristics Features

The Indo-Gangetic plain of India is one of the wealthiest areas holding a variety of flora. Since ancient times, use of plants as a source of medicine has been an inherent part of life. Some serine protease inhibitors that we have recently explored are described below:

3.1 **Eugenia jambolana** *(Jambul)*

Eugenia jambolana belongs to the family of Myrtaceae, is a large evergreen tree which thrives well under both tropical and subtropical climate. The seeds, leaves, bark wood of *Eugenia jambolana* provide practical and potential medicinal benefits. A trypsin-inhibiting protein EjTI of ~17.4 kDa was recovered from its seeds (Singh et al. 2014). The yield of EjTI was low which may be attributed to interference from a high level of phenols/and mucilaginous polysaccharides during purification as in the case of *Terminalia arjuna* (Rai et al. 2008). EjTI was sufficiently stable at a high temperature upto 80 °C and lost its trypsin inhibitory activity upto 90% at pH 2 and above pH 11. The high stability of EjTI over a wide range of pH and temperature is a remarkable feature that dictates its efficiency in controlling a variety of phytophagous insects. Insects have variations in their gut environment, which could range from highly acidic (pH 3.0) condition in midgut fluid of beetles to an alkaline state in most of the lepidopteron caterpillars (pH 12.0) (Jongsma et al. 1996). Therefore, the broad pH stability would be suitable to lock the critical digestive enzymes in the gut of several insects. Thermo-stability of inhibitor(s) due to their compact structure stabilized by some disulfide linkages makes it operational even under extremes of temperature.

The inhibition constant (K_i) value measures the affinity between the enzyme and inhibitor which is high for most of the explored plant trypsin inhibitor including EjTI (Macedo et al. 2000; Bhattacharyya et al. 2006). The inhibition kinetics of EjTI indicated a noncompetitive nature of inhibition for trypsin-like midgut proteases (Singh et al. 2014). In case of noncompetitive inhibition, the maximum catalytic rate that can be achieved by a particular enzyme (V_{max}) decreases whereas the substrate concentration at which half V_{max} is achieved (K_m) remains constant when compared to the response in the absence of inhibitor. The inhibition pattern of EjTI is as for other related inhibitors, such as *Acacia nilotica* trypsin inhibitor (Babu et al. 2012).

3.2 Madhuca indica *(Mahua)*

Madhuca indica belongs to the family of Sapotaceae. The tree, its flowers, and seeds have been useful in the Indian economy for a long time. The inhibitor from seed flour extracts of *Madhuca indica* (MiTI) is a single polypeptide chain with a molecular mass of ~19.8 kDa. MiTI inhibited to the total proteolytic and trypsin-like activities of the midgut protease of *H. armigera* larva by 87.51 and 76.12% respectively at a concentration of 5 μg/ml against total midgut proteases (Jamal et al. 2014). Inhibitory activity of MiTI on exposure to different pH was stable predominantly in alkaline conditions (pH 6–11). The optimal pH for digestive proteases among most lepidopteran larvae falls in the alkaline state with a maximal pH of 10 and 11 (Johnston et al. 1991). Trypsin inhibitor isolated from jackfruit and pigeon pea seeds were also static over a broad range of pH from 7 to 10 and 3 to 12 respectively (Annapurna et al. 1991). MiTI showed excellent stability over a wide range of temperature (20–60 °C), relatively lesser than EjTI. Trypsin inhibitors from oat were stable in a much wider range (0–100 °C) (Mikola and Mikkonen 1999). The high stability of PIs is an outcome of hydrophobic interactions of short stretches of H-bonded sheets (Bhattacharyya and Babu 2009)/disulfide linkages derogating their conformational entropy and in result enhance their stability (Ramasarma et al. 1995).

MiTI is a competitive inhibitor with a K_i value of 4.1×10^{-10} M for HGP. Serine protease inhibitors are capable of high-affinity inhibition by competitive, noncompetitive mechanism (Bhattacharyya and Babu 2009; Macedo et al. 2004). The inhibition constant of MiTI for HGP was 4.1×10^{-10} M, suggesting a high affinity of the MiTI for HGP. Similar K_i values were reported for other trypsin inhibitors from *Caesalpinia bonduc* seeds (2.75×10^{-10} M) (Bhattacharyya et al. 2007), *Derris trifoliate* (1.7×10^{-10} M) (Bhattacharyya and Babu 2009), *Dimorphandra mollis* (5.3×10^{-10} M) (Macedo et al. 2000) and *Putranjiva roxburghii* (1.4×10^{-10} M) (Sarate et al. 2012). Kinetic studies have shown competitive inhibition of trypsin by MiTI on *H. armigera* larvae fed with MiTI dosage inhibited the *H. armigera* larval growth and development.

3.3 Butea monosperma *(Flame of Forest)*

The protease inhibitor from seeds of *Butea monosperma* (BmPI) is a 14 kDa protein having an isoelectric point in the acidic region (PI-5.6). Multiple sequence analysis data confirms that the BmPI contains a sequence motif which is conserved in various trypsin and chymotrypsin inhibitors of Kunitz type (Jamal et al. 2015). The inhibitor shows trypsin inhibitory activity in a broad range of pH (4–10) and temperature (10–80 °C). The enzyme kinetic studies unveiled BmPI as a competitive inhibitor with a K_i value of 1.2×10^{-9} M.

BmPI turned out to be a Kunitz type inhibitor which caused a reduction of 90% activity of the midgut protease of *H. armigera* larvae and inhibited bovine trypsin. The digestive tract of *H. armigera* is rich in trypsin (95%) and chymotrypsin (5%) (Srinivasan et al. 2006). Therefore, serine protease inhibitors from natural non-host plant sources are assets in dealing with the insect pests. The IC_{50} of BmPI for *H. armigera* midgut proteases was calculated to be ~2.0 μg/ml. While measuring the instar-specific *Helicoverpa armigera* gut protease (HGP) activity; it was observed that the fourth instar larvae were most insensitive to inhibition. The inhibition achieved was ~35.6 and 90% of total HGP activities at 0.05 and 0.50% m/m dose of BmPI, respectively. As the fourth instar stage is the most greedy and devastating phase of *H. armigera* development in which maximum proteases are expressed, it is likely that intervention by using insecticidal protein molecules would adversely affect the insect pest and crop yield can be increased (Patankar et al. 2001; Damle et al. 2005).

3.4 Tamarindus indica

Another trypsin inhibitor of 21 kDa from the seeds of *Tamarindus indica* (TTI) showed inhibitory action against the total gut proteases of *H. armigera* (~87%) and bovine trypsin (~84%). Lethal doses which caused mortality and weight reduction by 50% were 1% w/w and 0.50% w/w, respectively. IC_{50} of TTI against trypsin-like *Helicoverpa* midgut proteases and bovine trypsin was ~2.10 μg/ml and 1.68 μg/ml respectively. This Kunitz type protein was found to retard growth and development, prolonged the larval-pupal development duration of the insect along with adversely affecting the fertility and fecundity of *H. armigera*. In artificial diet, at 0.5% w/w TTI, the efficiency of conversion of ingested food as well as of digested food, relative growth rate, growth index showed decline whereas approximate digestibility, metabolic cost, relative consumption rate, consumption index and total developmental period enhanced for *H. armigera* larvae. These results suggest that TTI has toxic and adverse effect on the developmental physiology of *H. armigera* and could be useful in controlling the insect pest *H. armigera* naturally (Pandey and Jamal 2014).

3.5 Pithecellobium dulce *(Manila Tamarind)*

A trypsin inhibitor from the seeds of *Pithecellobium dulce* (PDTI) is a 21 kDa protein which exhibited inhibitory activity towards total gut proteolytic enzyme of *H. armigera* larvae (~83%) (~1.5 μg/ml IC_{50}) and against bovine trypsin (~86%) (~1.33 μg/ml IC_{50}). PDTI exhibited stability in a wide range of pH. Annapurna et al. (1991) accounted that the trypsin inhibitor from jackfruit seed was stable over a broader range of pH (3–12). In contrast, the trypsin inhibitor from the seeds of pigeon pea maintained its full activity between pH 7.0 and pH 10.0, but when unwrapped to acidic pH from 3.0 to 5.0, 20% of the activity was doomed (Godbole et al. 1994). Johnston et al. (1991) observed that midgut of lepidopteran larvae is highly alkaline and the digestive peptidases have optimal activity at pH 10.0–11.0.

Generally, peptidase inhibitors from legumes are quite stable up to 80 °C although in several cases they lose their activity above this temperature (Rufino et al. 2013; Godbole et al. 1994). PDTI retained its activity at a temperature up to 85 °C. Heat stability of legume inhibitors may be attributed to their compact structure stabilized by some disulfide linkages. Rufino et al. (2013) observed that about 90% PI activity was maintained at 37–100 °C and pH 2–12 for 30 min incubation. Earlier studies on many of the Kunitz PI have also reported similar defiance to the extreme of the thermo-pH condition and have attributed this stability to the apparent ability of the inhibitor to reversible denaturation through a transient intermediate.

The covalent link of cysteine residues through disulfide bonds is vital for the folding, stability and function of many proteins (Zavodszky et al. 2001). With increasing concentration of dithiothreitol (DTT), the activity of PDTI decreased. Macedo et al. (2003) observed that a Kunitz type trypsin inhibitor from *Pithecellobium dulce* exhibited a loss in inhibitory activity after exposure with 100 mM DTT. Although intramolecular disulfide linkages have been ascribed to be the prime factor behind the functional stability of the inhibitor against denaturants (Oliveira et al. 2007), additional factors such as placement/arrangement of the disulphide linkages (buried, rather than surface exposed) within the molecule are also critical for the overall stability of the inhibitor (Yoshizaki et al. 2007). It was supposed that an abundance of β-sheets along with intramolecular disulfide bond might provide conformational stability to the active site of PDTI, which protect it against denaturants like DTT (causes structural modifications promoting inhibitory inactivity). Most Kunitz inhibitors have a direct relationship between the reduction of disulfide bonds and loss of their biological activity. Stability of this inhibitor to an extent against the different concentration of DTT suggests that stability of PDTI does not depend exclusively on the disulfide bonds. For instance, upon exposure to 100 mM DTT, the inhibitory activities of some of the Kunitz PIs were abolished (Garcia et al. 2004) whereas some others remained completely unaffected (Cruz et al. 2013; Chaudhary et al. 2008). However, studies on the susceptible (single polypeptide) soybean trypsin inhibitor (Tetenbaum and Miller 2001) or stable *Bauhinia* species inhibitors (devoid of any disulfide linkages), indicated that intramolecular disulfide bonds in Kunitz type PIs might be essential, but not exclusive to the stabilization of the reactive site loop. Another presumption is that the intramolecular disulfide bonds and their distance from the reactive site residues determine the functional stability of these inhibitors in the presence of denaturating agents (Garcia et al. 2004). It is interesting to note that under extreme DTT conditions the magnitude of loss in inhibitory activity of Kunitz PIs varies from species to species. Stability is an essential factor for PIs to overcome the physiologically stressful situation. As PDTI is a Kunitz type inhibitor, a relatively higher amount of hydrophobic interaction/H-bonded sheets contribute towards its high stability (Bhattacharyya and Babu 2009; Zhou et al. 2008).

The initial rates of reaction in the presence or absence of PDTI followed the Michaelis-Menten equation and Lineweaver-Burk double reciprocal plots created thereafter for inhibition mode of interaction of PDTI exhibited a similar in vitro mechanism of inhibition of the competitive type as those observed for other Kunitz inhibitor (Bhattacharyya and Babu 2009; de Oliveira et al. 2012). The inhibition constant (K_i) value establishes a high affinity between the enzyme and inhibitor, in concord with the data reported for other plant trypsin inhibitors (Rufino et al. 2013; Bhattacharyya et al. 2006; Bhattacharyya and Babu 2009). Dixon and Lineweaver-Burk double reciprocal plots analysis disclosed a competitive inhibition mechanism and a K_i of ~3.9×10^{-8} M. The inhibitory activity of PDTI was unaltered over a wide range of temperature, pH and in the presence of dithiothreitol.

A dose-dependent outcome on mean larval weight and a series of nutritional disturbances were observed when PDTI was incorporated in the artificial diet. At 0.25% w/w PDTI present in the artificial diet, the efficiency of conversion of ingested food, of digested food, growth index, and relative growth rate declines whereas relative consumption rate, approximate digestibility, consumption index, metabolic cost and total developmental period were increased in the larvae (Pandey et al. 2015). Thus PDTI is a strong antifeedant, with antimetabolic stress on midgut peptidases of *H. armigera*.

3.6 Cassia fistula *(Golden Shower)*

Cassia fistula L., (Leguminosae), a semi-wild Indian Laburnum (also known as the Golden Shower), is distributed in various countries including West Indies, Mexico, South Africa, China, East Africa and Brazil. The whole plant is used to treat diarrhea; seeds, flowers, and fruits are used to treat fever, abdominal pain, skin disease, and leprosy by traditional people (Perry 1980). It is useful in treating diabetes, leucoderma, and haematemesis (Asolkar et al. 1992). It is the national tree of Thailand and the state flower of Kerala in India. Trypsin inhibitors of 5.4 kDa (CFT-1) and ~7.1 kDa (CFT-2) extracted from the seeds of *Cassia fistula* was previously reported by Wijaya et al. (2000). CFTI-1 and CFTI-2 exhibited 92% and 78% inhibitory activity against midgut trypsin-like proteases, respectively. Dixon and Lineweaver-Burk double reciprocal plots analysis revealed a competitive inhibition pattern with the K_i values 2.9×10^{-10} M and 4.2×10^{-10} M for CFTI-1 and CFTI-2 respectively (Pandey et al. 2016). Lethal dose (0.34% m/m) and dosage for weight reduction by 50% (0.24% m/m) determined for CFTI-1; showed a dose-dependent effect on mean larval weight, a decline of the fertility and fecundity of the moths, and an extended total developmental duration of *H. armigera* life cycle (Pandey et al. 2016).

4 Inhibitor Proteins and Its Influence on *H. armigera*

H. armigera is a voracious eater of economically important crop plants. The protein substance of the different instar stages of *H. armigera* larvae parallels the high trypsin-like activity, and therefore the changes in protein concentration of different larval instar stages midgut administrate the changes in midgut proteolytic activity. It was apparent that a high inhibitory activity toward general proteolysis would give pronounced detrimental effects on larval growth and physiology. Such observations are supported through studies involving many other lepidopteran larvae (McManus and Burgess 1995; Gatehouse et al. 1999; Araujo et al. 2005).

4.1 In Vivo Influence of EjTI on H. armigera

EjTI inhibited 86% of HGP by 3.35 μg/ml at pH 7.8 and 10.0. The extent of inhibition demonstrated by EjTI on the activity of the gut enzyme of *H. armigera* larvae is profound. Telang et al. (2003) reported that nonhost PIs from bitter gourd inhibited more than 80% of *H. armigera* protease activity. Tomato PIs are also stable to insect proteases and inhibited about 50–80% proteinase activity of *H. armigera* larvae feeding on various nonhost and host plants. The IC_{50} of EjTI for midgut trypsin was approximately 2.00 μg/ml, but against bovine trypsin, it was slightly more (Babu and Subrahmanyam 2010). A lower concentration of EjTI is more effective against HGP activity, as a higher concentration perhaps causes molecular aggregation, and consequently, an increase in percent inhibition of trypsin was not favored. Being effectively toxic on *H. armigera* in low amount, such proteins would be an exciting candidate for incorporation in transgenic varieties.

PIs have been considered to decrease survival and delay growth and development of the *H. armigera* larvae (Charity et al. 1999). So the investigation into its insecticidal potential on the fourth instar *H. armigera* larvae indicated that EjTI has a profound and adverse effect on the growth and development of *H. armigera*. Similar to larval body weight, retardation pattern in the growth rate with increasing time was also evident among the EjTI-fed larvae. The survival pattern upon moderate to high doses of EjTI exhibited high to severe mortality rates.

While exploring its detrimental effects on developmental physiology, it was observed that EjTI caused a marked reduction in pupal growth, abnormality in molting, and subsequently died. Different degree of irregularities in the larval, pupal and adult stages of *H. armigera* development was noticed, and high level of malformed pupae and complete malformed adults were observed on the larvae fed on test diet. Moreover, EjTI adversely affected the moths emerging from fed larvae. PIs from *Capsicum annum* and *Madhuca latifolia* that adversely affects the larval growth and development, enhance the mortality rate, deformities, and highly significant in halting the adult emergence, fecundity and fertility of *H. armigera* (Tamhane et al. 2005; Pawar et al. 2004). These developmental anomalies that silently yet profoundly lead the larvae towards death are significant attributes of PIs that predisposes it for use in pest management in an eco-friendly manner.

Further, as opposed to a regular diet, EjTI-fed *H. armigera* adults drastically reduce its egg-laying capacity. The downfall in the ability of egg laying and egg hatching was dose-dependent. As against control, eggs hatching was also very low (19.3%) at the highest experimental concentration of EjTI. Thus, the PI from *E. jambolana* is determined for predators, which are yet to adapt to the inhibitory mechanism.

Another observation crucial for utilizing PIs is its stability to insect gut proteases. The food retention time in the midgut of the larva is 3 h, and on exposure with EjTI for this duration, there was no substantial decrease in its inhibitory potential. This is true even for potato, bitter gourd, winged bean, peanut, and another nonhost PIs against *H. armigera* gut proteases (Harsulkar et al. 1999; Telang et al. 2003).

4.2 *In Vivo Influence of MiTI on* H. armigera

Another fruitful protein inhibitor MiTI showed decreased proteolytic activities of the midgut proteases of *H. armigera* larvae. Inhibitor extracted from chickpea (P-256), bitter gourd, *Prosopis juliflora* inhibited 66%, 80%, and 83% of *H. armigera* midgut trypsin activity, respectively (Kansal et al. 2008; Telang et al. 2003; Sivakumar et al. 2005).

In vivo experiments with different concentration of MiTI in artificial diet expressed an effective loss in the larval body weight anda gain in the larval mortality. The concentration of MiTI in the artificial feed to get 50% mortality (LD_{50}) of larva was 1.5% w/w, and that to reduce the mass of larvae by 50% (ED_{50}) was 1.0% w/w. The consumption of PIs added stress on gut protease expression system to synthesize new and higher amount of proteases which may lead to the arrested growth and mortality of *H. armigera* larvae (Singh et al. 2014; Sarate et al. 2012). The inhibitors from other nonhost sources including chickpea, mungbean, and jambul also showed a significant reduction in the larval growth of *H. armigera* (Singh et al. 2014; Sudheendra and Mulimani 2002). MiTI halts larval and pupal developmental periods and causes several deformities in neonates. The duration of pupation increased due to the interference of MiTI in the molting process; consequently, the larvae are unable to go into the next developmental stages of their life cycle. In general, extended pupation times directly affected the survival of pupae.

Deformities of neonatal adults, low survival rate and reduction in fecundity (number of eggs laid down per female) were remarkable in MiTI treated larvae.

Interestingly, the MiTI was powerfully effective in reduction of fertility and fecundity which persisted for successive generations. The fecundity of lepidopteran adults are the most commonly used parameters to specify the influence of larval diet on the adult stage. In vivo study showed that the quality of food influences the effectiveness of the inhibitor (Pompermayer et al. 2003). Fecundity of *H. armigera* was severely affected by 0.33% concentration of winged bean PI in the diet (Gupta et al. 2002; Telang et al. 2003).

4.3 In Vivo Influence of BmPI on H. armigera

BmPI turns out to be a Kunitz type inhibitor which caused a reduction of 90% activity of the midgut proteases of *H. armigera* larvae and inhibits bovine trypsin. Srinivasan et al. (2006) have determined the abundance of trypsin (95%) and chymotrypsin (5%) in the digestive tract of *H. armigera.* The IC_{50} of BmPI for *H. armigera* midgut proteases was calculated ~2.0 µg/ml. The inhibition achieved in the fourth instar larvae was approximately 35.6 and 90% of total HGP activities at 0.05 and 0.5% m/m dose of BmPI, respectively. The fourth instar stage is the most voracious and devastating phase of *H. armigera* development in which maximum protease activity and protease isoforms are expressed (Patankar et al. 2001; Damle et al. 2005).

The larvae fed on BmPI supplemented diet expressed a drastic reduction in weight to 178 mg/larvae as compared to control (356 mg/larvae) at a dosage of 0.1% m/m. the weight loss was more marked at 0.5% m/m dosage of BmPI in comparison to 0.05 and 0.1% (m/m). These data are also in conformity with the result of McManus and Burgess (1995), where the mean weight of larvae fed on a diet containing 0.2 or 0.5% (m/v) SBTI (soybean trypsin inhibitor) was fruitfully lower than the mean weight of larvae fed on standard chickpea diet. The association of BmPI with *H. armigera* midgut proteases may hinder the transport of enzymes and hydrolyzed products and together in combination with limited proteolysis may restrict the accessibility of amino acids consequently leading to poor larval growth, development and high mortality.

The larval developmental period increased upon treatment with different concentration of BmPI. In comparison to control (~9.8 days) the BmPI treated *H. armigera* larvae exhibited an enhanced developmental duration of ~11.5 days and ~16.8 when reared at a concentration of 0.05% m/m and 0.50% m/m respectively. The pupal period of *H. armigera* development was ~15.9 days at the maximum level of BmPI (0.5% m/m). This increase in the developmental duration over the control group (~10.5 days) is an indicator of the physiological stress due to the presence of BmPI in the feeding assay. The increase in larval and pupal duration was due to the interference of the protease inhibitor in the molting processes; consequently, the larvae were not able to go into further developmental stages of their life cycle. In general, extended larval-pupal periods were directly in proportion to the addition in pupicidal activities. This observance was corroborated with different degrees of abnormalities in the larval, pupal and adult stages of *H. armigera* development and delay in molting.

The proteins placed in the insect cuticle are essential in the production of new adult tissues and enzymes (Missios et al. 2000). Hopkins et al. (2000) suggested that these proteins are synthesized primarily from free amino acids following a re-organization of the peptide from larval proteins without total degradation of amino acids. Franco et al. (2003) reported that larvae of *A. grandis* consuming soybean Kunitz trypsin inhibitor exhibited similar growth and developmental pattern. Such disruptions in the typical growth pattern may be attributed to the adverse effects of BmPI activity in inhibiting protein utilization and uptake, important in metamorphosis (Chapman 1982). Although, the real *modus operandi* of these inhibitors may be a fall out of anti-digestive effect, through proteolysis inhibition (Jongsma et al. 1995), or a consequence of toxic effect by inducing overproduction of protease, leading to a crisis in the availability of amino acids (Broadway and Duffey 1986). Even the BmPI is casting similar challenge in the gut of *H. armigera* or possibly follows a different mode of action of inhibitors against insects which is responsible for the deformity, larvicidal, pupicidal and retardation of metamorphosis.

The downfall in egg laying capacity was dose-dependent with reduction observed as 63.78 and 95.91% at BmPI concentration of 0.05% m/m and 0.5% m/m respectively. BmPI also exhibited an adverse effect on egg laying and hatching capacities of *H. armigera* female moth. At lower concentration of BmPI (0.05% m/m) the egg hatching was 48% which decline to 13% on increased concentration (0.5% m/m). Thus, BmPI too was like to PIs from other sources such as bitter gourd, winged bean, potato (PIN2), groundnut, etc., which exhibited adverse effects on egg laying capacity and hatchability of *H. armigera* moth egg (Telang et al. 2003; Damle et al. 2005). During the different larval developmental phases, significant accumulation proteins are critical and crucial for the process of vitellogenesis. Thus, it is likely that BmPI may be intervening in the process of vitellogenesis, thereby adversely affecting the growth and development of *H. armigera.*

4.4 *In Vivo Influence of TTI on* H. armigera

Another Kunitz type inhibitor protein of 21 kDa from *Tamarindus indica* seeds (TTI) demonstrated the in vitro bioinsecticidal activity of TTI against insect digestive enzymes of different orders including lepidopteron viz., *Plodia interpuncptella* (26.7%), *Alabama argillacea* (53.8%) and *Spodoptera frugiperda* (75.5%) (Araujo et al. 2005). Our data suggest that the toxic effect of TTI on *H. armigera* is much higher and remarkable and therefore, this inhibitor protein could be a better choice to control this pest effectively. TTI inhibited up to ~87% activity of the midgut trypsin-like enzymes of *H. armigera* larvae suggesting that this inhibitor is trypsin-like serine protease inhibitor. Further, the in vitro midgut protein assay of *H. armigera* larvae with continuous TTI feeding confirmed the remarkable reduction in the protease activities from the beginning towards the termination of the larval period.

The data have been substantiated using SKTI, a well-known Kunitz type inhibitor from soybean. Interestingly, TTI seems to be a more effective inhibitor as compared to those reported from bitter gourd and tomato. The IC_{50} of TTI for bovine trypsin and *H. armigera* midgut proteases was 1.68 µg/ml and 2.10 µg/ml respectively. This value of TTI against total HGP was close to inhibitors from AsPI (*Acacia senegal*), SKTI (soybean) and CaKPI (*Cicer arientinum*) (<2 µg/ml) against trypsin enzyme of *H. armigera* (Babu and Subrahmanyam 2010).

It was observed that there was a marginal decrease in the inhibitory potential of TTI after 3 h incubation suggesting that the protein TTI is not only highly stable to insect gut proteases but may remain stable in the insect gut too. Nonhost PIs from other sources like tomato, bitter gourd, winged bean, peanut, etc. show similar results against *H. armigera* gut proteases (Telang et al. 2003; Harsulkar et al. 1999). The stability is an essential factor in the role as PI and as TTI is a Kunitz type inhibitor like SKTI, a relatively higher amount of hydrophobic interaction/ hydrogen-bonded sheets contribute to its high stability (Rao and Gowda 2008; Bhattacharyya and Babu 2009).

Accumulating nutrients for energy and growth is very crucial during the larval stage which is subsequently expended for pupal and adult development, fertility and fecundity. Therefore arrested growth and mortality of *H. armigera* larvae could be an outcome of starvation synergized with added stress on gut protease expression system to synthesized new and higher amounts of proteases. Further, upon supplementing the artificial diet with purified PIs in three different doses, it was observed that *H. armigera* larvae not only developed slowly but were also lethargic. This effect has an added advantage because if the larval stage is prolonged and mobility diminished the chance exposure to environmental stress as well as predators in the field would also increase and such insects may not be able to protect themselves as vigorously as the normal ones.

Our observation in the conformity with the result of McManus and Burgess (1995), where the mean weight of larvae fed on standard chickpea diet was significantly higher than the mean weight of larvae on a diet containing 0.2% or 0.5% (w/v) soybean trypsin inhibitor the downfall in weight may be a consequence of non-utilization of amino acids essential for growth and development (Telang et al. 2003). PIs act in a straightforward way by significantly reducing protein digestion. By this, the possible association of TTI to *H. armigera* midgut may hinder the transport of enzymes and their hydrolysis products, which in combination with reduced proteolysis is a limiting factor in the availability of amino acids for larval growth, leading to the poor development and high mortality. It was also noteworthy that at an ED_{50} dosage of TTI the food consumption by the *H. armigera* larvae increased whereas there was no effect on the antifeedant activity implying that TTI act as a feeding deterrent only at higher doses. Thus such molecules working as antifeedant may have a significant impact on crop protection until slow acting biopesticides produce their effect (Murray et al. 1993). TTI clearly showed high antifeedant activity against *H. armigera* suggesting its greater susceptibility. Kuhar et al. (2013) had also reported the antifeedant activity of a double-headed inhibitor from *Dolichos biflorus* seeds.

An increase in the larval and pupal developmental periods suggests that the protein TTI has the potential to affect the developmental physiology of *H. armigera* adversely. The interference of protease inhibitor in the moulting process accounts for enhanced larval duration whereas the failure of larvae to sustain on TTI supplemented diet prevented it from going into further instars. It was also noteworthy that larvae fed on TTI were significantly small in size with various kinds of abnormalities. The presence of higher concentration of TTI in the diet caused several harmful effects at all stages of *H. armigera* development, including larvae, pupae, and adult insects. Due to the very low consumption of food, the larvae were unable to continue the normal physiological process. These effects were previously observed in larvae of *A. grandis* consuming SKTI (Franco et al. 2003). A conceivable explanation for these effects is that TTI activity may inhibit proteins essential in metamorphosis (Chapman 1982). Production of new adult tissues and enzyme system are dependent on proteins localized in the insect cuticle (Missios et al. 2000). These proteins are synthesized primarily from free amino acids following a re-organization of peptides from larval proteins without total degradation of amino acids (Hopkins et al. 2000). Proteolytic activeness including that of serine and cysteine proteases is necessary for this re-organization.

Further, this inhibitor protein had a deleterious effect on egg laying and egg hatching capacities of *H. armigera* female moth. As compared to control, significantly lower eggs were laid by female moths, and the effect is dose-dependent. Gupta et al. (2002) reported that the fecundity of *H. armigera* was severely affected at 0.33% concentration of winged bean protease inhibitor in the diet. Damle et al. (2005), showed that tomato PIs caused adverse effects on various developmental parameters of *H. armigera,* most notably on fecundity at a concentration of inhibitor used in this study. Telang et al. (2003) reported that ingestion of PIs from non-host such as BGPIs adversely affected protein uptake at the larval stage, which not only caused developmental abnormalities but also reduced fertility and fecundity of the adult. Thus, our observations are in agreement with these studies that accumulation of proteins during the larval stage is critical to vitellogenesis and that the in vivo assays results indicated that TTI adversely affects the growth and development of *H. armigera.*

4.5 *In Vivo Influence of PDTI on* H. armigera

In the seeds of several *Pithecellobium* species peptidase inhibitory activity has been registered. In *P. keyense* the inhibitory activity against trypsin was higher than that for soybean (Garcıa-Carreno et al. 1996). Delgado-Vargas et al. (2004) only demonstrated the inhibitory specificity of PDTI against purified commercial enzyme (bovine trypsin) and suggested the potential of PDTI as either an anti-carcinogen or an insect deterrent. Further, they did not perform any antifeedant and antimetabolic study on the voracious feeder *H. armigera*/other pests. We reported the influence of the 21 kDa inhibitory protein (PDTI) from *Pithecellobium dulce* seeds on the developmental physiology of *H. armigera*. Generally, bioassays with PIs use artificial diet concentrations ranging from 0.1 to 1% (Alves et al. 2009). Similar concentration range was used in studies with *Adenanthera pavonina* seed protease inhibitor against *Diatraea saccharalis* larvae (da Silva et al. 2014).

Further, the biochemical analysis showed that the inhibitor is refractory to the digestion by midgut peptidases in all evaluated time intervals. There are two possible explanations for this fact: first, the newly synthesized *H. armigera* trypsin is insensitive, and the inactivation of PI ceases to be a necessary adaptation. Thus PIs would represent no more risk for insect digestive process. And second, PIs are known for their resistance to digestion by midgut peptidases for some insects, and this would be another characteristic that needs to be studied with special attention (de Oliveira et al. 2013). A prerequisite for toxicity is that the inhibitor should be able to survive the hostile proteolytic environment of the insect midgut. It is possible to conclude that insecticidal effects of PDTI are related to its resistance to digestion by *H. armigera* midgut peptidases. A similar finding was also observed by Macedo et al. (2011) and Oliveira et al. (2011). Our data confirm the effects on larval physiology, reaffirming that the inhibitor appears to have maintained its structure to exert its insecticidal activity.

4.6 In Vivo Influence of CFTI on H. armigera

Similar finding on viability with *Albizia lebbeck* seeds against *H. armigera* larvae shown by Hivrale et al. (2013) strengthen our toxic result on *H. armigera* (Singh et al., 2014). The life cycle extension is desirable in an insect pest since any delay in its development reduces the emergence ratio of new generations; thereby reducing the damage(s) caused by the species in the targeted crop (Oliveira et al. 2011). Once in the digestive tract, the inhibitor form complexes with the trypsin enzymes, it starts hindering the digestive process. Beyond the impairment of proteolysis through inhibition of trypsin, it is possible that more chymotrypsin is being synthesized in an attempt to re-establish the average level of proteolysis. In this way, part of the amino acids that would be used in larval development are used in the enzymes synthesis, and are associated to the impairment of digestion exerted by PDTI; the delay in larval development is re-established, with consequence on adult emergence and longevity.

The extension of the larval and the pupal stage combined with the reduction of larval and pupal mass may be a signal of the use of essential amino acids in the synthesis of midgut peptidases (Broadway and Duffey 1986) and other proteins related to metabolism, detoxification and stress tolerance (Petek et al. 2012). In this way, the extension of the larval stage would be a compensatory effect, because during the adaptive process there is a considerable energy expense. Therefore, the larvae should accumulate the energy necessary to complete the next stage of development (de Oliveira et al. 2013; Singh et al. 2014). Moreover, control strategies interfering with the growth rate of pest populations without having a high impact on insect mortality minimize the risk of insect resistance. A decrease in population growth rates could be a result of delayed larval development and reproduction, lower fecundity, high deformity, or a combination of these factors (de Oliveira et al. 2014). The growth index (GI) measures the effects of food substrate, such as a dietary inhibitor, on both survival and developmental time (Howe 1971). The principle behind this parameter is that adverse conditions, such as inappropriate food substrate, may prolong the developmental period, while fewer individuals survive, resulting in low GI value. Conversely, a suitable food substrate should result in high GI value (Machuka et al. 2000).

5 Antifeedant Activity of PI(s)

Antifeedant are "ecologically safe pesticides" and can be described as allomone substances which inhibit feeding and do not kill the insect pest directly. They limit their developmental potential considerably and act as a phagodeterrent or phagorepellent over test as well as permanent insect pest feeding on the plant. Quantification of the antifeedant effect of botanicals is of great importance in the field of insect pest management. From an ecological point of view, antifeedants are crucial since they never kill the target insect directly but allow them to be available to their natural enemies and help in the maintenance of natural balance. For most antifeedants, the modality of action is directed at the taste cells. A distinctive gustatory sensillum in an insect contains receptors selective for deterrents and other for stimulants (sugars and amino acids). Although most of the antifeedants act by stimulating a deterrent receptor, that in the response indicates a signal ("do not feed") to the feeding center in the insect's central nervous system. More or fewer antifeedants are thought to block or otherwise interfere with the perception of feeding stimulants, while others may cause an erratic burst of electrical impulses in the nervous system. It prevents the insect from acquiring appropriate taste information on which it may choose a proper feeding behavior. Recently antifeedant activity of a double-headed inhibitor was reported from *Dolichos biflorus* and *Pithecelobium dumosum* seeds by Kuhar et al. (2013) and Oliveira et al. (2007) respectively support our findings. The antifeedant activity with EjTI increased with all concentrations. Thus, such proteins molecules serving as antifeedant could be used to protect crops until slow acting biopesticides manifest their effects (Murray et al. 1993).

6 Nutritional Analysis of PIs fed *H. armigera*

A study of nutritional indices revealed the anti-nutritional attribute of MiTI. ECI is in general a measure of an insect's caliber to exploit the food ingested for growth (Abdel-Rahman and Al-Mozini 2007). Along with nutritional indices, ECI may alter with the digestibility of food and the equivalent sum of the digestible part of the food, which is modified to body mass and metabolized for energy, crucial for vital activities (Abdel-Rahman and Al-Mozini 2007). In this study, change in ECD also points out the increased/decreased proportion of digested meal metabolized for energy (Nathan et al. 2005). A close or no change in ECI and ECD values as compared to control reveal that ingested biochemicals such as PIs do not show any chronic toxicity (Koul et al. 2003). However, the nutritional indices of both the fourth and fifth instar larvae of *H. armigera* were substantially dissimilar in comparison to the control. Therefore the data incurred for the fourth and fifth instar larvae are not dependable with each other. This fact explains that the nutritional need for insect changes during development and such changes are dependent on the consumption of food and feeding behavior (Browne 1995). The higher ECI value of *H. armigera* for fourth instar larvae compared to control suggests that they were more capable of converting the ingested food into biomass.

The nutritional analysis suggested that these PIs exert toxic effects on larvae by changing its physiological and metabolic performance. ECD was negatively affected, whereas CM enhanced on the treatment of the *H. armigera* with PI. The ECD decreases as the ratio of digested food metabolized for energy increases (Coelho et al. 2007). The reduction in ECD value suggests that PIs exhibits toxicity in insects while increasing the CM of these larvae. A decrease in the ECD indicates that more food is being metabolized for energy and less is being converted to bodily substances, most likely to circumvent the anti-nutritional effect of this protein. Percent ECI may differ with the digestibility of food and the proportional amount of digestible portion of food that is converted into body mass and metabolized for energy for principal activity (Shazly 1993). The increase in AD observed for PI-fed larvae was most likely a result of the lower output of feces by these larvae relative to the control group. A greater AD would help to meet the increased demand for nutrients (Pandey et al. 2014; Jamal et al. 2014) and compensate for the deficiency in foodstuff conversion (reduction in ECI and ECD), perhaps by diverting energy from biomass production into detoxification (Macedo et al. 2011). Mordue and Blackwell (1993) and Rayapuram and Baldwin (2006) reported similar findings. The growth index (GI) measures the effects of the food substrate, such as a dietary inhibitor on both survival and developmental times (Machuka et al. 2000). The principle behind this parameter is that in adverse conditions including unsuitable food substrate, the developmental period increases, and fewer individuals survive, resulting in a low GI value. Conversely, a suitable food substrate should result in a high GI value (Machuka et al. 2000). Thus, this study would be helpful in dealing with the complexity of *H. armigera* gut proteases.

7 Conclusion and Future Direction

Bt transgenic crops have been much successful in controlling insect infestation and helping to reduce the use of chemical pesticides. Bt Cry proteins are presently the proteins of significance in the management of insect pests. Although a potent insecticidal protein, the resistance to Bt insects has evolved in which the critical cause seem to be an alteration in the midgut receptor protein. Bt resistance is also associated with reduced activity of digestive enzymes engaged in the solubilization and activation of Cry1 Ac and Cry1 Ab. Another mechanism reported is overexpression of specific proteases involved in the degradation of the Cry toxin. The phenomenon of resistance development is dynamic and threatens the continued success of any particular protein. Therefore, researchers need to hunt new insecticides that have attributes of green technologies.

With increasing report of insect adaptation against PIs bioassays evaluating the insecticidal potential of novel PIs on larval developmental physiology must be performed before encouraging further investigations. The growth inhibitory potential of PIs towards individual insect is a function of (1) feeding behavior of insect, (2) nature of PIs, and (3) stability of the inhibitor in insect gut (Jongsma and Bolter 1997). Depending on the developmental stages and diet composition; *H. armigera* regulates gut protease expression which not only encourages polyphagy but also enables the insect to sustain the challenges posed by dietary PIs (Srinivasan et al. 2006). By minimizing the digestion and assimilation of nutrients necessary for the growth and development of larvae, one can minimize its detrimental effects on food crops.

Under these circumstances, the generation of insecticides that are less harmful to beneficial insects and environment is necessary. Plants-originated proteins such as PIs, lectins, or bacterial vegetative insecticidal proteins could serve as the potential insect controlling agents shortly. Novel gene pools from wide varieties of plants with more critical and versatile properties must be explored so that these genes/gene product(s) can be exploited for developing crops that can be efficiently managed and the use of dangerous pesticides can be restricted. Interestingly, in a coevolving plant-insect interaction choosing nonhost plants represents one of the best sources of identifying active PIs. In conclusion, the action of some of these nonhost protease inhibitors on the growth and development of *H. armigera* was remarkable. Performed bioassay experiments (both qualitative and quantitative) and nutritional indices analysis leads to a solid conclusive remark that PI protein may have the tremendous biotechnological potential to combat with the pest *H. armigera*/other phytophagous insects. Additional studies of these PIs in other insects are required to confirm the biotechnological potential of PIs as an agent against phytophagous insects.

In comparison to total eradication approaches of pest control, plant PIs do not lead to high selection pressure. This restricts the possibility of developing resistance in the insect population against PIs. Another merit of this approach lies in the fact that PIs are a plant's natural defense response against phytophagous insects. It is also suggestive that better crop protection can be achieved using a combination of inhibitors and for such reasons exploring novel protease inhibitors will hold great promise in controlling corn earworm using either independently or in combination(s).

Acknowledgments The authors express their gratitude to Ministry of Science and Technology, Govt. of India and University Grants Commission, New Delhi for providing financial support to the department under the DST-FIST scheme for developing infrastructure to carry out work on such aspects. Prabhash K. Pandey gratefully acknowledges University Grant Commission, New Delhi for providing the financial assistance (UGC-DSKPDF No.F.4-2/2006 BSR/BL/17-18/0442). The corresponding author is thankful to the Council of Science and Technology, U.P. India for providing recurring financial support vide project CST/SERPD/D-1136.

References

Abdel-Rahman HR, Al-Mozini RN (2007) Antifeedant and toxic activity of some plant extracts against larvae of cotton leafworm *Spodoptera littoralis* (Lepidoptera: Noctuidae). Pak J Biol Sci 10:4467–4472

Alves DT, Vasconcelos IM, Oliveira JTA, Farias LR, Dias SC, Chiarello MD, Maria-Neto S, Franco OL (2009) Identification of four novel members of Kunitz-like alpha-amylase inhibitors family from *Delonix regia* with activity toward Coleopteran [sic] insects. Pestic Biochem Physiol 95:166–172

Annapurna SS, Ramadoss CS, Prasad DS (1991) Characterization of a trypsin/chymotrypsin inhibitor from jack fruit (*Artocarpus integrifolia*) seeds. J Sci Food Agric 54:605–618

Araujo CL, Bezerra IW, Oliveira AS, Moura FT, Macedo LL, Gomes CE, Barbosa AE, Macedo FP, Souza TM, Franco OL, Bloch JC, Sales MP (2005) In vivo bioinsecticidal activity toward *Ceratitis capitata* (fruit fly) and *Callosobruchus maculatus* (cowpea weevil) and in vitro bioinsecticidal activity toward different orders of insect pests of a trypsin inhibitor purified from tamarind tree (*Tamarindus indica*) seeds. J Agric Food Chem 53:4381–4387

Asolkar LV, Kakkar KK, Chakre OJ (1992) Second supplement to glossary of Indian medicinal plant with active principles. Publication and Information Directorate, CSIR, I, New Delhi, p 177

Babu SR, Subrahmanyam B (2010) Bio-potency of serine proteinase inhibitors from *Acacia senegal* seeds on digestive proteinases, larval growth and development of *Helicoverpa armigera* (Hubner). Pestic Biochem Physiol 98:349–358

Babu SR, Subrahmanyam B, Srinivasan SIM (2012) In vivo and in vitro effect of *Acacia nilotica* seed proteinase inhibitors on *Helicoverpa armigera* (Hubner) larvae. J Biosci 37:269–276

Bhattacharyya A, Babu CR (2009) Purifcation and biochemical characterization of a serine proteinase inhibitor from *Derris trifoliata* Lour. seeds: insight into structural and antimalarial features. Phytochemistry 70:703–712

Bhattacharyya A, Mazumdar S, Babu CR (2006) A Kunitz proteinase inhibitor from *Archidendron ellipticum* seeds: purification, characterization, and kinetic properties. Phytochemistry 67:232–241

Bhattacharyya A, Rai S, Babu CR (2007) A trypsin and chymotrypsin inhibitor from *Caesalpinia bonduc* seeds: isolation, partial characterization and insecticidal properties. Plant Physiol Biochem 45:69–177

Broadway RM, Duffey SS (1986) Plant proteinase inhibitors: mechanism of action and effect on the growth and digestive physiology of larval *Heliothis zea* and *Spodoptera exigua*. J Insect Physiol 32:827–834

Browne B (1995) Ontogenetic changes in feeding behaviour. In: Chapman RF, de Boer G (eds) Regulatory mechanisms in insect feeding. Chapman & Hall, New York, pp 307–342

Browse J, Howe GA (2008) New weapons and a rapid response against insect attack. Plant Physiol 146:832–838

Chapman RF (1982) The insects: structure and function, 3rd ed. Macmillan, India

Charity JA, Anderson MA, Bittisnich DJ, Whitecross M, Higgins TJV (1999) Transgenic tobacco and peas expressing a proteinase inhibitor from *Nicotiana alata* have increased insect resistance. Mol Breed 5:357–365

Chaudhary NS, Shee C, Islam A et al (2008) Purification and characterization of a trypsin inhibitor from *Putranjiva roxburghii* seeds. Phytochemistry 69:2120–2126

Chen H, Gonzales-Vigil E, Wilkerson CG, Howe GA (2007) Stability of plant defense proteins in the gut of insect herbivores. Plant Physiol 143:1954–1967

Christeller J, Liang W (2005) Plant serine protease inhibitors. Protein Pept Lett 12:439–447

Christeller JT, Farley PC, Ramsay RJ, Sullivan PA, Laing WA (1998) Purification, characterization and cloning of an aspartic proteinase inhibitor from squash phloem exudate. Eur J Biochem 254:160–167

Coelho MB, Marangoni S, Macedo MLR (2007) Insecticidal action of *Annona coriacea* lectin against the flour moth *Anagasta kuehniella* and the rice moth *Corcyra cephalonica* (Lepidoptera: Pyralidae). Comp Biochem Physiol C 146:406–414

Cruz ACB, Massena FS, Migliolo L, Macedo LLP, Monteiro NKV, Oliveira AS et al (2013) Bioinsecticidal activity of a novel Kunitz trypsin inhibitor from Catanduva (*Piptadenia moniliformis*) seeds. Plant Physiol Biochem 70:61–68

da Silva DS, de Oliveira CFR, Parra JRP, Marangoni S, Macedo MLR (2014) Short and long-term antinutritional effect of the trypsin inhibitor ApTI for biological control of sugarcane borer. J Insect Physiol 61:1–7

Damle MS, Giri AP, Sainani MN, Gupta VS (2005) Higher accumulation of proteinase inhibitors in flowers than leaves and fruits as a possible basis for differential feeding preference of *Helicoverpa armigera* on tomato (*Lycopersicon esculentum* Mill, Cv. Dhanashree). Phytochemistry 66:2659–2667

De Leo F, Gallerani R (2002) The mustard trypsin inhibitor 2 affects the fertility of *Spodoptera littoralis* larvae fed on transgenic plants. Insect Biochem Mol Biol 32:489–449

de Oliveira CFR, Luz LA, Paiva PMG, Coelho LCBB, Marangoni S, Macedo MLR (2011) Evaluation of seed coagulant *Moringa oleifera* lectin (cMoL) as a bioinsecticidal tool with potential for the control of insects. Process Biochem 46:498–504

de Oliveira CFR, Vasconcelos IM, Aparicio R, Freire MGM, Baldasso PA, Marangoni S et al (2012) Purification and biochemical properties of a Kunitz type trypsin inhibitor from *Entada acaciifolia* (Benth.) seeds. Process Biochem 47:929–935

de Oliveira CFR, Souza TP, Parra JRP, Marangoni S, Silva-Filho MC, Macedo MLR (2013) Insensitive trypsins are differentially transcribed during *Spodoptera frugiperda* adaptation against plant protease inhibitors. Comp Biochem Physiol B Biochem Mol Biol 165:19–25

de Oliveira CFR, Marangoni S, Macedo MLR (2014) The trypsin inhibitor from *Entada acaciifolia* seeds affects negatively the development of Mediterranean flour moth, *Anagasta kuehniella*. Pestic Biochem Physiol 108:74–79

Delgado-Vargas F, Lopez-Valdes HE, Valdes-Rodriguez S, Blanco-Labra A, Chagolla-Lopez A, Lopez-Valenzuela EJ (2004) Isolation and properties of a Kunitz-type protein inhibitor obtained from *Pithecellobium dulce* seeds. J Agric Food Chem 52:6115–6121

Fan S-G, Wu G-J (2005) Characteristics of plant proteinase inhibitors and their applications in combating phytophagous insects. Bot Bull Acad Sin 46:273–292

Franco OL, Dos-Santos RC, Batista JAN et al (2003) Effects of black-eyed pea trypsin/chymotrypsin inhibitor on proteolytic activity and on development of *Anthonomus grandis*. Phytochemistry 63:343–349

Garcia VA, Freire MDGM, Novello JC et al (2004) Trypsin inhibitor from *Poecilanthe parviflora* seeds: purification, characterization, and activity against pest proteases. Protein J 23:343–350

Garcıa-Carreno FL, Navarrete-del-Toro MA, Diaz-Lopez M, Hernandez-Cortes MP, Ezquerra JM (1996) Proteinase inhibition of fish muscle enzymes using legume seed extracts. J Food Prot 59:312–318

Gatehouse AMR, Norton E, Davison GM, Babbe SM, Newell CA, Gatehouse JA (1999) Digestive proteolytic activity in larvae of tomato moth, *Lacanobia oleraceae*: effects of plant protease inhibitors in vitro and in vivo. J Insect Physiol 45:545–558

Godbole SA, Krishna TG, Bhatia CR (1994) Further characterisation of protease inhibitors from pigeon pea (*Cajanus cajan* L. millsp) seeds. J Sci Food Agric 64:331–335

Greenblatt HM, Ryan CA, MNG J (1989a) Structure of the complex of *Streptomyces griseus* proteinase B and polypeptide chymotrypsin inhibitor-1 from Russet Burbank potato tubers at 2.1 Å resolution. J Mol Biol 205:201–228
Greenblatt HM, Ryan CA, MNG J (1989b) Structure of the complex of *Streptomyces griseus* proteinase B and polypeptide chymotrypsin inhibitor-1 from Russet Burbank potato tubers at 2.1 Å resolution. J Mol Biol 205:201–228
Gupta GP, Ranjekar PK, Birah A, Rani S, Gupta V, Sainani MN (2002) Effective inhibition of *Helicoverpa armigera* by non-host plant proteinase inhibitors. ICAR News 8:1–2
Haq SK, Atif SM, Khan RH (2004) Protein proteinase inhibitor genes in combat against insects, pests, and pathogens: natural and engineered phytoprotection. Arch Biochem Biophys 431:145–159
Harsulkar AM, Giri AP, Patankar AG, Gupta VS, Sainani MN, Ranjekar PK, Despande VV (1999) Successive use of non-host plant proteinase inhibitors required for effective inhibition of *Helicoverpa armigera* gut proteinases and larval growth. Plant Physiol 121:497–506
Hivrale VK, Lomate PR, Basaiyye SS, Kalve ND (2013) Compensatory proteolytic responses to dietary proteinase inhibitors from *Albizia lebbeck* seeds in the *Helicoverpa armigera* larvae. Arthropod Plant Interact 7:259–266
Hopkins TL, Krichma LJ, Ahamad SA, Kramer KJ (2000) Pupal cuticle proteins of *Manduca sexta*: characterization and profiles during sclerotization. Insect Biochem Mol Biol 30:19–27
Howe RW (1971) A parameter for expressing the suitability of an environment for insect development. J Stored Prod Res 7:63–65
Hung CH, Huang CC, Tsai WS, Wang HL, Chen YL (2003) Purification and characterization of a trypsin inhibitor from *Brassica campestris* seeds. J Yuanpei Univ Sci Technol 10:13–22
Jamal F, Singh D, Pandey PK (2014) Negative effects of a nonhost proteinase inhibitor of ~19.8 kDa from *Madhuca indica* seeds on developmental physiology of *Helicoverpa armigera* (Hübner). BioMed Res Int 2014:202398
Jamal F, Pandey PK, Singh D, Wareed A (2015) A Kunitz-type serine protease inhibitor from *Butea monosperma* seed and its influence on developmental physiology of *Helicoverpa armigera*. Process Biochem 50:311–316
Johnson SL, Pellecchia M (2006) Structure and fragment-based approaches to protease inhibition. Curr Top Med Chem 6:317–329
Johnston KA, Lee MJ, Gatehouse JA, Anstee JH (1991) The partial purification and characterisation of serine protease activity in midgut of larval *Helicoverpa armigera*. Insect Biochem 21:389–397
Jongsma MA, Bolter C (1997) The adaptation of insects to plant protease inhibitors. J Insect Physiol 43:885–895
Jongsma MA, Bakker PL, Peters J, Bosch D, Stiekema WJ (1995) Adaptation of *Spodoptera exigua* larvae to plant proteinase inhibitors by induction of gut proteinase activity insensitive to inhibition. Proc Natl Acad Sci USA 92:8041–8045
Jongsma MA, Stiekema WJ, Bosch D (1996) Combating inhibitor-insensitive proteases of insect pest. Trends Biotechnol 14:331–333
Kansal R, Gupta RN, Koundal KR et al (2008) Purification and characterization of trypsin inhibitor from *Cicer arietinum* L. and its efficacy against *Helicoverpa armigera*. Braz J Plant Physiol 20:313–322
Koul O, Daniewski WM, Multani JS, Gumulka M, Singh G (2003) Antifeedant effects of the limonoids from *Entandrophragma candolei* (Meliaceae) on the Gram Pod Borer, *Helicoverpa armigera* (Lepidoptera: Noctuidae). J Agric Food Chem 51:7271–7275
Kuhar K, Kansal R, Subrahmanyam B, Koundal KR, Miglani K, Gupta VK (2013) A Bowman-Birk protease inhibitor with antifeedant and antifungal activity from *Dolichos biflorus*. Acta Physiol Plant 35:1887–1903
Lawrence PK, Koundal KR (2002) Plant protease inhibitors in control of phytophagous insects. Electron J Biotechnol 5:93–109
Macedo MLR, de Matos DGG, Machado OLT, Marangoni S, Novello JC (2000) Trypsin inhibitor from *Dimorphandra mollis* seeds: purification and properties. Phytochemistry 54:553–558

Macedo MLR, Freire MGM, Cabrini EC et al (2003) A trypsin inhibitor from *Peltophorum dubium* seeds active against pest proteases and its effect on the survival of *Anagasta kuehniella* (Lepidoptera: Pyralidae). Biochim Biophys Acta 1621:170–182

Macedo MLR, de Sa CM, Freire MDGM, Parra JRP (2004) A Kunitz-type inhibitor of coleopteran proteases, isolated from *Adenanthera pavonina* L. seeds and its effect on *Callosobruchus maculatus*. J Agric Food Chem 52:2533–2540

Macedo MLR, Freire MGM, Franco OL, Migliolo L, de Oliveira CFR (2011) Practical and theoretical characterization of *Inga laurina* Kunitz inhibitor on the control of *Homalinotus coriaceus*. Comp Biochem Physiol B Biochem Mol Biol 158:164–172

Machuka JS, Okeola OG, Chrispeels MJ, Jackai LE (2000) The African yam bean seed lectin affects the development of the cowpea weevil but does not affect the development of larvae of the legume pod borer. Phytochemistry 53:667–674

McManus MT, Burgess EPJ (1995) Effects of the soybean (Kunitz) trypsin inhibitor on growth and digestive proteases of larvae of *Spodoptera litura*. J Insect Physiol 41:731–738

Mikola M, Mikkonen A (1999) Occurrence and stabilities of oat trypsin and chymotrypsin inhibitors. J Cereal Sci 30:227–235

Missios S, Davidson HC, Linder D, Mortimer L, Okobi AO, Doctor JS (2000) Characterization of cuticular proteins in the red flour beetle *Tribolium castaneum*. Insect Biochem Mol Biol 30:47–56

Mordue AJ, Blackwell A (1993) Azadirachtin: an update. J Insect Physiol 39:903–924

Murray KD, Alford AR, Groden E, Drummond FA, Storch RH, Bentley MD, Sugathapala PM (1993) Interactive effects of an antifeedant used with *Bacillus thuringiensis* var. San diego delta endotoxin on Colorado potato beetle (Coleoptera: Chrysomelidae). J Econ Entomol 86:1793–1801

Nathan SS, Chung PG, Murugan K (2005) Effect of biopesticides applied separately or together on nutritional indices of the rice leaffolder *Cnaphalocrocis medinalis*. Phytoparasitica 33:187–195

Neurath H (1989) Proteolytic processing and physiological regulation. Trends Biochem Sci 14:268–271

Oliveira AS, Migliolo L, Aquino RO et al (2007) Purification and characterization of a trypsin–papain inhibitor from *Pithecelobium dumosum* seeds and its in vitro effects towards digestive enzymes from insect pests. Plant Physiol Biochem 45:858–865

Pandey PK, Jamal F (2014) Bio-potency of a 21 kDa Kunitz-type trypsin inhibitor from *Tamarindus indica* seeds on the developmental physiology of *H. armigera*. Pestic Biochem Physiol 116:94–102

Pandey PK, Singh D, Singh S, Khan MY, Jamal F (2014) A nonhost peptidase inhibitor of ~14 kDa from *Butea monosperma* (Lam.) Taub. seeds affects negatively the growth and developmental physiology of *Helicoverpa armigera*. Biochem Res Int 2014:361821

Pandey PK, Singh D, Jamal F (2015) The influence of a 21 kDa Kunitz-type trypsin inhibitor from non-host Madras thorn, *Pithecellobium dulce* seeds on *H. armigera* (Hübner) (Lepidoptera: Noctuidae). Arch Insect Biochem Physiol 89:18–34

Pandey PK, Singh D, Singh R, Sinha MK, Singh S, Jamal F (2016) *Cassia fistula* seed's trypsin inhibitor(s) as antibiosis agent in *Helicoverpa armigera* pest management. Biocatal Agric Biotechnol 6:202–208

Patankar AG, Giri AP, Harsulkar AM, Sainani MN, Deshpande VV, Ranjekar PK et al (2001) Complexity in specificities and expression of *Helicoverpa armigera* gut proteinases explains polyphagous nature of the insect pest. Insect Biochem Mol Biol 31:453–464

Pawar P, Joseph M, Tungikar V, Joshi S (2004) Bioactivity of nonedible oil seed extracts and purified extracts against *Helicoverpa armigera* (Hubner). Indian J Exp Biol 42:91–95

Perry LM (1980) Medicinal plants of East and South East Asia. MIT Press, Cambridge

Petek M, Turnsek N, Gasparic MB, Novak MP, Gruden K, Slapar N, Popovic T, Strukelj B, Gruden K, Strukelj B, Jongsma MA (2012) A complex of genes involved in adaptation of *Leptinotarsa decemlineata* larvae to induced potato defense. Arch Insect Biochem Physiol 79:153–181

Pompermayer P, Falco MC, Parra JRP, SilvaFilho MC (2003) Coupling diet quality and Bowman-Birk and Kunitz type soybean proteinase inhibitor effectiveness to *Diatraea saccharalis* development and mortality. Entomol Exp Appl 109:217–224

Rai S, Aggarwal KK, Babu CR (2008) Isolation of a serine Kunitz trypsin inhibitor from leaves of *Terminalia arjuna*. Curr Sci 94:1509–1512

Ramasarma PR, Appu Rao AG, Rao DR (1995) Role of disulfide linkages in structure and activity of proteinase inhibitor from horsegram (*Dolichos biflorus*). Biochim Biophys Acta 1248:35–42

Rao DH, Gowda LR (2008) Abundant class III acidic chitinase homologue in tamarind (*Tamarindus indica*) seed serves as the major storage protein. J Agric Food Chem 56:2175–2182

Rawlings ND, Tolle DP, Barrett AJ (2004) MEROPS: the peptidase database. Nucleic Acids Res 32:D160–D164

Rayapuram C, Baldwin IT (2006) Using nutritional indices to study LOX3-dependent insect resistance. Plant Cell Environ 29:1585–1594

Richardson MJ (1991) Seed storage proteins: the enzyme inhibitors. In: Richardson MJ (ed) Methods in plant biochemistry. Academic, New York, pp 259–305

Rufino FPS, Pedroso VMA, Araujo JN, Franca AFJ, Rabelo LMA, Migliolo L, Kiyota S, Santos EA, Franco OL, Oliveira AS (2013) Inhibitory effects of a Kunitz-type inhibitor from *Pithecellobium dumosum* (Benth) seeds against insect-pests' digestive proteinases. Plant Physiol Biochem 63:70–76

Ryan CA (1990) Protease inhibitors in plants: genes for improving defenses against insects and pathogens. Annu Rev Phytopathol 28:425–449

Sarate PJ, Tamhane VA, Kotkar HM et al (2012) Developmental and digestive flexibilities in the midgut of a polyphagous pest, the cotton bollworm, *Helicoverpa armigera*. J Insect Sci 12:1–16

Schluter D, Marchinko KB, Barrett RDH (2010) Natural selection and the genetics of adaptation in three spine. Philos Trans R Soc B 365:2479–2486

Shazly NM (1993) Quantitative evaluation of food intake and assimilation by *Spilostethus pandurus* (Scopoli) (Hemiptera: Lygaeidae). Bull Entomol Soc Egypt 71:109–117

Singh D, Jamal F, Pandey PK (2014) Kinetic assessment and effect on developmental physiology of a trypsin inhibitor from *Eugenia jambolana* (Jambul) seeds on *Helicoverpa armigera* (Hübner). Arch Insect Biochem Physiol 85:94–113

Sivakumar S, Franco OL, Tagliari PD, Bloch CJ, Mohan M, Tayumanavan B (2005) Screening and purification of a novel trypsin inhibitor from *Prosopis juliflora* seeds with activity toward pest digestive enzymes. Protein Pept Lett 12:561–565

Srinivasan A, Giri AP, Gupta VS (2006) Structural and functional diversities in lepidopteran serine proteases. Cell Mol Biol Lett 11:132–154

Sudheendra K, Mulimani VH (2002) Effect of feeding legume proteinase inhibitors on *Helicoverpa armigera*gut proteinase activity. Proceedings of the 2nd international conference on precision engineering and nanotechnology (ICPN '02) 9:51–53

Tamhane VA, Chougule NP, Giri AP, Dixit AR, Sainani MN, Gupta VS (2005) In vitro and vivo effects of *Capsicum annum* proteinase inhibitors on *Helicoverpa armigera*gut proteinases. Biochim Biophys Acta 1722:155–167

Telang M, Srinivasan A, Patankar A, Harsulkar A, Joshi V, Damle A, Deshapande V, Sainani M, Ranjekar P, Gupta G, Birah A, Rani S, Kachole M, Giri A, Gupta V (2003) Bitter gourd proteinase inhibitors: potential growth inhibitors of *Helicoverpa armigera* and *Spodoptera litura*. Phytochemistry 63:643–652

Tetenbaum J, Miller LM (2001) A new spectroscopic approach to examining the role of disulfide bonds in the structure and unfolding of soybean trypsin inhibitor. Biochemistry 40:12215–12219

Umezawa H (1982) Low molecular weight enzyme inhibitors of microbial origin. Annu Rev Microbiol 36:75–99

Wijaya R, Neumann GM, Condron R, Hughes AB, Polya GM (2000) Defense proteins from seed of *Cassia fistula* include a lipid transfer protein homologue and a protease inhibitory plant defensin. Plant Sci 159:243–255

Yoshizaki L, Troncoso MF, Lopes JLS, Hellman U, Beltramini LM, Wolfenstein-Todel C (2007) *Calliandra selloi* Macbride trypsin inhibitor: isolation, characterization, stability, spectroscopic analyses. Phytochemistry 68:2625–2634

Zavodszky M, Chao-Wei C, Huang JK, Zolkiewski M, Wen L, Ramaswamy K (2001) Disulfide bond effects on protein stability: designed variants of *Cucurbita maxima* trypsin inhibitor-V. Protein Sci 10:149–160

Zhou JY, Liao H, Zhang NH et al (2008) Identification of a Kunitz inhibitor from *Albizzia kalkora* and its inhibitory effect against pest midgut proteases. Biotechnol Lett 30:495–1499

Printed by Printforce, the Netherlands